Probability
and Random Processes:
A First Course with Applications

A. BRUCE CLARKE
College of Arts and Sciences
Western Michigan University

RALPH L. DISNEY
Gordon Professor
of Industrial Engineering and Operations Research
Virginia Polytechnic Institute and State University

John Wiley & Sons

New York • Chichester • Brisbane • Toronto • Singapore

Library of Congress Cataloging in Publication Data:

Clarke, A. Bruce, 1927–
    Probability and random processes.

    Rev. ed. of: Probability and random processes for
engineers and scientists. 1970.
    Bibliography: p.
    Includes index.
    1. Probabilities.   2. Stochastic processes.   I. Disney,
Ralph, 1928–   . II. Clarke, A. Bruce, 1927–
Probability and random processes for engineers and
scientists. III. Title.

QA273.C597 1985   519.1   84–15312
ISBN 0-471-08535-9

Printed in the United States of America

10 9 8 7 6 5 4 3 2 1

Probability and Mathematical Statistics (Continued)

SEBER • Linear Regression Analysis
SEBER • Multivariate Observations
SEN • Sequential Nonparametrics: Invariance Principles and Statistical Inference
SERFLING • Approximation Theorems of Mathematical Statistics
TJUR • Probability Based on Radon Measures
WILLIAMS • Diffusions, Markov Processes, and Martingales, Volume I: Foundations
ZACKS • Theory of Statistical Inference

Applied Probability and Statistics

ABRAHAM and LEDOLTER • Statistical Methods for Forecasting
AGRESTI • Analysis of Ordinal Categorical Data
AICKIN • Linear Statistical Analysis of Discrete Data
ANDERSON, AUQUIER, HAUCK, OAKES, VANDAELE, and WEISBERG • Statistical Methods for Comparative Studies
ARTHANARI and DODGE • Mathematical Programming in Statistics
BAILEY • The Elements of Stochastic Processes with Applications to the Natural Sciences
BAILEY • Mathematics, Statistics and Systems for Health
BARNETT • Interpreting Multivariate Data
BARNETT and LEWIS • Outliers in Statistical Data, _Second Edition_
BARTHOLOMEW • Stochastic Models for Social Processes, _Third Edition_
BARTHOLOMEW and FORBES • Statistical Techniques for Manpower Planning
BECK and ARNOLD • Parameter Estimation in Engineering and Science
BELSLEY, KUH, and WELSCH • Regression Diagnostics: Identifying Influential Data and Sources of Collinearity
BHAT • Elements of Applied Stochastic Processes, _Second Edition_
BLOOMFIELD • Fourier Analysis of Time Series: An Introduction
BOX • R. A. Fisher, The Life of a Scientist
BOX and DRAPER • Evolutionary Operation: A Statistical Method for Process Improvement
BOX, HUNTER, and HUNTER • Statistics for Experimenters: An Introduction to Design, Data Analysis, and Model Building
BROWN and HOLLANDER • Statistics: A Biomedical Introduction
BROWNLEE • Statistical Theory and Methodology in Science and Engineering, _Second Edition_
CHAMBERS • Computational Methods for Data Analysis
CHATTERJEE and PRICE • Regression Analysis by Example
CHOW • Analysis and Control of Dynamic Economic Systems
CHOW • Econometric Analysis by Control Methods
CLARKE and DISNEY • Probability and Random Processes: A First Course with Applications
COCHRAN • Sampling Techniques, _Third Edition_
COCHRAN and COX • Experimental Designs, _Second Edition_
CONOVER • Practical Nonparametric Statistics, _Second Edition_
CONOVER and IMAN • Introduction to Modern Business Statistics
CORNELL • Experiments with Mixtures: Designs, Models and The Analysis of Mixture Data
COX • Planning of Experiments
DANIEL • Biostatistics: A Foundation for Analysis in the Health Sciences, _Third Edition_
DANIEL • Applications of Statistics to Industrial Experimentation
DANIEL and WOOD • Fitting Equations to Data: Computer Analysis of Multifactor Data, _Second Edition_
DAVID • Order Statistics, _Second Edition_
DAVISON • Multidimensional Scaling
DEMING • Sample Design in Business Research
DILLON and GOLDSTEIN • Multivariate Analysis: Methods and Applications
DODGE and ROMIG • Sampling Inspection Tables, _Second Edition_
DOWDY and WEARDEN • Statistics for Research

continued on back

**Dedicated
to the memory of
Edward**

*"(He) should have died hereafter.
There would have been time for such a thing."*
Macbeth

"Good night sweet prince."
Hamlet

PREFACE

In the first edition of this book (then titled *Probability and Random Processes for Engineers and Scientists*) we tried to do something rather new for the times by presenting an undergraduate textbook that introduced the student to probability and random processes instead of the more common probability-statistics sequence. This attempt was based on our feeling that in the world of the engineer and scientist one builds models from which to learn. Topics such as Markov processes provide the beginner with a rich and often-used class of stochastic models. Thus, this material should be in the background of every beginner in such fields.

Our effort was moderately successful. The book was used not only in areas that we expected and wrote for, but in areas we had not anticipated. This result has encouraged us to prepare a second edition and at the same time to broaden its intended audience. Thus the change of title.

At the same time we found some topics, of interest to unexpected users, were more important than we had foreseen. Other topics, our intuition tells us, were obtrusive and really not as important to the intended audience as we had originally thought. Still, based on the feedback that we received from reviewers and users, the basic concepts proved sound.

With these thoughts in mind we have revised the first edition. Most noteworthy of the changes are: the elimination of Chapter 14, "Further Properties of Stochastic Processes"; the omission of certain material on combinatorial probability; the introduction of expanded material on functions of random variables; a new section on laws of large numbers and the central limit theorem; and the addition of many new problems, exercises, and examples, some of which are less engineering oriented. Some of these are "real-life" applications drawn from the literature.

We hope that such changes better meet the needs of our readers, including those who previously used the book but were neither engineers nor physical scientists. We hope, furthermore, that we have not thrown out the baby with the bath water and that engineers and physical scientists will continue to find the book useful.

The literature of applied probability is vast. We reviewed some of it and tried to cull interesting problems which had real-life content and which contained experimental data. We eschewed purely theoretical studies no matter how interesting the topic was to us. We also sought breadth of coverage rather than depth because we want the beginning student to see how widely used probability is.

We deliberately have attempted to organize the book around a series of real-world "case studies" that are continued from chapter to chapter to provide practical motivation for the concepts.

Each application, as we have written it, contains a reference to the authors from whom we got the example, and a date. Anyone interested can consult the original literature to go deeper into the particular application. In this connection we have used both primary sources and secondary sources. Some of our applications were originally found in foreign language journals.

In such cases we deliberately chose to quote the English source, not the original source. This decision should make the literature more accessible to the undergraduate student.

In pursuing these applications we have occasionally bent the author's data by presenting it in summary form or a form different from his presentation. This was done, where it was done, to illustrate the use of probability, not the basic problems that might be present with the data. Where we have done this, we have tried to state clearly that we have manipulated the data. Distortions caused by our manipulations, of course, are our fault, not those of the original author.

One final point needs to be made concerning these applications. We have presented one or two tools often used in data analysis (e.g., test of independence, Kolmogorov-Smirnov test of good fit). We "pull these out of the air." The student should be told that. We have assiduously tried to avoid introducing statistical methods that might be used better on specific data, or long discussions of the strengths and weaknesses of the methods we do use. The idea here is to hurry the student into using probability on real problems. We want the student to play, probabilistically, with real problems. For this purpose we have chosen statistical methods that seem least obtrusive yet let the student get from reality to probabilistic analysis with a minimum of background or sidetracks. This cavalier approach to statistics should do three things. First, we hope the student sees, at an elementary level, that there is an important interface between experimentation and probability used to model the experiment. Second, we hope that the student sees that probability is widely useful and used. Third, we hope that the student is better motivated to subsequent studies of statistics through questioning how to better or differently look at our data.

We made two points in the preface to the first edition that we continue to think are important. In writing an applied mathematics book one always fears stepping over the line into too much mathematics or too much application for the use of the audience. It is not our intention to do either, but only the user can determine if we have. We also noted in the preface to the first edition that we are firmly convinced that topics in probability and random processes are just as important to the modern student's background as many of the courses presently given in many curricula, including engineering. We are also convinced that pre-college education does not prepare a student to feel at home in these probability-based courses. Therefore, the student must be helped into this world. Our text, as viewed by us, is simply a vehicle to help the student. It is not a substitute for competent teaching (as is no textbook).

As in the first edition, there are too many people to thank for the preparation of this second edition to do an honest job of it. Certainly everyone who helped us on the first edition deserves our thanks again. The two authors of the first edition have both changed affiliations since the first edition appeared. This means many new people have been imposed upon to help us with the second edition. These include Bob Foley, whose advice helped to keep us from departing too much from important principles, as well as Cynthia Will and Rosemary Lyon who helped put our rough efforts into usable form. Our wives, Florence and Lois, were unreasonably supportive in assisting us in completing an activity that inevitably requires more of a time and effort commitment than is ever anticipated.

A. Bruce Clarke

Ralph L. Disney

Robert D. Foley

CONTENTS

1 · Sample Spaces and Events

1.1 Some Background

Probability theory has a strange history. There is little chance of finding "the beginning."

Some writers ascribe its beginnings to the earliest ages of man. Small cubical bones from the legs of horses were found in prehistoric temples, suggesting that dice may have been used for early religious rites. By the fifteenth century, marine insurance had developed in Italy, indicating that there was at least a rudimentary knowledge of probability.

But the earliest carefully documented use of probability (and therefore the point often taken as the beginning of probability) occurred during the seventeenth century in France and centered on games of chance. Unfortunately, this emphasis has continued in textbooks and has given the beginner a feeling that games are the major use of probability. It is hoped that the many other examples now available, some of which appear in this book, will dissuade the reader from this feeling.

Throughout most of its history, indeed until the latter part of the nineteenth century, probability was without a home and among many potential users was looked at askance. It was a curiosity devoid of content except possibly to win a few hands of cards or dice games.

At least two major developments occurred near the end of the nineteenth century that had a major impact on the use of probability.

In physics, which was undergoing a revolution at about this time, attempts were made to explain the quite erratic behavior of molecules suspended in a liquid. The behavior had been observed more than 100 years earlier and was named Brownian motion. A great step occurred when a probabilistic model was developed that satisfactorily described the molecules' behavior.

The second major impetus to probability usage occurred in the then new field of telephony. The telephone had been invented in the latter part of the nineteenth century and by 1900 was becoming ubiquitous. It became clear that it would be physically impossible to connect every telephone to every other telephone. Rather one needed a system whereby one could interconnect many, but not all, phones at once. This led early telephone engineers into a host of questions, some of which are still under active study. It was clear almost from the beginning that such questions could only be answered with the help of probabilistic models.

Throughout the early twentieth century new uses were found for probabilistic thinking. But even with this burgeoning growth of the uses of probability, a well-defined, accepted theory of probability did not develop until about 1930.

One should remember that while this burst of applications of probability models was taking place there was a revolution occurring in mathematics. In particular the Cantor theory of sets and the final emergence of the Lebesgue theory of functions and integration were changing the face of mathematics. By about 1930 much of this development was complete. Probability was overripe for a rigorous underpinning, and the mathematics necessary to provide a substructure of probability was at hand. The merger came about during the four years 1928 to 1932.

Our purpose in this book is to study, insofar as possible, some of the principles of probability theory and to point out some of its uses. We do not assume that the student has a background in all of the related mathematics (e.g., Lebesgue integrals) so we will try to make some of our arguments plausible rather than rigorous. The student should be aware of this and recognize that there is a large body of knowledge beyond the scope of this book. The person wishing to use probability may find that the topics covered here allow one to understand the applied literature and to start using probability in solving real work problems.

1.2   An Overview

A theory of probability (not "the" theory, for there are several) is based on four rather simple ideas. One starts with a primitive notion called a *sample space*. Such a space is nothing more than a set whose elements are interpreted as possible experimental outcomes. We discuss the idea in the remainder of this chapter.

Then from this sample space one chooses subsets. These chosen subsets with their complements and unions then form a new set called *the set of events*. This set is the major working set of probability. Unfortunately, the construction of a set of events takes some care. One would like this set to contain every conceivable subset of the sample space, but if the set of events becomes too large it is not possible to go to the next step. By keeping the sample space rather simple (e.g., countable) this problem never arises for us and we usually can take the set of events to be composed of the set of all subsets of the sample space.

The third major element we need to develop a theory of probability is the concept of *probability* itself. Here also we must exercise a degree of care. Because of its historical antecedents in things such as games one would hope to develop the concept of probability in such a way that it includes these classical, intuitive ideas. A simple way to do this is to define a probability as a function (called a "measure") from the set of events onto the unit interval. In this way each *event* is assigned a number between 0 and 1, termed the *probability* of the event. Now not every function will have the intuitive properties used in early probability applications so we will have to put some restrictions (called axioms) on the class of functions that give probabilities. These concepts are developed in Chapter 2.

The fourth major element is the *random variable*. A random variable is simply (for us) a function, but not just any function. We want a function that assigns real numbers *to points in the sample space*. But here we have a bit of a problem. Probabilities are assigned to elements in the event space, that is, subsets of the sample space. Real numbers (not just those in the unit interval) are assigned to elements in the sample space by the random variable. There is no direct connection between probability and the range of the random variable. To obtain a useful connection, not just any function will do for a random variable. We will define our random variables in a way such that a useful connection exists between the range of the random variable and the set of events. This concept will be elucidated in Chapter 3.

In Chapter 3, we will learn that the random variable has to be a function that allows us to reverse direction and to go from the real line back into an event and then to a probability—certainly a circuitous route. Fortunately, life is not that devious. Rather in Chapter 3 we will see that we can go directly from the real line to a probability with a third function called a *distribution function*. Indeed, in many applications one collects data to estimate this distribution function (Chapter 8). Then probabilistic studies are made from the distribution function, and the basic building blocks of sample space, event set, and probability function fade into the background.

After defining the idea of a conditional probability in Chapter 2, basic probability is about complete. One learns, further, how to compute probabilities, how to obtain some useful information even if the entire probability distribution is not known, and how to use certain probability distributions that occur so commonly that they are given special names (Chapter 7).

Just as in basic calculus one proceeds from the behavior of functions of one variable to functions of several variables (i.e., multivariate calculus), so too we will have to progress to two, many, countably many, and finally uncountably many random variables. In all instances, however, the basic sample space, event set, and probability function will remain unchanged. This step introduces some new concepts of vital importance to probabilistic modeling. These concepts are developed in Chapter 4. The entire set of Chapters 8 to 13 then examines this class of problems in special, but extremely useful, cases.

1.3　Random Experiments and Sample Spaces

Probability theory is rooted in the real-life situation in which a person performs an experiment the outcome of which may not be certain. Such an experiment is called a *random experiment*. Mathematically, the concept of a random experiment is considered a primitive notion that is not otherwise defined.

Examples of random experiments are plentiful in real-life situations. The strength of a manufactured part is uncertain prior to its manufacture. The process of manufacturing the part can be construed as the performance of a random experiment. Games of chance are classic examples of random experiments; the outcome of the game is not known with certainty prior to playing the game. Firing a rocket is an example of performing a random experiment, as is conducting a sales campaign.

In each of the examples it is clear that the outcome of the performed experiment is not certain a priori. The exact strength of the manufactured part is not certain *before* the part is manufactured. The face appearing on the cast die is not certain prior to casting the die. The increased sales are not a priori known. Success, failure, or partial success of a rocket firing is not certain prior to firing.

Thus the study of probability can be initiated with the primitive idea of a random experiment. Associated with the experiment are the possible results, or outcomes, of the experiment; for example, associated with the die-tossing experiment are the possible outcomes: an ace turns up, a two turns up, et cetera. Associated with the manufacture of a part are the possible strengths of the completed part. The important concept here is that associated with any random experiment there is a set, S, of all possible outcomes. This set S is called a *sample space*.

We must point out that the sample space is not determined completely by the experiment. It is partially determined by the purpose for which the experiment is carried out. If a coin is tossed twice, for some purposes it might be sufficient to consider only three possible outcomes: two heads, two tails, a combination of one head with one tail. These three outcomes would then constitute the sample space S. However, one might be playing a game in which the exact *order* of appearance of the sides of the coin is important. In this case the sample space S must be considered as containing *four* possible outcomes: head–head, tail–tail, head–tail, tail–head. Other possible sample spaces would be obtained if we took into account such things as the exact locations of the fall of the coins, the number of times they spin in the air, et cetera. Frequently we use a larger sample space than is strictly necessary because it is easier to use. For instance, in the preceding illustration the sample space with four possible outcomes is easier to discuss, since for an "honest" coin those four outcomes are all "equally likely," although this is not true in the sample space with three outcomes.

EXAMPLE 1.1

Let us begin our examples with a very simple one where the underlying experiment has only two outcomes. These outcomes could be "rain" or "dry" as has been used in an early attempt to model weather. They could be "good" or "defective" as is often used to model the quality of produced parts. They could be "left" or "right" as has been used to model spin directions of electrons. To avoid long discussions let us take these outcomes as "heads" or "tails" as is common when modeling the outcomes of tossing a coin. In this case the sample space, S, is simply

$$S = \{\text{Heads, Tails}\}$$

If we now flipped *two* coins (or looked at *two* consecutive days of weather or *two* produced parts or *two* electrons) then

$$S = \{(\text{Heads, Heads}), (\text{Tails, Tails}), (\text{Heads, Tails}), (\text{Tails, Heads})\}$$

The generalization to many coins is now apparent.

EXAMPLE 1.2

A lot consists of five manufactured articles labeled $1, 2, 3, 4, 5$. An experiment consists of selecting three of these articles for testing. If order is disregarded, by direct enumeration we find that the sample space consists of the following 10 possible outcomes:

$$123, 124, 125, 134, 135, 145, 234, 235, 245, 345$$

If order is taken into account, there will be 60 possible outcomes. For instance, the sample consisting of articles 1, 2, and 3 can be selected in the following six orders:

$$123, 132, 213, 312, 231, 321$$

Which one of these possible sample spaces should be used depends on the method of testing and evaluating. For example, if the articles are tested sequentially under the rather loose rule that the lot is rejected only if at least two of the articles tested are defective, then clearly the second sample space should be used. In this case notice that if the first two articles selected and tested were found to be either both good or both defective it would be unnecessary even to select a third article in order to reach a decision.

EXAMPLE 1.3

In anthropological studies one "draws a person at random" and measures his height and weight. To use the resulting information to design an elevator a designer might be concerned only with the weight load on the elevator. Thus, the sample space would consist of the set of all possible weights of the people drawn at random. On the other hand, the same elevator designer might be concerned only with the head room design of the elevator. If this is true, the sample space could be taken to be the set of all possible heights of the people. In most real-life design problems the designer at some time would be concerned with both the set of all possible heights and the set of all possible weights of users of the elevator. More specifically the designer would usually be interested in the outcomes that simultaneously gave the height and weight of the possible users. Thus, the sample space could be the set of all possible *pairs* (height, weight).

We can categorize three types of sample spaces for most applications.

(a) In the experiment of tossing a single coin there may be only two possible outcomes. More generally, the sample space may contain only a finite number of outcomes. We say that the sample space associated with the experiment is a *finite sample space*.

(b) If the outcomes of a sales campaign are given in terms of the *number* of units sold, we can take as an idealization of the sample space the set of all nonnegative integers. Certainly no real sales campaign will result in an arbitrarily large number of units sold, but the set of nonnegative integers could serve as a mathematical idealization, or *model*, of the campaign. Sets of outcomes in these cases can be put into a one-to-one correspondence with the counting numbers. Hence, such a sample space is said to be *countably infinite*.

(c)   The random experiment of manufacturing a part and measuring its strength conceivably could have an entire *interval* of real numbers as possible values. Since an interval of real numbers cannot be enumerated in a sequence (even an infinite sequence) such a sample space is said to be *noncountable* or *nondenumerable*. Certainly no real experiment conducted using real measuring instruments can ever yield such a sample space, since there is a limit to the fineness to which any instrument can measure. Nonetheless, such a sample space frequently can be taken as an idealization of, an approximation to, or a model of a real-life situation, which is easier to discuss than a more exact model.

In many cases it is not necessary to distinguish between finite and countably infinite sample spaces. Therefore, if a sample space is either finite or countably infinite we say that it is a *countable sample space*. In some usages we speak of a countable sample space as a *discrete sample space* and of a noncountable sample space as a *continuous sample space*.

1.4   Set Notation

Since we have identified a sample space S as the *set* of all possible outcomes of a random experiment, it is convenient to review and to consolidate some *set* ideas that will be useful.

If s is a possible outcome of the experiment, then s is one of the members of S. We indicate the statement "s is a member of S" or "s belongs to S" or "s is an element of S" by the notation

$$s \in S$$

Elements of S are termed *sample points*.

If s is not a possible outcome of the experiment, then s is not an element of S and we write

$$s \notin S$$

Several notations are used to specify a set in terms of its elements.

If the set S is finite, it may be convenient to merely list the elements of S (Example 1.1). Thus, for the experiment of tossing a single coin there were two possible outcomes, *heads* or *tails*, and the sample space was

$$S = \{\text{Heads, Tails}\}$$

In a similar fashion

$$S = \{1, 2, 3, 4, 5, 6\}$$

would denote the set S whose elements are the counting numbers $1, 2, 3, 4, 5, 6$. This set could serve as the sample space for a die-tossing experiment.

For infinite sets, the listing of the elements is impossible; for large finite sets, the listing is inconvenient. Thus, we seek another notation for these sets. We first note that any set is uniquely specified once its elements are known. For a given set, we could state a rule, an inclusion condition, that could be applied to any object to determine whether the object is or is not to be included in the set. Thus, a set is uniquely defined in terms of a category of possible objects and an inclusion condition.

Suppose that the experiment consists of manufacturing a part and of measuring its strength, and that each manufactured part must have a strength between a and b, with all such numbers being possible strengths. The sample space S then consists of elements that are real numbers that lie in the interval from a to b. Notationally, then, S is defined by

$$S = \{x : a < x < b\}$$

This notation is read as "S is a set whose elements x consist of all numbers in the interval from a to b." To make the notation more explicit, we could say that "S is the set (the braces indicate this) of elements x, so that (the colon is used to denote these words) x satisfies some statement (the statement is explicit and forms the inclusion condition)." In this case the sample points are real numbers. This type of interval is termed an *open* interval. For further discussion, see Exercise 14 at the end of this chapter.

As a further example of set notation, we might be interested in a set of points in a plane that lie on the same given line. Then, S can be considered as a set of points whose coordinates are given by ordered pairs (x, y). The inclusion condition for S is that a point belongs to S if it satisfies an equation of the form $ax + by = c$ for given a, b, and c. Symbolically, then,

$$S = \{(x, y) : ax + by = c\}$$

says "S is the set whose elements are those ordered pairs (x, y) that satisfy the condition $ax + by = c$."

Notice that we might consider the pair (x, y) above to be a vector emanating from the origin with terminus at (x, y). Then, if we consider an element $\mathbf{x} = (x, y)$ as such a vector, the set S can be taken as a set of vectors. The particular set of vectors of interest in the example would be denoted by

$$S = \{\mathbf{x} : \mathbf{a}\mathbf{x}^T = c; \ \mathbf{a} = (a, b)\}$$

[Here $\mathbf{x}^T$ is the column vector $\begin{pmatrix} x \\ y \end{pmatrix}$ called the *transpose* of the vector $\mathbf{x} = (x, y)$.]

In the elevator design (Example 1.3), if the designer is interested in both the height and the weight of the users, a pair of numbers (height, weight) taken in that order would be a point (a sample element) in the sample space consisting of all such pairs. Hence, S can be taken to be a set of vectors. A vector is one of the ordered pairs (height = x, weight = y).

1.5   Combinations of Sets

In many applications we are given more than one set, for instance, $A_1, A_2, \ldots,$ and we wish to combine them to arrive at new sets that are of importance. For example, in most applications we have the sample space S, but we are interested in only some of the elements of S. In other words, we are concerned with some elements of S that share a common property as distinguished from other elements of S that do not have this property. Thus, if S is the sample space for the strength of a manufactured part and if

$$S = \{x : a < x < b\}$$

we may be particularly concerned with the outcomes for which $a < x < c$ where c is some number between a and b. For example, c might be the least strength that a part could have to serve a given function. The new set, for example, A, would be given by

$$A = \{x : a < x < c\}$$

and could be considered in this example as the "set of defects."

Other kinds of combinations are also of interest. Suppose that we conduct the experiment of measuring the resistance x (in ohms) of an electrical part and suppose that

$$S = \{x : a < x < b\}$$

Suppose further that

$$A_1 = \{x : a < x < c\}$$

represents the strengths of the parts whose resistance is "too low" for use in a television set. Also suppose that

$$A_2 = \{x : d < x < b\}$$

represents the strengths of the parts whose resistance is "too great" for use in a television set. Then the set

$$A_3 = \{x : x \in A_1 \text{ or } x \in A_2\}$$

represents the set of defective strengths, that is, the resistances that are either too low or too great for use in a television set. Certainly the manufacturer of television sets is interested in A_3, which is a set formed by "combining" sets A_1 and A_2.

Hence, clearly, many questions of interest require that one be able to take given sets and operate on them to produce new sets. We discuss some of these operations in this section.

(a) Equal Sets   Two sets A and B are *equal sets*, $A = B$, if they contain the same sample points. Thus, the sets $A = \{1, 2\}$ and $B = \{2, 1\}$ are equal sets. Notice that $C = \{1, 2, 1, 2, 1\}$ is a set that is also equal to A, since it contains the same elements. However, $D = \{1, 2, 3, 1, 2, 1, 2\}$ is not equal to A, B, or C.

(b) Subsets   A set A is said to be a *subset* of B, $A \subset B$, if every sample point, or element, of A is also an element of B. If, in addition, every element of B is also an element of A, then by the above definition $A = B$.

(c) Unions   Suppose A and B are two sets. The *union* of A and B, symbolically $A \cup B$, is the set whose elements belong to either A or B (or both). That is,

$$A \cup B = \{ s : s \in A \text{ or } s \in B \}$$

Notice that s may belong to both A and B, in which case it is still a member of $A \cup B$.

If A and B are characterized by verbal phrases, then $A \cup B$ is characterized by placing these phrases together with the word "or" between them. For example, when tossing a coin twice, if the sets A and B are defined by

A: "the first coin is a head" = {Head–Head, Head–Tail}

B: "the coins match" = {Head–Head, Tail–Tail}

then the set $A \cup B$ is defined by

$A \cup B$: "the first coin is a head or the coins match"

= {Head–Head, Head–Tail, Tail–Tail}

Observe that the word "or" is always used in the mathematical or nonexclusive sense of "one or the other or both." The set A_3 in the introduction to this section is another example of a set that is the union of the two sets A_1 and A_2.

(d) Intersections   Suppose that A and B are two sets. The *intersection* of A and B, symbolically $A \cap B$, is the set whose elements belong to both A and B. That is,

$$A \cap B = \{ s : s \in A \quad \text{and} \quad s \in B \}$$

If A and B are characterized by verbal phrases, then $A \cap B$ is characterized by placing these phrases together with the word "and" between them. In the preceding illustration we have

$A \cap B$: "the first coin is a head and the coins match"

= {Head–Head}

Suppose that in testing parts for use in a satellite it is required that a part be "strong enough" and yet "light enough." Thus, suppose that any part whose yield strength exceeds a is strong enough and that any part whose weight is less than b is light enough. Then let A be the set of all parts whose strength exceeds a and let B be the set of parts whose weight is less than b. Then $A \cap B$ would represent the set of parts that is acceptable for use in the satellite.

(e) Complements   Suppose A is a subset of S. The *complement* of A (symbolically $\bar{A}$) is the set whose elements belong to S but not to A. Since we are concerned only with the points that do belong to S (the sample space), the set $\bar{A}$ is just those points that are not members of A. Symbolically

$$\bar{A} = \{ s : s \in S \text{ and } s \notin A \}$$

Verbally, the phrase characterizing A is changed into the phrase characterizing $\overline{A}$ by the insertion of the word "not." In the coin-tossing illustration

$\overline{A}$: "the first coin is not a head" = {Tail–Head, Tail–Tail}

Again, in testing electronic parts, if S is the set of all parts whose resistance x is between a and b, if a part is "good" when its resistance exceeds c (that is, $c < x < b$), and if A represents the set of "good" parts in S, then $\overline{A}$ represents the set of defective parts.

Although the definitions are given for two sets, we immediately determine that they extend to any finite number of sets. However, it is convenient to change notation somewhat. Thus we define

$$\bigcup_{i=1}^{n} A_i = A_1 \cup A_2 \cup A_3 \cup \cdots \cup A_n$$

$$= \{s : s \in A_1 \text{ or } s \in A_2 \text{ or } s \in A_3 \text{ or } \ldots s \in A_n\}$$

$$\bigcap_{i=1}^{n} A_i = A_1 \cap A_2 \cap A_3 \cap \cdots \cap A_n$$

$$= \{s : s \in A_1 \text{ and } s \in A_2 \text{ and } \ldots s \in A_n\}$$

Obviously these definitions can be extended to the union and intersection of an infinite number of sets.

Two particular sets require further discussion. The sample space S itself is a set. By any reasonable definition of the term *subset*, we must conclude that S is a subset of itself because otherwise we would be called on to produce an element of the left side of the inclusion $S \subset S$, which is not an element of the right side, and this is clearly impossible. Thus, $S \subset S$. Let us also postulate the existence of a set $\varnothing$, which contains no elements whatsoever (the *empty set*). Necessarily, $\varnothing$ must be a subset of *every* set, since for the formula $\varnothing \subset A$ to be false we would have to produce an element of $\varnothing$ that is not an element of A, and clearly this is impossible because $\varnothing$ has no elements. Hence, in particular, $\varnothing \subset S$.

We notice that the following obvious identities always hold true.

$$\overline{S} = \varnothing$$
$$\overline{\varnothing} = S$$
$$S \cup A = S$$
$$S \cap A = A$$
$$A \cup \overline{A} = S$$
$$A \cap \overline{A} = \varnothing$$

The statement that two sets A and B have no elements in common can be symbolized by the formula $A \cap B = \varnothing$. In this case, we say that A and B are *mutually exclusive* (or *disjoint*) sets.

Notice also that for sets of numbers $\varnothing \neq \{0\}$. The right side is a set containing one element 0 and the left side is a set containing no elements. Thus they cannot be the same set.

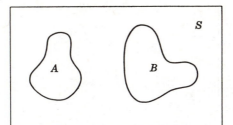

FIGURE 1.1   Venn diagram for sample
space S and two sets A and B.

1.6   Venn Diagrams

A convenient way to illustrate some set relations is to draw a Venn diagram of
the sets of interest. Thus, for a given sample space S and two subsets A and B we
have the Venn diagram shown in Figure 1.1.

In this figure the points constituting the sample space are symbolically denoted
by the ones within the rectangle. The subsets A and B are represented by certain
regions within S.

The set $A \cup B$ is represented by the set of points lying in either A or B. $A \cup B$
is represented by the shaded region in Figure 1.2. Notice that in that figure
$A \cap B = \emptyset$, that is, A and B are mutually exclusive events.

More generally, if A and B are arbitrary subsets of S that can overlap, they
might be represented by a Venn diagram like Figure 1.3. The region representing
$A \cap B$ is indicated on this diagram and $A \cup B$ is represented by the shaded
region. Again notice $A \cap B \neq \emptyset$. Here the events A and B are not mutually
exclusive.

For any set A the complement $\overline{A}$ consists of all points in S that do not lie in
A and, hence, $\overline{A}$ is represented by the unshaded region in Figure 1.4.

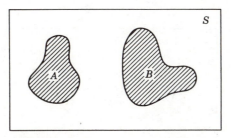

FIGURE 1.2   Venn diagram for sample
space S and the set $A \cup B$ when
$A \cap B = \emptyset$.

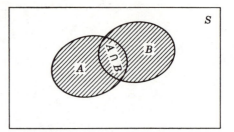

FIGURE 1.3   Venn diagram for sample space S and the set $A \cup B$ when $A \cap B \neq \emptyset$.

By reference to Figures 1.3 and 1.4, the following identities are observed to hold true for any subsets A and B of S:

$$A \cap S = A$$

$$A \cup S = S$$

$$\left(\overline{\overline{A}} \right) = A$$

$$\left(\overline{A \cup B} \right) = \overline{A} \cap \overline{B}$$

$$\left(\overline{A \cap B} \right) = \overline{A} \cup \overline{B}$$

If we wish to discuss combinations of three sets, A, B, and C, a Venn diagram like Figure 1.5 is useful.

By shading appropriate regions in a diagram like Figure 1.5, the reader is urged to check these identities:

$$A \cap B \cap C \subset A \cap B \subset A$$

$$A \subset A \cup B \subset A \cup B \cup C$$

$$A \cap (B \cup C) = (A \cap B) \cup (A \cap C)$$

$$A \cup (B \cap C) = (A \cup B) \cap (A \cup C)$$

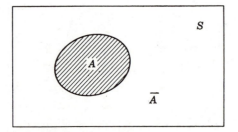

FIGURE 1.4   Venn diagram for sample space S and sets A and $\overline{A}$.

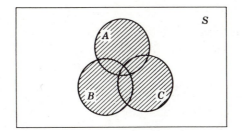

FIGURE 1.5   Venn diagram for sample space S and the sets A, B, C, and $A \cup B \cup C$.

1.7   Events

We have seen that a subset of a sample space S can be determined either by a direct listing of its elements or by a statement or formula giving the condition that elements must satisfy in order to belong to the subset. In probability terminology the term *event* is used to describe any subset that is of interest in the discussion or, equivalently, any statement of conditions that defines this subset is also termed an *event*.

For example, in the die-tossing experiment having the sample space $S = \{1, 2, 3, 4, 5, 6\}$, the subset $A = \{2, 4, 6\}$ is defined by the statement "the toss yields an even number." We use the term *event* interchangeably to describe the subset or the statement. Thus, we say that A is an event, and that the statement "the toss yields an even number" defines the same event.

Vaguely speaking, an event can be defined as a statement whose truth or falsity is determined *after* the experiment. (Not until after the die is tossed does one know whether an even number comes up.) The corresponding subset of S is simply the set of all experimental outcomes for which the statement is true.

No error will be introduced if the student interprets the term *event* simply to mean *subset of the sample space*. All subsets of S considered in this book are events.

Suppose that we consider an event to be defined in the die-tossing experiment by the statement B: "the toss yields the number 2." Thus the event B is a subset of S consisting of just one element:

$$B = \{2\}$$

We must exercise some notational care at this point. Notice, for example, that $2 \in S$ and $2 \in B$ but 2 is not B. B is a subset of S, while 2 is an *element* of S. Hence, $B \neq 2$. Clearly B is not an element of S, since the elements of S are numbers not sets of numbers.

In the coin-tossing example the sample space is $S = \{\text{Head, Tail}\}$. Events from this space are: S, $\varnothing$, $A_1 = \{\text{Head}\}$, $A_2 = \{\text{Tail}\}$.

EXAMPLE 1.4

The sequence of times between breakdowns of a certain piece of machinery in a factory is observed for a one-month period.

Here the sample space S would consist of elements (sample points) each of which is itself a finite sequence $(T_1, T_2, \ldots)$ of positive numbers, the times between breakdowns. The event B: "breakdowns occur at least 24 hours apart" would be represented by the subset of S consisting of those sequences $(T_1, T_2, \ldots)$ in S with each $T_i \geq 24$.

Exercises 1

1. A penny is tossed until the first head appears, then the experiment is stopped. Define a sample space for this experiment.

2. Two dice are rolled once. Construct a sample space whose elements involve the number of spots on each of the two dice.

3. Illustrate the following properties by means of appropriate shadings on Venn diagrams for any sets A, B, and C.

   (a) $A \cap B \subset A \subset A \cup B$.

   (b) If $A \subset B$, then $A \cap B = A$.

   (c) If $A \subset B$, then $A \cup B = B$.

   (d) If $A \cap B = \varnothing$, then $A \subset \bar{B}$.

   (e) If $A \subset B$ and $B \subset C$, then $A \subset C$.

4. The *relative complement* of a set A relative to B is usually denoted by $B \backslash A$. This new set is sometimes called the *difference set*. Formally,

$$B \backslash A = \{ x : x \in B, \ x \notin A \}$$

   Use Venn diagrams to illustrate the identities

$$B \backslash A = \bar{A} \cap B; \qquad B \backslash A \subset B; \qquad A \subset \overline{B \backslash A}; \qquad A \backslash \bar{A} = A$$

5. The *symmetric difference* of two sets A, B is defined as

$$A \vartriangle B = (A \backslash B) \cup (B \backslash A)$$

   (a) Give an interpretation for the symmetric difference in words.

   (b) Write the symmetric difference in terms of unions, intersections, and complements.

6. If $\{ A_i; \ i = 1, 2, \ldots n \}$ is a collection of subsets of S with $A_i \cap A_j = \varnothing$ for every $i \neq j$ and $\bigcup A_i = S$, then this collection is termed a *partition of S*. If $S = \{ a, b, c, d, e \}$, determine three different partitions of S.

7. Let A, B, and C be three events. Write out expressions for the following events in terms of A, B, and C, using set notation.

(a) A occurs but neither B nor C occurs.

(b) A and B occur, but not C.

(c) A or B occurs, but not C.

(d) None of A, B, and C occurs.

(e) A, B, or C occurs.

(f) Exactly one of A, B, and C occurs.

(g) Exactly two of A, B, and C occur.

(h) Two or more of A, B, and C occur.

(i) Less than two of A, B, and C occurs.

(j) Either A occurs and not B, or B occurs and not A.

8. Describe a possible sample space for each of the following experiments.

   (a) A coin is tossed four times and observed to be either a head or a tail each time.

   (b) A large lot of machine screws is known to contain a number of defectives. Three screws are chosen at random and tested.

   (c) A box of ten screws is known to contain one defective and nine good screws. Three screws are chosen at random from this box and tested.

9. An electronic system contains three components, any one of which may fail during operation. A random experiment now consists of a field test of the system. Define three events A_1, A_2, and A_3 as follows:

   $$A_i: \text{"component } i \text{ does not fail in the test,"} \qquad i = 1, 2, 3$$

   (a) Express each of the following events in words.

   $$A_1 \cup A_2, \; A_1 \cup A_2 \cup A_3, \; A_1 \cap A_2, \; A_1 \cap A_2 \cap A_3,$$
   $$\overline{A_1}, \; \overline{A_1} \cap A_3, \; \overline{A_1} \cup \overline{A_2}, \; \overline{A_1} \cap \overline{A_2}, \; \overline{A_3}, \; \overline{A_1 \cap A_2}, \; \overline{A_1 \cup A_2}$$

   (b) The test is termed "successful" if no more than one component fails. It is termed "highly successful" if no components fail. Denote these two events by B and C. Express B and C in terms of A_1, A_2, and A_3.

10. Let the sample space S consist of all real numbers and define the following events (subsets).

    $$A = \{x : x > 0\}$$
    $$B = \{x : -1 < x < 2\}$$
    $$C = \{x : x < 0\}$$

    (a) Describe and sketch on a number line (real axis) the following subsets of S.

    $$A \cup C, \; A \cap C, \; \overline{A}, \; B \cap C, \; A \cap \overline{B}, \; A \cup B$$

    (b) Express the following subsets of S in terms of A, B, and C.

    $$D = \{x : 0 < x < 2\}$$
    $$E = \{x : x = 0\}$$
    $$F = \{x : x \leq -1\}$$

11. Using a Venn diagram illustrate the following set relation. Assume that A_1, A_2, A_3, A_4, and A_5 form a *partition* of the sample space S (see Exercise 6) and that E is any event, $E \subset S$. Then,

$$(E \cap A_1) \cup (E \cap A_2) \cup (E \cap A_3) \cup (E \cap A_4) \cup (E \cap A_5) = E$$

12. Consider the series electrical circuit shown below.

Suppose that components R_1 and R_2 can be either Good or have Failed (G or F). Let the sample space S consist of the four possible combinations of Good and Failed components, $S = \{(G,G), (G,F), (F,G), (F,F)\}$. Let E_1 be the event "the entire circuit is operative" and E_2 be the event "at least one of R_1, R_2 has failed." Are E_1 and E_2 mutually exclusive? Interpret $\overline{E}_1$ and $\overline{E}_2$. What can we say about the relation between $\overline{E}_1$ and E_2?

13. Consider the following series-parallel assembly consisting of five components.

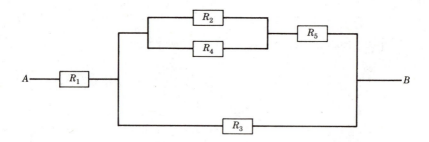

Each component is either good or fails under load. However, the entire assembly fails only if the path from A to B is broken. Let the sample space S consist of the 32 possible combinations of good or failed components: $S = \{(G,G,G,G,G),\ldots,(F,F,F,F,F)\}$ where, for example, (G,F,G,F,F) means components R_1 and R_3 are good and components R_2, R_4, and R_5 have failed. Let E_1 be the event "the assembly is operative," let E_2 be the event "R_2 has failed but the assembly is operative," and let E_3 be the event "R_3 has failed but the assembly is operative."

(a) List the elements of S, E_1, E_2, and E_3.

(b) Are the events E_1 and E_2 mutually exclusive?
   Are the events E_1 and E_3 mutually exclusive?
   Are the events E_2 and E_3 mutually exclusive?

14. If a and b are numbers, $a \leq b$, then the set of all real numbers between a and b is termed an *interval*. Different kinds of intervals are defined depending on

whether the end points are included in the set or not. The following notation and terminology are standard.

$$[a, b] = \{x : a \le x \le b\}, \qquad \text{the } \textit{closed} \text{ interval from } a \text{ to } b$$

$$(a, b) = \{x : a < x < b\}, \qquad \text{the } \textit{open} \text{ interval from } a \text{ to } b$$

$$[a, b) = \{x : a \le x < b\}, \qquad \text{the } \textit{left-closed} \text{ interval from } a \text{ to } b$$

$$(a, b] = \{x : a < x \le b\}, \qquad \text{the } \textit{right-closed} \text{ interval from } a \text{ to } b$$

(i) Let

$$S = [0, 100], \qquad A = [10, 100], \qquad B = [5, 10), \qquad C = [0, 50],$$
$$D = [0, 25), \qquad E = (90, 100]$$

and describe the following sets using interval notation wherever possible. Sketch each on a number line (real axis).

(a) $A \cup B$ (f) $\tilde{A} \cap B$

(b) $A \cap C$ (g) $A \cap \bar{B}$

(c) $C \cap D$ (h) $\overline{A \cap B}$

(d) $A \cup E$ (i) $[(C \cap \bar{D}) \cap D] \cap E$

(e) $A \cap B$ (j) $B \cup E$

(ii) If S is the interval $(0, 1)$ give a partition of S into three intervals (see Exercise 6).

2 · Probability

2.1 Introduction

In the previous chapter we learned that the physical behavior of random experiments can be modeled rather naturally using concepts such as sets, subsets, and classes of sets. In addition, we would like to assign numbers to events that, in some sense, tell us what the chances are of a particular event occurring. To this end, we must assign some measure for events termed the *probability of the event A* and symbolized by Pr[A].

Since much of the earliest development of the theory of probability arose in games of chance, it seemed, at least historically, quite natural to interpret the *probability of the event A* in the following sense.

Consider a sequence of repetitions of the same experiment under identical conditions. Let f_n denote the number of occurrences of the event A in the first n repetitions of the experiment. The ratio f_n/n then gives the *proportion* of "successes" among the first n trials. For example, if the experiment consists of tossing a coin and if the event A corresponds to "heads," then f_n/n gives the fraction of heads in the first n tosses. Intuitively, we feel that as n increases the ratio f_n/n should stabilize and approach some fixed number that would measure the likelihood of the event A occurring. Thus, we would like to assign probabilities to events in our model in such a way that

$$\Pr[A] = \lim_{n \to \infty} \frac{f_n}{n} \tag{2.1}$$

Of course the existence of the limit in (2.1) could never be checked in any real situation because no *infinite* sequence of real experiments is possible. Nonetheless, it is intuitively so attractive that we would hope to have a mathematical model in which this formula holds true.

In repeated tosses of an "honest" coin we expect the proportion of heads to approach $\frac{1}{2}$, so that by (2.1) we assign the value $\frac{1}{2}$ to Pr["Heads"]. In fact, we could *define* an "honest" coin as one in which $\lim_{n \to \infty} (f_n/n) = \frac{1}{2}$. Figure 2.1 gives one possible graph of f_n/n for n up to 100 tosses of a supposedly honest quarter. Although this graph neither proves the existence of the limit of f_n/n nor guarantees a numerical value of $\frac{1}{2}$ for this limit if it exists, it does illustrate the behavior that we expect from a relative frequency sequence.

Clearly, conceptually at least, we could model repetitive experiments such as coin tossings, card playing, and games of chance in the sense described above. Furthermore, we could model probabilities for more complicated random experi-

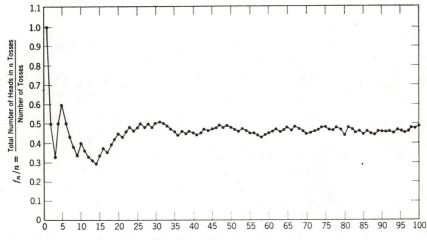

FIGURE 2.1 Proportion of heads in 100 tosses of a quarter.

ments in this sense, for example, the sales campaign mentioned in Chapter 1. By taking some liberty with existing measuring instruments we could use the foregoing approach to probability to model probabilities for the manufacturing operation that measures the strength of a part. Hence, for a considerable amount of engineering and scientific application, this *relative frequency* sense of probability serves to give an adequate model of the probability measure. That is, for many engineering problems we can imagine repeating an experiment many times and can imagine measuring the number of "successes." Therefore, any mathematical construction of a probability measure must allow us to interpret the probability in this relative frequency sense.

Although a relative frequency approach to probability is satisfactory for many engineering applications and games of chance, such an approach is inadequate for many uses. The mathematical construction of a probability measure must be independent of any envisioned application but must be usable for many applications. To this end the modern theory of probability begins with the construction of a set of *axioms of probability* that, we will observe, do not refer to how the probabilities are actually obtained. How we determine the actual probability of a given event depends strongly on the random experiment that is conducted, and, hence, the determination of a real probability is an extramathematical problem. The theory of probability properly begins with the assumption that probabilities can be assigned that satisfy the following *axioms of probability*. Whether, in a particular case, these probabilities are determined by repetition of a random experiment or are based on personal judgment, on communion with nature, or on a visitation with the local soothsayer is irrelevant to the study of probability theory.

2.2 Axioms of Probability

If probabilities are to be assigned to events in a way consistent with (2.1), certain conditions are obviously necessary.

(a) Since $0 \leq f_n/n \leq 1$ for every n, then we should require that $0 \leq \Pr[A] \leq 1$ for every A.

(b) If $A = S$, then $f_n = n$ and $f_n/n = 1$. Thus, events certain to occur should have probability 1.

(c) Suppose that A and B are mutually exclusive events, and let $f_n(A)$, $f_n(B)$, and $f_n(A \cup B)$ be the number of occurrences of the events A, B, and $A \cup B$, respectively, in the first n trials. Clearly $f_n(A \cup B) = f_n(A) + f_n(B)$ and, thus, $f_n(A \cup B)/n = f_n(A)/n + f_n(B)/n$. Hence we would expect $\Pr[A \cup B]$ to be the sum of $\Pr[A]$ and $\Pr[B]$.

In view of these properties, we postulate that each event, A, is assigned a number, $\Pr[A]$, in such a way that the following *axioms* are satisfied.

(i) $0 \leq \Pr[A] \leq 1$.

(ii) $\Pr[S] = 1$, where $S =$ the entire sample space.

(iii) If $A_1, A_2, A_3, \ldots$ is a sequence (finite or infinite) of *mutually exclusive* events ($A_i \cap A_j = \varnothing$ for $i \neq j$), and if

$$B = A_1 \cup A_2 \cup A_3 \cdots$$

is the event "at least one of the A_i's occurs," then

$$\Pr[B] = \Pr[A_1] + \Pr[A_2] + \Pr[A_3] + \cdots$$
$$= \sum_i \Pr[A_i]$$

Certainly these axioms are consistent with our intuitive ideas of how probabilities behave. Furthermore, we can show that if these axioms are satisfied formula (2.1) holds true, with appropriate definitions of the meaning of the terms "experiments repeated under identical conditions" and "converges" (see Section 6.11).

To avoid certain mathematical difficulties we must place some restrictions on the types of subsets of the sample space that we term "events" and for which we define probabilities. These restrictions do not affect any part of our discussion. For our purposes every subset constructed here can be considered as an event having a probability.

A random experiment with its associated sample space, events, and probability assignment is termed a *probability system*.

EXAMPLE 2.1

Let us return to Example 1.1 to see how we can build a probability system for the problem discussed there. In Example 1.1 where we were flipping a penny once we can construct four events very simply. $A_1 = S$, $A_2 = \varnothing$, $A_3 = \{\text{Heads}\}$, $A_4 = \{\text{Tails}\}$. Then by our axioms of probability we must have

$$\Pr[S] = 1$$

Since

$$\{\text{Heads}\} \cup \{\text{Tails}\} = S$$

and

$$\{\text{Heads}\} \cap \{\text{Tails}\} = \varnothing$$

we also must have

$$1 = \Pr[\{\text{Heads}\} \cup \{\text{Tails}\}] = \Pr[\{\text{Heads}\}] + \Pr[\{\text{Tails}\}]$$

Since $\{\text{Heads}\} \cap \{\text{Tails}\} = \varnothing$, these two events are mutually exclusive. Thus

$$\Pr[\{\text{Tails}\}] = 1 - \Pr[\{\text{Heads}\}]$$

But what should we choose for Pr[{Heads}]? Here probability theory itself is of no help. We must go outside of probability theory to get a number for this probability. If we think the coin is evenly weighted and will be flipped in such a manner as to give heads and tails equal chances of occurring we might choose

$$\Pr[\{\text{Heads}\}] = 1/2$$

and hence

$$\Pr[\{\text{Tails}\}] = 1/2$$

In scientific or engineering work one might think of performing an experiment several times and then using a relative frequency definition of probability. In a sense this is a "good" way to get an *estimate* of a probability, however, it is not the only way. In some cases one uses educated guesses or one might provide a probability based on geometric reasoning or a counting method under an assumption that all events are "equally likely." We will spend quite a bit of time doing such computations.

2.3 Elementary Properties of Probability

A number of properties of probability follow directly from axioms (i), (ii), and (iii).

(a) For any event A, $\Pr[\overline{A}] = 1 - \Pr[A]$.

PROOF A and $\overline{A}$ clearly are mutually exclusive and $S = A \cup \overline{A}$. Hence, by axioms (ii) and (iii),

$$1 = \Pr[S] = \Pr[A] + \Pr[\overline{A}]$$

from which the assertion follows.

(b) If $\varnothing$ is the impossible event, then $\Pr[\varnothing] = 0$.

PROOF   Since $\varnothing \cup \varnothing = \varnothing$ and $\varnothing = \varnothing \cap \varnothing$, it follows from axiom (iii) that $\Pr[\varnothing] = \Pr[\varnothing] + \Pr[\varnothing] = 2\Pr[\varnothing]$. Hence, $\Pr[\varnothing] = 0$.

There is an important and sometimes confusing point that we must eliminate. The empty set $\varnothing$ as shown above must be assigned a probability of 0 to be consistent with our axioms of probability. We call $\varnothing$ the *impossible event*. One must be careful not to reverse this and assume that since $\varnothing$ has probability 0 then any event A with probability 0 is "impossible." We suppose that what one has in mind by "impossible" is that it can never happen, but one must be wary of the logic here. The empty set $\varnothing$ has probability 0 and is called the impossible event. That does not mean that an event A that has probability 0 necessarily is "impossible" or equal to $\varnothing$. We will see examples later.

(c)   If A_1 and A_2 are any events, not necessarily mutually exclusive, then
$$\Pr[A_1 \cup A_2] = \Pr[A_1] + \Pr[A_2] - \Pr[A_1 \cap A_2]$$

PROOF   Clearly (see Figure 2.2)
$$A_1 \cup A_2 = A_1 \cup (\bar{A_1} \cap A_2) \qquad \text{and} \qquad A_2 = (A_1 \cap A_2) \cup (\bar{A_1} \cap A_2)$$
where in each equation the unions on the right side are of mutually exclusive events.

By axiom (iii)
$$\Pr[A_1 \cup A_2] = \Pr[A_1] + \Pr[\bar{A_1} \cap A_2]$$
$$\Pr[A_2] = \Pr[A_1 \cap A_2] + \Pr[\bar{A_1} \cap A_2]$$
The second equation gives $\Pr[\bar{A_1} \cap A_2] = \Pr[A_2] - \Pr[A_1 \cap A_2]$, which, when substituted into the first equation, yields the given assertion.

The formula in (c) can be generalized as follows.

(d)   If $A_1, A_2, \ldots, A_k$ are any events, then
$$\Pr\left[\bigcup_{i=1}^{k} A_i\right] = \Pr[A_1 \cup A_2 \cup \cdots \cup A_k] = \sum_i \Pr[A_i] - \sum_{i<j} \Pr[A_i \cap A_j]$$
$$+ \cdots + (-1)^{k+1}\Pr[A_1 \cap A_2 \cap \cdots \cap A_k]$$

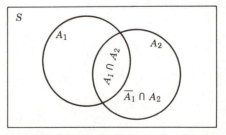

FIGURE 2.2   Venn diagram for the set relation $A_1 \cup A_2 = A_1 \cup (\bar{A_1} \cap A_2)$.

where the successive sums are over all possible events, pairs of events, triples of events, et cetera.

PROOF   The proof follows from (c) by induction. The details are omitted. A student familiar with induction arguments should have no difficulty adding them.

2.4   Discrete Sample Spaces; Equally Likely Sample Points

A probability system is particularly simple when the sample space consists of only a finite number N of possible experimental outcomes, sample points, or *elementary events*. To determine the probability of all events in this case we only need to know the probabilities $p_1, p_2, \ldots, p_N$ of the elementary events, $\sum p_i = 1$. Since any event A consists of a certain collection of these sample points, $\Pr[A]$ can be found, using axiom (iii), by adding up the probabilities of the separate sample points that make up A.

EXAMPLE 2.2
Three subjects in a psychological learning experiment are each given a sequence of digits to remember for a fixed time. The probabilities that these subjects will remember correctly are .4, .6, and .8, respectively. Now consider the experiment of giving the sequence of digits and of determining how many of the subjects remember them correctly after the fixed time. Thus, the sample space for this experiment contains four sample points: 0 subjects remember correctly, 1 remembers correctly, 2 remember correctly, and 3 remember correctly. The probabilities that can be assigned to these elementary events are: .048, .296, .464, and .192, respectively (see Exercise 6). The event "at least one subject remembers incorrectly" = {0 correct, 1 correct, 2 correct} thus will have probability $.048 + .296 + .464 = .808$. The event "no more than one remembers incorrectly" = {2 correct, 3 correct} will have probability $.464 + .192 = .656$.

The situation is even simpler when the N individual sample points all have equal probabilities (*equally likely sample points*),

$$p_1 = p_2 = \cdots = p_N = \frac{1}{N}$$

Calculation of probabilities in this case reduces simply to counting. If the event A consists of k sample points, then

$$\Pr[A] = \frac{1}{N} + \frac{1}{N} + \cdots + \frac{1}{N} = \frac{k}{N}$$

EXAMPLE 2.3
A group of four radios consists of two good radios, labeled g_1 and g_2, and two defective radios, labeled d_1 and d_2. If three radios are selected at random from

this group, what is the probability of the event A: "two of the three selected are defective"?

By direct enumeration we observe that there are four ways of selecting three radios (if order is disregarded): .

$$g_1 g_2 d_1, \qquad g_1 g_2 d_2, \qquad g_1 d_1 d_2, \qquad g_2 d_1 d_2$$

The phrase "selected at random" can be interpreted here to mean that each of these four sample points is equally likely and, therefore, each has probability $\frac{1}{4}$. Since the event A can occur in two ways, $A = \{ g_1 d_1 d_2 \} \cup \{ g_2 d_1 d_2 \}$, we have

$$\Pr[A] = \tfrac{2}{4} = \tfrac{1}{2}$$

Systems with equally likely sample points occur frequently when we discuss various games of chance. Because of their simple structure, the discussion of these systems occupies much space in many elementary works on probability. Unfortunately, most applications of interest in engineering, business, or science do not fall into this category.

The discussion of finite sample spaces immediately extends to random experiments that have an infinite sequence of possible sample points (*countable sample spaces*). Again, only the probabilities of the elementary events need be given; the probabilities of other events can be obtained by addition. Of course the equally likely case cannot occur here, since the sum of infinitely many equal numbers cannot add up to one.

EXAMPLE 2.4

Consider the experiment of tossing an honest coin repeatedly and of counting the number of tosses required until the first "head" occurs. Clearly the terms of the sequence $1, 2, 3, \ldots$ are all possibilities for the number of tosses required. Suppose that the corresponding probabilities are $p_1, p_2, p_3 \ldots$. That is,

$$p_i = \Pr[\text{first "head" occurs on the } i\text{th toss}], \qquad \text{for } i = 1, 2, 3, \ldots$$

Since there are 2^i equally likely ways of tossing an honest coin i times, only one of which consists of $i - 1$ "tails" followed by a "head," we observe that

$$p_i = 1/2^i, \qquad i = 1, 2, 3, \ldots$$

As a check on this formula, notice that

$$\sum_{i=1}^{\infty} p_i = \sum_{i=1}^{\infty} 1/2^i = \frac{1/2}{1 - 1/2} = 1$$

by using the formula for the sum of a geometric series (Appendix A.3).

If A is the event "the first head occurs on an even-numbered toss," then

$$\Pr[A] = p_2 + p_4 + p_6 + \cdots$$
$$= \sum_{k=1}^{\infty} 1/2^{2k} = \frac{1/4}{1 - 1/4} = \frac{1}{3}$$

2.5 Independent Events

Axiom (iii) for a probability system stated that for *mutually exclusive* events there is a connection between *unions* of events and *sums* of probabilities, that is, for such events the probability of their union is the sum of the probabilities of the events. For certain kinds of events a similar connection holds between their *intersection* and the *product* of their probabilities.

We *define* two events, A and B, to be *independent* (or *statistically independent*) if

$$\Pr[A \cap B] = \Pr[A]\Pr[B] \tag{2.2}$$

It will turn out (Section 2.6) that this property is equivalent to asserting that if one knows that one of the events has indeed occurred, then the probability of the other occurring is unchanged.

Important cases exist where knowledge of the occurrence of an event B does *not* change the probability of an event A, and this fact is important in a discussion of the two events. If a coin is tossed twice, we would reasonably suppose that the probability of the event "a head on the second toss" would not be influenced if we knew that the event "a head on the first toss" had occurred. In life-testing of sets of transistors it seems reasonable to suppose that the lifetime of a transistor in no way depends on the lifetime of any other transistor.

EXAMPLE 2.5

An electronic assembly consists of a transistor and a condenser. The transistor is selected from a lot of 100, 10 of which are defective, and the condenser is selected from a lot of 300, 15 of which are defective. Let A be the event "the transistor part of the assembly is defective," ($\Pr[A] = 10/100 = 0.10$), and let B be the event "the condenser part of the assembly is defective," ($\Pr[B] = 15/300 = 0.05$). Since the choices are from different lots, it seems reasonable to expect that knowledge of whether one of the events has occurred will not affect the other, that is, that the events are independent. This can be checked. There are a total of $100 \cdot 300$ possible assemblies, of which $10 \cdot 15$ have both parts defective. Thus

$$\Pr[A \cap B] = \frac{10 \cdot 15}{100 \cdot 300} = 0.005 = (.10)(.05) = \Pr[A]\Pr[B]$$

and, thus, A and B satisfy (2.2) and *are* independent.

By using the concept of independence, we are able to construct many complicated probabilities from simpler ones.

The preceding definition extends to sets of more than two events by generalizing formula (2.2). For k events $A_1, A_2, \ldots, A_k$ to be *independent*, we require not only that

$$\Pr[A_1 \cap A_2 \cap \cdots \cap A_k] = \Pr[A_1]\Pr[A_2] \cdots \Pr[A_k] \tag{2.3}$$

but also that a similar multiplicative formula holds true for every subcollection of the k events.

If events $A_1, A_2, \ldots, A_k$ are such that every pair is independent, then they are said to be *pairwise independent*. It does not follow in this case that the entire set is independent.

EXAMPLE 2.6

Suppose two honest coins are tossed. Let A be the event "the first coin is a head," let B be the event "the second coin is a head," and let C be the event "the coins match" (both heads or both tails). Assume A and B are independent events, and also A and C, as well as B and C, are independent events. (Why?) The events A, B, and C are *pairwise independent*. But observe that $\Pr[A \cap B \cap C] = \frac{1}{4}$, while

$$\Pr[A]\Pr[B]\Pr[C] = \left(\tfrac{1}{2}\right)\left(\tfrac{1}{2}\right)\left(\tfrac{1}{2}\right) = \tfrac{1}{8} \neq \Pr[A \cap B \cap C]$$

Thus the events A, B, and C are not *independent*, although they are *pairwise independent*.

One obvious and useful result of independence is the following.

Theorem

If events A and B are independent, then so are events A and $\overline{B}$, events $\overline{A}$ and B, and events $\overline{A}$ and $\overline{B}$.

PROOF $A \cap B$ and $A \cap \overline{B}$ are mutually exclusive events whose union is A. Hence

$$\Pr[A] = \Pr[A \cap B] + \Pr[A \cap \overline{B}]$$
$$= \Pr[A]\Pr[B] + \Pr[A \cap \overline{B}]$$

since A and B are independent. It follows that

$$\Pr[A \cap \overline{B}] = \Pr[A](1 - \Pr[B]) = \Pr[A]\Pr[\overline{B}]$$

and the independence of A and $\overline{B}$ follows from (2.2).

The independence of $\overline{A}$ and $\overline{B}$ follows immediately if the events A and B are replaced by $\overline{B}$ and A, while the independence of $\overline{A}$ and B follows by interchanging the roles of A and B.

This theorem generalizes in obvious fashion to sets of more than two events.

2.6 Conditional Probability

In many situations we are concerned with two (or more) events whose occurrences are in some way connected. Roughly speaking, we wish to consider a situation where knowledge of the occurrence of one event makes the occurrence of another event more (or less) likely than if this information were not known. For example, consider the experiment of choosing a word at random in an English dictionary. Let A be the event "the letter u occurs in the word" and let B be the event "the letter q occurs in the word." Certainly event A has a probability but if we know that B occurs we are much surer that A must occur too, since q rarely occurs in a word without being followed by u. To consider another example, if buses are known to pass a certain corner at approximately 10-minute intervals, the probability of a bus passing in the next minute would be strongly influenced by the knowledge of whether a bus had passed during the past 5 minutes.

Thus, if A and B are events, we wish to define a quantity termed the *conditional probability of the event A given that the event B occurs*, or the *conditional probability of A given B*, in symbols

$$\Pr[A|B]$$

Various interpretations of the situation described here are possible. We might consider the experiment to be only partially completed, sufficiently so for the occurrence of B to be known but not far enough for the occurrence of A to be determined. The above example of the bus passing a corner is of this type. Another interpretation is simply to ignore all experiments in which B fails and to only consider the ones for which B succeeds. Still another interpretation is to assume that the occurrence of A cannot be directly observed even after the experiment, although B can be directly observed. This would be the case if the experiment consisted of manufacturing an item, B referred to a certain test measurement on the item, and A referred to the existence of some flaw that could only be shown with 100% accuracy by destructive testing.

Since, knowing that B occurs, A can occur only in conjunction with B, it seems reasonable to define the conditional probability as in some way proportional to $\Pr[A \cap B]$. Furthermore, by any reasonable definition we should have $\Pr[B|B] = 1$ (the conditional probability of B occurring given that B occurs should be 1). Therefore, it seems reasonable to define $\Pr[A|B]$ to be the relative "size" of the event $A \cap B$ relative to the "size" of the event B, with "size" measured by "probability." Thus we make the following *definition*:

$$\Pr[A|B] = \frac{\Pr[A \cap B]}{\Pr[B]} \tag{2.4}$$

This definition only makes sense in the case $\Pr[B] \neq 0$. If $\Pr[B] = 0$, we leave $\Pr[A|B]$ undefined.

Formula (2.4) can also be written in the form

$$\Pr[A \cap B] = \Pr[A|B]\Pr[B] \tag{2.5}$$

Notice here that if A and B are *independent*

$$\Pr[A \cap B] = \Pr[A]\Pr[B]$$

by (2.2). Thus

$$\Pr[A|B] = \Pr[A] \tag{2.6}$$

whenever A and B are independent. This justifies the interpretation in our discussion of independence in Section 2.5 that knowledge of the occurrence of B does not affect the probability of the occurrence of A.

EXAMPLE 2.7

A company owns two factories that produce similar items. Factory 1 produces 1000 items, 100 of which are defective. Factory 2 produces 4000 items, 200 of which are defective. An item is chosen at random from the production of the company and found to be defective. What is the probability that it came from factory 1?

Let B be the event "the item chosen is defective" and let A be the event "the item chosen came from factory 1." Thus we wish to find $\Pr[A|B]$.

We can reason directly as follows. There are a total of 300 defective items, 100 of them from factory 1; hence, the probability that a defective comes from factory 1 is $100/300 = \frac{1}{3}$.

If we apply definition (2.4) we find that $\Pr[A \cap B] = \Pr[\text{item chosen is defective and from factory 1}] = 100/5000$, and that $\Pr[B] = \Pr[\text{item chosen is defective}] = 300/5000$; therefore, by (2.4)

$$\Pr[A|B] = \frac{100/5000}{300/5000} = \frac{100}{300} = \frac{1}{3}$$

in agreement with our direct reasoning.

This type of example assures us that (2.4) is a *reasonable* definition of conditional probability, in accord with what our intuition tells us this phrase should mean.

EXAMPLE 2.8

A box contains two pennies—one honest and one two-headed. A penny is chosen at random, tossed, and observed to come up heads. What is the probability that the other side is also a head?

We wish to compute $\Pr[A|B]$ where B is the event "the coin comes up heads" and A is the event "the coin is two-headed." We easily determine that $\Pr[B] = \frac{3}{4}$ and $\Pr[A \cap B] = \frac{2}{4}$. (Of the four sides that can come up, three are heads, and two are heads with the two-headed penny.) Thus, by (2.4)

$$\Pr[A|B] = \frac{2/4}{3/4} = \frac{2}{3}$$

On first thought a person might reason that, since each coin is known to have at least one head, the toss gives no information and, therefore, that the penny has a 50–50 chance of being either of the given coins and that the probability of a head on the other side is $\frac{1}{2}$. The student should convince himself of the fallaciousness of this reasoning.

Formulas (2.4) and (2.5) can be generalized as follows. If A_i, $i = 1, 2, 3, \ldots, n$, are n events, then the conditional probability of A_n, given that $A_1, A_2, \ldots, A_{n-1}$ occur, is given by

$$\Pr[A_n|A_1, A_2, A_3, \ldots A_{n-1}] = \frac{\Pr[A_1 \cap A_2 \cap \cdots \cap A_{n-1} \cap A_n]}{\Pr[A_1 \cap A_2 \cap \cdots \cap A_{n-1}]} \quad (2.7)$$

It follows directly that

$$\Pr[A_1 \cap A_2 \cap A_3 \cap \cdots \cap A_n] = \Pr[A_1]\Pr[A_2|A_1]\Pr[A_3|A_1 \cap A_2]$$
$$\cdots \Pr[A_n|A_1 \cap A_2 \cap \cdots \cap A_{n-1}] \quad (2.8)$$

An interesting and useful result occurs when the sequence of events has the property that each depends only on the preceding one, that is, that A_{j+1} depends on A_j but, once A_j is given, A_{j+1} does not depend on $A_{j-1}, A_{j-2} \ldots A_1$. Then the right-hand side of (2.8) can be simplified under this assumption to

$$\Pr[A_1 \cap A_2 \cap \cdots \cap A_n] = \Pr[A_1]\Pr[A_2|A_1]\Pr[A_3|A_2] \ldots \Pr[A_n|A_{n-1}] \quad (2.9)$$

Now one must be careful here. Formula (2.9) is not always true, while (2.8) is. As we shall see, starting in Chapter 9, this seemingly strong assumption is often realized in many real-life applications and forms a useful assumption for modeling very diverse behavior.

To see one way that this assumption can be made plausible one can consider a very simple structure that is used in combat models.

EXAMPLE 2.9

We are interested in the probability of destroying a target. If we are to destroy the target several consecutive things must occur. First, there must be a target available to us. Second, we must detect that target. Third, we must fire at the target. Fourth, we must destroy the target with our fire. All of these things must occur if the target is to be destroyed.

> Let A_1 be the event "target present,"
>
> A_2 be the event "target detected,"
>
> A_3 be the event "target fired on,"
>
> A_4 be the event "target destroyed."

Since clearly each of these events is included in the preceding one we can write

$$A_4 = A_1 \cap A_2 \cap A_3 \cap A_4$$

So

$$\Pr[A_4] = \Pr[A_1 \cap A_2 \cap A_3 \cap A_4]$$
$$= \Pr[A_1]\Pr[A_2|A_1]\Pr[A_3|A_1 \cap A_2]\Pr[A_4|A_1 \cap A_2 \cap A_3]$$

The argument now proceeds as follows. A_2 clearly depends on A_1 because we assume the probability of detection is zero unless the target is present and is nonzero if the target is present. So A_1 and A_2 are not independent events. However, if we know the target has been detected then A_3 depends only on A_2. (We assume we do not fire on undetected targets and we may fire on detected ones.) Therefore, the probability of firing depends only on whether the target is detected or not. So

$$\Pr[A_3|A_1 \cap A_2] = \Pr[A_3|A_2]$$

Another way of saying this is that all of the necessary information about whether the target is present or not that is needed to determine whether we fire or not is included in what we know about A_2. Given A_2, the added information (if there is any) included in A_1 is of no value in determining whether or not to fire (A_3). By analogous reasoning

$$\Pr[A_4|A_1 \cap A_2 \cap A_3] = \Pr[A_4|A_3]$$

Therefore, the probability of destroying the target is simply

$$\Pr[A_4] = \Pr[A_1 \cap A_2 \cap A_3 \cap A_4]$$
$$= \Pr[A_1]\Pr[A_2|A_1]\Pr[A_3|A_2]\Pr[A_4|A_3]$$

$\Pr[A_2|A_1]$ is determined by the characteristic of the detecting devices. If one is using eyesight this probability might be small. If one is using sounding devices or more sophisticated detection devices (e.g., sonar or radar) this probability might be larger.

$\Pr[A_3|A_2]$ can depend on many conditions. If ammunition is low we may not fire even at a detected target. If the target is not exposed to our weapons (we do not have a line-of-sight) we may not fire. Weapons take time to aim at a target. (How long depends on the weapon.) We may not fire at a detected target simply because we do not have time to aim the weapon.

$\Pr[A_4|A_3]$ also depends on many things. It depends, obviously, on the weapon being used, the target being fired at, evasive tactics of the target, point of hit, and the like.

In the study of battlefield tactics, weapon systems evaluation, et cetera, considerable time is spent trying to estimate conditional probabilities as above. Some of these probabilities can be computed from a relative frequency estimate but some of them, especially in a design phase of development, must be based on the best estimate of the designer.

2.7 Total Probability and Bayes' Formula

The concept of conditional probability leads us to another formula that is very useful for our subsequent work.

Suppose that $B_1, B_2, B_3, \ldots$ is a sequence of *mutually exclusive* and *exhaustive* events, that is, no two of the B_i's can occur at the same time, and one of them must occur. Such a collection of sets is a *partition* of S. Let A be any event. If A

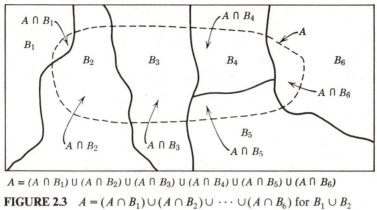

$$A = (A \cap B_1) \cup (A \cap B_2) \cup (A \cap B_3) \cup (A \cap B_4) \cup (A \cap B_5) \cup (A \cap B_6)$$

FIGURE 2.3 $A = (A \cap B_1) \cup (A \cap B_2) \cup \cdots \cup (A \cap B_6)$ for $B_1 \cup B_2$
$\cup \cdots \cup B_6 = S$ and $B_i \cap B_j = \varnothing$, for i, j.

occurs, then it must occur in conjunction with one of the B_i's. This statement implies the formula (see Figure 2.3)

$$A = [A \cap B_1] \cup [A \cap B_2] \cup [A \cap B_3] \cup \cdots$$

By axiom (iii) of Section 2

$$\Pr[A] = \Pr[A \cap B_1] + \Pr[A \cap B_2] + \Pr[A \cap B_3] + \cdots$$

and by formula (2.5)

$$\Pr[A] = \Pr[A|B_1]\Pr[B_1] + \Pr[A|B_2]\Pr[B_2] + \cdots$$
$$\Pr[A] = \sum_i \Pr[A|B_i]\Pr[B_i] \tag{2.10}$$

This formula is called the *total probability* formula. It will be quite useful.

EXAMPLE 2.10

A company producing transistor radios has three assembly plants producing 15%, 35%, and 50%, respectively, of its output. Suppose the probabilities that a radio produced by these plants is defective are .01, .05, and .02, respectively. If a radio is chosen at random from the output of the company, what is the probability that it is defective?

By formula (2.10)

$$\Pr[\text{defective}] = \sum_{i=1}^{3} \Pr[\text{defective}|\text{from plant } i] \, \Pr[\text{from plant } i]$$
$$= (.01)(.15) + (.05)(.35) + (.02)(.50)$$
$$= .029$$

Let us go back to the general situation and suppose that after the experiment we observe that the event A actually occurs, but that it is not known directly which of the mutually exclusive and exhaustive events $B_1, B_2, B_3, \ldots$ holds true. We then might ask for the conditional probability, $\Pr[B_j|A]$, that one of these events B_j occurs, given that A occurs. By using (2.10) and (2.5), we find that

$$\Pr[B_j|A] = \frac{\Pr[B_j \cap A]}{\Pr[A]}$$

$$= \frac{\Pr[A|B_j]\Pr[B_j]}{\sum\limits_{i} \Pr[A|B_i]\Pr[B_i]} \tag{2.11}$$

Formula (2.11) is termed *Bayes' formula* and is useful in many applications.

Referring to the preceding example, let us assume that a radio is chosen at random and is found to be defective. What then is the probability that it came from plant number 2?

By (2.11)

$\Pr[\text{from plant no. 2}|\text{defective}]$

$$= \frac{\Pr[\text{defective}|\text{from plant no. 2}] \cdot \Pr[\text{from plant no. 2}]}{\Pr[\text{defective}]}$$

$$= \frac{(.05)(.35)}{(.029)}$$

$$= .603$$

Thus the knowledge that the radio is defective increases the chance that it came from plant no. 2 from 35% to more than 60%.

2.8   Further Examples

EXAMPLE 2.11

Suppose that $(100p)\%$ of the output of a manufacturing process is defective. What is the probability that three randomly chosen parts will have no more than one defective among them?

The probability of no defectives among the three is observed to be

$$(1-p)^3$$

provided we assume that occurrences of defectives do not depend on defectives appearing in previously chosen parts (that is, the events "the ith chosen part is defective" are independent), since

Pr[no defectives] = Pr[part 1 is good]Pr[part 2 is good]Pr[part 3 is good]

$$= (1 - p)^3$$

Similarly the probability that exactly one is defective is $3p(1 - p)^2$. [1]

It follows that the probability of no more than one defective is

$$(1 - p)^3 + 3p(1 - p)^2$$

EXAMPLE 2.12

In Example 2.11 the mutually exclusive cases that make up the required event could be enumerated directly. In many situations this is not so and other methods must be considered. There are many ways of doing some of these counts, but the following argument is direct up to a point (and beyond if one has the proper background), easy to generalize, and quite easy to visualize. To set up our arguments reconsider Example 2.11 now with n parts randomly chosen. Suppose we want to find the probability of exactly k defective parts among these n. Let us define the event

$E_{n,k}$ = there are exactly k defects among n randomly chosen parts

How could this event come about? Well, of course, there are many ways but if someone told us the number of defects found after looking at the first $n - 1$ parts, then finding the probability of $E_{n,k}$ would be easy. For if we set $q = 1 - p$, ($p + q = 1$), then

$$Pr[E_{n,k}|E_{n-1,k}] = q$$
$$Pr[E_{n,k}|E_{n-1,k-1}] = p$$

(Any other conditional probabilities, e.g., $Pr[E_{n,k}|E_{n-1,k-2}]$, would be zero since we have only one more item to choose.) By using our conditional probability formula (2.5) and formula (2.10) we can write immediately

$$Pr[E_{n,k}] = p \, Pr[E_{n-1,k-1}] + q \, Pr[E_{n-1,k}] \qquad (2.12)$$

This equation is called a *difference equation*. It relates an unknown function (the probabilities) with argument n, k to the same function but with arguments

---

[1] This fact could be derived formally by using the notations and theory of this section as follows. Let $\bar{A}_i$ be the event "part i is good" and, thus, A_i is the event "part i is defective." Then the probability of exactly one defect is

$$Pr[(A_1 \cap \bar{A}_2 \cap \bar{A}_3) \cup (\bar{A}_1 \cap A_2 \cap \bar{A}_3) \cup (\bar{A}_1 \cap \bar{A}_2 \cap A_3)]$$

Clearly the events $[A_1 \cap \bar{A}_2 \cap \bar{A}_3]$, $[\bar{A}_1 \cap A_2 \cap \bar{A}_3]$, and $[\bar{A}_1 \cap \bar{A}_2 \cap A_3]$ are mutually exclusive. Also the events

$$A_1, A_2, A_3$$

are mutually independent. Hence the probability of exactly one defect is

$$p(1 - p)^2 + (1 - p)p(1 - p) + (1 - p)^2 p = 3p(1 - p)^2$$

using the theorem at the end of Section 2.5.

$n-1, k$ or $n-1, k-1$. The difference equation can be solved by formal argument, however, at this point we simply note that the solution is

$$\Pr[E_{n,k}] = \frac{n!}{k!(n-k)!} p^k q^{n-k}, \qquad k = 0, 1, 2, \ldots, n \qquad (2.13)$$

if we take $\Pr[E_{0,0}] = 1$. (This solution can be checked by formal substitution.)

Formula (2.13) is enormously important so it is given a name. It is called *the binomial probability* because the right-hand side is simply the general term of the binomial expansion of $(p+q)^n$. The coefficient $n!/[k!(n-k)!]$ is often abbreviated as $\binom{n}{k}$.

EXAMPLE 2.13

The trick that allowed us to solve Example 2.12 can be extended in many ways to get many useful probabilities. Here is another example of exactly the same logic. But again we will have to be slippery when it comes to solving our equation. To set up the problem similar to Example 2.12 let us assume again that we choose n parts at random. The new idea is that instead of the parts being effective or defective, let us now assume they can be effective, defective, *or repairable*. Then let $E_{n,k,j}$ be the event that of these n items k are effective and j are defective, and then of course $n-k-j$ will be repairable. Let us use precisely the same logic as before. If we knew the number of effective and defective items after the $(n-1)$st item had been inspected computing the probability for $E_{n,k,j}$ would be easy because

$$\Pr[E_{n,k,j} | E_{n-1,k,j}] = q, \qquad \text{say}$$

$$\Pr[E_{n,k,j} | E_{n-1,k-1,j}] = p, \qquad \text{say}$$

$$\Pr[E_{n,k,j} | E_{n-1,k,j-1}] = r, \qquad \text{say}$$

where $p+q+r = 1$. Any other conditional probabilities (e.g., $\Pr[E_{n,k,j}| E_{n,k-1,j-1}]$) would be 0 since we only have one more item to inspect. Then we can again use our formulas of conditional probability (2.5) and total probability (2.10) to write

$$\Pr[E_{n,k,j}] = p \Pr[E_{n,k-1,j}] + q \Pr[E_{n,k,j}]$$
$$+ r \Pr[E_{n,k,j-1}] \qquad (2.14)$$

This equation is a bit tougher to solve but it can be done. Its solution is

$$\Pr[E_{n,k,j}] = \frac{n!}{k!j!(n-k-j)!} p^k q^j r^{n-k-j} \qquad (2.15)$$

This result is called *the trinomial probability*. The right-hand side is just the general term of the expanded form of $(p+q+r)^n$.

EXAMPLE 2.14

In the preceding two examples the events "the ith part chosen is defective" were independent. However, this is not always the case in practice. Suppose that we have a bin containing N parts, K of which are defective. If n parts are selected at random and if the first part chosen is defective, this affects the proportion of defectives left in the bin and, hence, affects the conditional probability that the second part chosen will also be defective. The events "the ith part is defective" are dependent, not independent.

The probability of exactly k defectives among n parts chosen can be calculated using the methods of this section.

Let D_n = the total number of defective parts in the n examined. If D_{n-1} is known then D_n can be determined easily. For if $D_{n-1} = k - 1$ then $D_n = k$ if the nth part chosen is a defect. On the other hand, $D_n = k$ if $D_{n-1} = k$ and the nth part chosen is not a defect. Since there is only one part to be chosen and it is either a defect or not, these are the only two ways that $D_n = k$.

Furthermore, if $D_{n-1} = k$ we have chosen $n - 1$ parts from the original N and of these k are defects from the original K. Thus, $N - (n - 1)$ parts remain of which $K - k$ are defects. The ratio $(K - k)/(N - n + 1)$ is the proportion of defects left to pick from. Thus, this ratio is the probability that the nth chosen part will now be defective. Of course $1 - (K - k)/(N - n + 1)$ is the probability that the nth chosen part will now not be defective. Therefore, we can write the equation

$$\Pr[D_n = k] = \left[1 - \frac{K-k}{N-n+1}\right]\Pr[D_{n-1} = k] + \left[\frac{K-k+1}{N-n+1}\right]\Pr[D_{n-1} = k-1]$$

$$(2.16)$$

This equation has as its solution

$$\Pr[D_n = k] = \frac{\binom{K}{k}\binom{N-K}{n-k}}{\binom{N}{n}} \qquad (2.17)$$

Formulas (2.13) and (2.16) occur often in quality control procedures. We say that formula (2.13) gives "the probability of exactly k defects in a sample of size n when sampling is done *with replacement* or when sampling from a 'large' population." We say that formula (2.16) gives "the probability of exactly k defects in a sample of size n when sampling is done *without replacement* from a 'finite' population." Here the important distinction is that the events "the ith part chosen is defective" are independent if the sampling procedure does not deplete the population as the sample is drawn (sampling from an infinite population or sampling with replacement) as compared to the case in which the sampling procedure does deplete the population (sampling without replacement from a finite population).

Formula (2.17) is termed the *hypergeometric formula*.

2.9 Estimating Probabilities

In Section 2.1 we noted that the theory of probability tells us how to use probabilities to make useful statements but that theory does not tell us how to determine these probabilities. Indeed, as noted in that section, where the probabilities came from is irrelevant to the ensuing theory. However, to use probability theory in applications one must be able to determine the probabilities somehow.

Undoubtedly the most common means of determining these is to estimate them from data using the relative frequency arguments of Section 2.1. This is the classical means and in application is the most often used means. Of course one is sure that all of the conditions required by the relative frequency approach to determining probabilities cannot be satisfied exactly. A phrase such as "repeat the experiment under identical conditions" is meaningless in many real applications. In automobile emissions tests, to repeat an experiment under identical conditions would require control of temperature, barometric pressure, and humidity, at least. Although this might be done in specially constructed labs, these are expensive. But the emissions output in these labs relies on a host of other variables such as driver characteristics, quality of tune-up, dynamometer behavior, et cetera, and some of these simply cannot be held constant to allow the replication of the experiment under absolutely identical conditions.

Furthermore, the relative frequency definition defines a probability as the limit of a ratio. In no physical experiment does this limit operation make sense. One may repeat the experiment 10 times or 100 times or even, in some cases, 10^6 times. But this still is not "in the limit."

In spite of these shortcomings the relative frequency is an often used means of estimating probabilities. Relative frequencies are practical to obtain in many cases, which is a strong argument in their favor. As we shall see later (Section 6.11) these relative frequencies have some nice properties. They often converge relatively rapidly to the probabilities and one can give some estimates about how far the relative frequency is from the "true" probability.

EXAMPLE 2.15

In a famous physics experiment, performed at about the turn of the century, Westgren observed a small fraction of a liquid in which he had suspended gold particles. He was attempting to verify a theory of Smoluchowski concerned with Brownian motion. For his verification he needed to count the number of particles in a small portion of liquid at many widely separated points. Table 2.1 lists 73 values from one of his experiments, as reported in Chandresekhar (1943).[2]

We would like to estimate the probability of the events E_i = "exactly i particles observed," for $i = 0, 1, 2, \ldots, 5$. We can use a relative frequency estimate based on Westgren's data.

i:	0	1	2	3	4	5
$\Pr[E_i] \approx$	12/73	19/73	28/73	10/73	3/73	1/73

---

[2] Chandrasekhar, S., (1943), "Stochastic Problems in Physics and Astronomy," *Reviews of Modern Physics*, **15**, pp. 2–89. Reprinted with permission of the author and the American Institute of Physics.

TABLE 2.1
Counts of Particles in Westgren's Experiment

2	1	1	1	1	1	0	2	2	1	1	1	2	3	2	3
0	0	0	0	0	0	1	1	0	0	1	0	1	1	1	2
2	3	3	4	5	3	4	2	2	1	2	1	3	2	0	2
2	1	0	2	2	2	1	2	3	2	2	2	3	2	2	2
2	2	2	1	3	3	4	2	2							

EXAMPLE 2.16

One of the famous examples of using a relative frequency definition of probability occurs in Weldon's dice example. In this example Weldon, an English biologist, threw a die 315,672 times and recorded 106,602 occurrences of either a 5 or a 6. Based on this data the probability of the event "a 5 or 6 turns up on a throw of a die" can be estimated as

$$\frac{106,602}{315,672} = .3377$$

From physical considerations (two out of six sides on a die) we would expect this event to have probability $1/3$. Thus, the Weldon results appear at first glance to be nearly what they should be (only .0044 away from $1/3 = .3333$). In fact, with 315,672 repetitions, Weldon's results are poor. Either he did not repeat his 315,672 experiments "under identical conditions" or his die was biased. With this many trials the relative frequency should have been within about .002 of $1/3$.

EXAMPLE 2.17

Hall (1969)[3] collected data on 5140 telephone calls arriving to a police dispatcher. Of these 415 were calls for an ambulance. Based on these data one concludes that .081 is a good estimation of the probability that a call to the dispatcher is for an ambulance. One must have some basis for assuming that such data is collected "under identical conditions." Hall provides a lengthy argument to defend this assumption.

Exercises 2

1. If $A \subset \varnothing$ prove $\Pr[A] = 0$.

2. Prove $\Pr[A \cap B] \le \Pr[A]$.

3. Is $\Pr[A \triangle B] \le \Pr[A]$ necessarily? Justify your answer.

[3] W. K. Hall (1969), *A Queueing Theoretic Approach to the Allocation and Distribution of Ambulances in an Urban Area*, Ph.D. dissertation, School of Business Administration, University of Michigan, Ann Arbor, Michigan. Reprinted with permission of the author.

4. Suppose $\Pr[A]$, $\Pr[B]$, and $\Pr[C]$ are known. Give a condition under which the probability for the event "either A, B, or C occurs" can be determined in terms of these probabilities and the axioms of probability. Determine this probability.

5. In Example 2.2 determine the probabilities of the four sample points under the assumption that the degrees of correctness of the memories of the three subjects are independent of each other.

6. Four digits are transmitted as a word. In a "noisy" transmission circuit each digit can be transmitted incorrectly with probability .2, independent of the transmission of any other digit. Show that each of the events "the ith and jth digits are transmitted correctly and the kth and lth digits are transmitted incorrectly" is equally probable with probability .0256, where i, j, k, l is any permutation of $1, 2, 3, 4$.

7. Prove that if A, B, and C are events, then

$$\Pr[A \cup B \cup C] \le \Pr[A] + \Pr[B] + \Pr[C]$$

by using the axioms and the properties developed in Sections 2.2 and 2.3.

8. Prove the following special case of property (d), Section 2.3.

$$\Pr[A_1 \cup A_2 \cup A_3] = \Pr[A_1] + \Pr[A_2] + \Pr[A_3]$$
$$- \Pr[A_1 \cap A_2] - \Pr[A_1 \cap A_3] - \Pr[A_2 \cap A_3]$$
$$+ \Pr[A_1 \cap A_2 \cap A_3]$$

by using property (c).

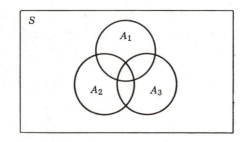

9. Show that for two events, A and B,

$$\Pr[A \triangle B] = \Pr[(A \cap \bar{B}) \cup (\bar{A} \cap B)] = \Pr[A] + \Pr[B] - 2\Pr[A \cap B]$$

10. A number x is formed by choosing six digits in order, each being either a 0 or a 1 and equally likely.

   (a) Describe the sample space and count the number of sample points it contains.

(b) Compute the probability of each of the following events.

A: "the number $x \geq 100,000$"

B: "the number x is even"

C: "no pair of successive digits in x are identical"

11. A number x is formed by arranging the six digits $0, 0, 1, 1, 2, 2$ in a random order. Answer the questions in Exercise 10 about this number.

12. An honest coin is tossed until the *second* head appears.

(a) Describe the same space S and compute the probabilities $\{p_i\}$ of the various sample points.

(b) Compute the probability that an even number of tosses is required. *Hint.* If

$$f(x) = \sum_{k=1}^{\infty} x^{2k-1} = \frac{x}{1-x^2}$$

then

$$f'(x) = \sum_{k=1}^{\infty} (2k-1)x^{2k-2} = \frac{d}{dx}\left(\frac{x}{1-x^2}\right)$$

13. An "honest" coin is tossed $2n$ times. Show that the probability of exactly n heads is $\binom{2n}{n}/2^{2n}$.

14. A pocket is known to contain two regular pennies and a two-headed penny. A coin is chosen at random and tossed. If it comes up heads, what is the probability that its other side is also heads?

15. Four urns each contain 3 balls.

urn 1 contains 3 white balls

urn 2 contains 2 white and 1 black balls

urn 3 contains 1 white and 2 black balls

urn 4 contains 3 black balls

An urn is chosen at random and a ball is chosen from it. If the ball turns out to be white, compute the conditional probability that urn 2 was the urn chosen.

16. In a noisy communication system transmitting binary digits (zeros and ones) the sender may send a digit (for example, 1) and the receiver may receive the other digit (for example, 0). Let A be the event "a 1 is sent" and let B be the event "a 1 is received." Hence $\bar{A}$ is the event "a 0 is sent" and $\bar{B}$ is the event "a 0 is received." Assume that

$$\Pr[\bar{B}|A] = .01$$
$$\Pr[B|\bar{A}] = .01$$

Suppose further that $\Pr[A] = .5$. Give a physical interpretation of the foregoing conditional probabilities. Determine the probabilities $\Pr[A|B]$, $\Pr[\overline{A}|B]$, $\Pr[A|\overline{B}]$, and $\Pr[\overline{A}|\overline{B}]$. These conditional probabilities are the ones that the receiver might want to know to help in interpreting a message.

17. In the design of automobile dashboards, warning lights have replaced gauges. These lights are intended to warn the driver of impending engine troubles. Hence we would like to have each light turn on when trouble is imminent, with a high probability, and to have the light not turn on when there is no trouble. Suppose T is the event "engine trouble" and O is the event "light turns on." Let $\Pr[O|T] = p$ and $\Pr[\overline{O}|\overline{T}] = q$. Suppose that $\Pr[T] = .01$. By using Bayes' rule determine the following in terms of p and q.

    (a) $\Pr[T|O]$.

    (b) $\Pr[T|\overline{O}]$.

    (c) $\Pr[T \cap O]$.

    (d) $\Pr[T \cap \overline{O}]$.

    How large should p and q be if the designer wants $\Pr[T|O] \geq .99$ and $\Pr[T|\overline{O}] \leq .02$?

18. One of three possible hypotheses H_1, H_2, or H_3 can "explain" the occurrence of a road accident. The probability of the accident if H_1 is correct is .75. The probability of the accident if H_2 is correct is .80. The probability of the accident if H_3 is correct is .85. H_1 has probability .4 of occurring, H_2 has probability .5 of occurring, and H_3 has probability .1 of occurring. Determine the "most probable" hypothesis to explain an observed accident.

19. Suppose that a supplier of material sends raw material that is 1% defective. If we sample 10 parts from a new batch of material and find 2 defective parts, could we reasonably doubt that this batch is more than 1% defective? Discuss.

20. Suppose that two events, A and B, are mutually exclusive and $\Pr[B] > 0$. Under what conditions will A and B be independent?

21. Suppose that $\Pr[A] > 0$ and $\Pr[B] > 0$ and that A and B are independent. Can A and B be mutually exclusive?

22. Show that every event A is independent of the impossible event $\varnothing$ and of the event S.

23. Show that in general $\Pr[A_1|A_2] + \Pr[A_1|\overline{A}_2] \neq 1$.

24. Each of four coins has probability p of turning up heads on any toss and probability q of turning up tails. An experiment consists of tossing the four coins. Enumerate the elements of the sample space. Let event E_i be defined as "exactly i heads appear," $i = 0, 1, 2, 3, 4$. Enumerate each of the events E_i in terms of the sample points. What probabilities would we assign to the events E_i if the coins are considered to be independently tossed?

25. An electronic component consists of three parts. Each part has probability .99 of performing satisfactorily. The component fails if two or more of the parts do not perform satisfactorily. Assuming that the satisfactory performance of one part in no way depends on the performance of the other parts, determine the probability that the component does not perform satisfactorily.

26. Manufacturing Station I produces 10 nondefectives and 3 defectives. Manufacturing Station II produces 3 nondefectives and 5 defectives. Two products from Station I's production bin are randomly chosen and sent to Station II's production bin. One product is then drawn from Station II's bin. Find the probabilities of

   (a) transferring 2 nondefectives

   (b) transferring 1 of each kind

   (c) drawing a defective from Station II's bin.

27. Show by substitution that equation (2.13) satisfies equation (2.12).

28. Show by substitution that equation (2.15) satisfies equation (2.14).

29. Show by substitution that equation (2.17) satisfies equation (2.16).

30. For large values of n an approximation for $n!$ is given by *Stirling's approximation*

$$n! \sim \sqrt{2\pi n}\, n^n e^{-n}$$

   ($\sim$ means that as $n \to \infty$ the ratio of the two sides approaches 1.) For $n = 2, 5, 10$ show that the ratio $n! / [\sqrt{2\pi n}\, n^n e^{-n}]$ approaches 1 but that the difference $n! - \sqrt{2\pi n}\, n^n e^{-n}$ grows.

31. Using Stirling's approximation estimate

$$\frac{\Pr[E_{n,k}]}{\Pr[E_{n,k+1}]}$$

   for $n = 20$, $k = 8$, and $p = q = \frac{1}{2}$ for the binomial probabilities given in Example 2.12.

32. In Exercise 13 use Stirling's approximation to show that the probability is approximately $1/\sqrt{\pi n}$.

3 · Random Variables and Their Distributions

3.1 Introduction

At this point in our construction of probability theory we have a complete probability structure. We have discussed a random experiment, a set of outcomes defining the sample space, certain subsets of the outcomes defining events, and an assignment of numbers to each event, which satisfy the axioms of probability. One troublesome point remains that we must attack. Thus far we have treated the sample space as the set of all possible outcomes of an experiment. Consequently, we have dealt with sample spaces such as

$$S = \{1, 2, 3, 4, 5, 6\}$$
$$S = \{\text{Heads, Tails}\}$$

That is, we have observed that some experiments yield sample spaces whose elements are numbers, but that other, quite reasonable, experiments do not naturally yield numerically valued elements. Other examples yielding such non-numerical sample spaces are:

(1) The rocket-firing experiment, where outcomes might be termed "success" or "failure."

(2) The quality check experiment whose outcomes might be "good" or "defective."

The important point, of course, is that nothing in our construction requires the sample space to be numerically valued; on the other hand, there is nothing that forbids numerically valued outcomes. For mathematical purposes this is inconvenient. It is desirable to have numbers associated with the outcomes. Hence, we need one more construction to extend the basic ideas of probability. The construction we now start leads to what is called a *random variable*.

3.2 The Definition of a Random Variable

Suppose that to each experimental outcome or sample point s we apply some rule of measurement or counting (X) that yields a *number*; for example, choose a person at random from a group and *measure* his or her height, or choose a sample from a production lot and *count* the number of defective items it contains. Such a *rule* for assigning numbers to sample points is termed a *random variable*.

42

The terminology used here is traditional, and clearly a random variable is not a variable at all in the usual sense. A more closely related concept is the concept of a *function* in elementary mathematics. Recall that an equation such as

$$y = y(x) = x^2 + 3$$

is said to define y as a function of x, meaning that it specifies a *rule* that for every *number* x assigns a *number* for y. The situation is exactly comparable when we say that the measurement of height is a *rule* that assigns a *number* to each *person*. Thus, we have a more precise *definition*.

A RANDOM VARIABLE X *is a function that to each sample point in the sample space S assigns a number* (usually a *real* number).

The use of a single letter, such as X, for this function, in place of $X(s)$ or $X(.)$, exactly corresponds to the use of a single letter, such as y or f, in place of $y(x)$ or $f(x)$, for a function in calculus. The use of a single letter is logically superior and, in any case, is traditional. If it is necessary to specify the value of the random variable for a *particular* sample point s we will use the notation $X(s)$. Thus, if X is the height function or random variable, the statement "Joe is 70 inches tall" might be written as

$$\text{"}X(\text{Joe}) = 70\text{"}$$

The sample space S is termed the *domain* of the random variable or function and the collection of all numbers that are *values* for X is termed the *range* of the random variable. Thus the range is a certain subset of the set of all real numbers.

Notice, in particular, that two or more *different* sample points might give the *same* value for X (two different people might have the same height), but two different numbers in the range cannot be assigned to the same sample point (one person cannot have two different heights) (see Figure 3.1).

For a coin-tossing experiment with only two sample points, we might define the function X as (see Figure 3.2)

$$X(\text{Heads}) = 0$$

$$X(\text{Tails}) = 1$$

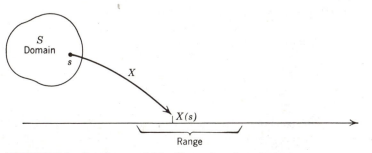

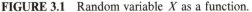

FIGURE 3.1   Random variable X as a function.

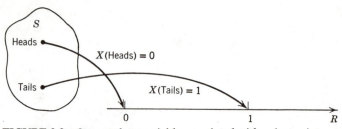

FIGURE 3.2   One random variable associated with coin tossing.

Observe that, usually, no unique rule or measurement must be applied. In general, many rules can be assigned to the same sample points. The choice of a random variable for use in a particular study is not fixed within the framework of probability, and the study itself dictates which one of the possible random variables is most appropriate. There is nothing to keep us from defining X as in Figure 3.2 if this is a useful function. On the other hand, we could just as well choose other functions, say, Y, Z, or U, with

$$Y(\text{Heads}) = 1$$
$$Y(\text{Tails}) = 0$$

or

$$Z(\text{Heads}) = 7 \times 10^6$$
$$Z(\text{Tails}) = 4$$

or even

$$U(\text{Heads}) = 0$$
$$U(\text{Tails}) = 0$$

The student should be aware that, in practice, the motivation for the choice of a random variable is often one of convenience alone and is dictated by the kind of questions we want to answer.

EXAMPLE 3.1

In many experiments the sample point s itself is a number. In such cases it is often convenient to use the *identity function*, $X(s) = s$, as the random variable. For example, in the toss of a die an outcome is a 1 spot, 2 spot,..., 6 spot. So we could take $X(s) = s$ as $X(1) = 1$, $X(2) = 2,..., X(6) = 6$.

EXAMPLE 3.2

In Example 2.17 we considered telephone calls arriving at a police dispatcher. If s is a call, then we can set $X(s) = 0$ if the call is for an ambulance and $X(s) = 1$ if the call is not for an ambulance. Then S is the set of all calls coming to the dispatcher and for each call s, $X(s) = 0$ if that call is for an ambulance and $X(s) = 1$ otherwise.

EXAMPLE 3.3

Remember the Westgren data (Example 2.15)? There the experimenter saw at most five particles in his series of observations. Thus, one would be tempted to choose the sample space to be the six integers $0, 1, \ldots, 5$. But remember the sample space is chosen as the set of *all possible* outcomes. Just because Westgren never saw six particles or seven or any number bigger than five in his data does not mean that there could not have been such things. For that reason it is reasonable to assume that any number of particles could have been seen. Then, because the experimenter was interested in the number of particles and because each sample point is the number of particles in the experimental volume, it seems natural to choose the following *identity function* for a random variable for that experiment: $X(0) = 0$; $X(1) = 1$; $\ldots$; $X(n) = n$; $\ldots$.

Henceforth, we shall begin with a given random variable and shall work with numbers in its range. We shall not often make explicit the sample space from which this random variable arises. This point should not be confusing; and we must be aware that some sample space lies behind each random variable.

3.3   Events Defined by Random Variables

In Chapter 2 *probabilities* were defined as numbers assigned to subsets of the sample space (events). More formally, *probability* is a set-function—a rule that assigns a number between 0 and 1 to a set of points of S. Its domain is the *set of events* of a random experiment and its range is contained in the interval [0, 1]. We have learned that a random variable X is also a function whose range is a set of numbers, but whose domain is the set of sample points s making up the sample space S, not subsets of S.

If X is a random variable and x is a fixed real number, we can define the event A_x to be the subset of S consisting of all sample points s to which the random variable X assigns the number x,

$$A_x = \{ s : X(s) = x \}$$

As an event, A_x will have a probability (see Figure 3.3)

$$p = \Pr[A_x]$$

Thus we see that for every number x we can use the random variable X to construct an event or subset A_x of the sample space containing all sample points whose X-value is x. This event, A_x, then has a probability $p = \Pr[A_x]$. Generally, we might interpret p as the probability that the random variable X takes on the value x.

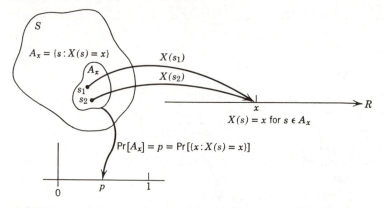

FIGURE 3.3   Relation between a random variable X and its probability function Pr.

More specifically, suppose that S consists of five sample points $S = \{s_1, s_2, s_3, s_4, s_5\}$, with individual probabilities $p_1 = \Pr[\{s_1\}]$, $p_2 = \Pr[\{s_2\}]$, $p_3 = \Pr[\{s_3\}]$, $p_4 = \Pr[\{s_4\}]$, $p_5 = \Pr[\{s_5\}]$, $\Sigma p_i = 1$. Define the random variable X by $X(s_1) = X(s_3) = 0$, $X(s_2) = X(s_4) = 1$, $X(s_5) = 2$. Thus, $A_0 = \{s : X(s) = 0\} = \{s_1, s_3\}$, and $p = \Pr[A_0] = \Pr[\{s_1, s_3\}] = p_1 + p_3$, et cetera (see Figure 3.4). Similarly, $A_1 = \{s_2, s_4\}$, $\Pr[A_1] = p_2 + p_4$ and $A_2 = \{s_5\}$, $\Pr[A_2] = p_5$.

For simplicity we introduce the notation $[X = x]$ for the set A_x. Thus

$$[X = x] = \{s : X(s) = x\} \tag{3.1}$$

This is a better notation than A_x since it displays the random variable X under consideration as well as the value x. The probability of this set is then

$$\Pr[X = x] = \Pr[\{s : X(s) = x\}] \tag{3.2}$$

This probability is termed *the probability that the random variable X equals x.*

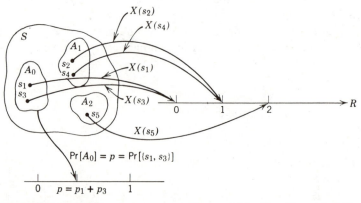

FIGURE 3.4   Relation between a random variable X and its probability function Pr— a specific example.

In the preceding example

$$\Pr[X=0]=p_1+p_3$$
$$\Pr[X=1]=p_2+p_4$$
$$\Pr[X=2]=p_5$$

Thus, by means of the random variable, we have assigned probabilities to the numbers 0, 1, 2 on the real line, although the original probabilities were only defined for subsets of the abstract set S. We say that the random variable has *induced* probabilities on the real line.

The events $[X=x]$ were defined by (3.1). This convenient notation obviously can be extended to define other types of events in terms of a random variable. For fixed numbers, x, a, b define

$$[X \le x] = \{s: X(s) \le x\}$$
$$[X > x] = \{s: X(s) > x\}$$
$$[a < X < b] = \{s: a < X(s) < b\}, \quad \text{et cetera} \tag{3.3}$$

In each case we have defined an event or subset of the abstract sample space S. These events have probabilities that are denoted by

$$\Pr[X \le x] = \text{``the probability that } X \text{ takes a value} \le x \text{''}$$
$$\Pr[X > x] = \text{``the probability that } X \text{ takes a value} > x \text{''}$$
$$= 1 - \Pr[X \le x]$$

$$\Pr[a < X < b] = \text{``the probability that } X \text{ takes a value}$$
$$\text{strictly between } a \text{ and } b\text{,''} \quad \text{et cetera}$$

EXAMPLE 3.4

An honest coin is tossed three times. The sample space then consists of eight equally likely sample points $S = \{HHH, \ldots, TTT\}$. Let X be the random variable that counts the number of heads in each sample point. Thus, X has range $\{0, 1, 2, 3\}$. Table 3.1 lists the value of X for each sample point and for the various events $[X=x]$. Notice that

$$[X \le 1] = \{TTT, HTT, THT, TTH\} = [X=0] \cup [X=1]$$
$$\Pr[X \le 1] = \Pr[X=0] + \Pr[X=1] = \tfrac{1}{8} + \tfrac{3}{8} = \tfrac{1}{2}$$
$$[X \le 1] \cup [X > 1] = S$$
$$\Pr[X \le 1] + \Pr[X > 1] = 1$$
$$\Pr[X > 1] = 1 - \Pr[X \le 1] = \tfrac{1}{2}$$

In each of the above unions the events are mutually exclusive.

$$[0 < X < 3] = [1 \le X \le 2] = \overline{[X=0] \cup [X=3]}$$

TABLE 3.1
Counting Heads when Tossing Three Honest Coins

s	$X(s)$	EVENTS DEFINED BY $X(s)$	PROBABILITIES INDUCED BY $X(s)$
HHH	3	$[X = 0] = \{TTT\}$	$\Pr[X = 0] = \frac{1}{8}$
HHT	2	$[X = 1] = \{HTT, THT, TTH\}$	$\Pr[X = 1] = \frac{3}{8}$
HTH	2	$[X = 2] = \{HHT, HTH, THH\}$	$\Pr[X = 2] = \frac{3}{8}$
THH	2	$[X = 3] = \{HHH\}$	$\Pr[X = 3] = \frac{1}{8}$
HTT	1		
THT	1	$[X = x] = \varnothing$ for $x \neq 0, 1, 2, 3$	$\Pr[X = x] = 0$
TTH	1		for $x \neq 0, 1, 2, 3$
TTT	0		

Hence

$$\Pr[0 < X < 3] = 1 - \Pr([X = 0] \cup [X = 3]) = 1 - \left(\tfrac{1}{8} + \tfrac{1}{8}\right) = \tfrac{3}{4}$$

If we tabulate $[X \leq x]$ and $\Pr[X \leq x]$ for $x = -1, 0, 1, 2, 3, 4$, we observe that each event is included in the following one and, hence, that the probabilities increase (or at least do not decrease) with increasing x (Table 3.2).

$$[X \leq -1] \subset [X \leq 0] \subset [X \leq 1] \subset [X \leq 2] \subset [X \leq 3] \subset [X \leq 4]$$

$$\Pr[X \leq -1] \leq \Pr[X \leq 0] \leq \Pr[X \leq 1] \leq \Pr[X \leq 2] \leq \Pr[X \leq 3] \leq \Pr[X \leq 4]$$

TABLE 3.2
Probability Distribution Function for Tossing Three Honest Coins

x	$[X \leq x]$	$\Pr[X \leq x]$
-1	$\varnothing$	0
0	$\{TTT\}$	$\frac{1}{8}$
1	$\{TTT, TTH, THT, HTT\}$	$\frac{4}{8}$
2	$\{TTT, TTH, THT, HTT, HHT, HTH, THH\}$	$\frac{7}{8}$
3	$\{TTT, TTH, THT, HTT, HHT, HTH, THH, HHH\} = S$	1
4	S	1

3.4 Distribution Functions

If X is a random variable and x is a number, we have defined the event $[X \leq x] = \{s : X(s) \leq x\}$ and its probability $\Pr[X \leq x]$. Clearly, for a given random variable, this event and its probability depend on the value used for x. Thus, $\Pr[X \leq x]$ is a number whose value depends on x. In other words, it is a *function* of x. This function is termed the *distribution function* (or *cumulative distribution function*) of the random variable and is denoted by $F(x)$ or $F_X(x)$.

$$F(x) = \Pr[X \leq x], \qquad -\infty < x < \infty \tag{3.4}$$

The distribution function is an ordinary function of the type familiar to every student. Its domain is the set of all real numbers and its range is a set of numbers between 0 and 1 (since every value of $F(x)$ is a probability). This is in contrast to the random variable X, which is a function whose domain is the sample space, or the probability function Pr, whose domain is a set of events.

Most of the important information about a probability system that refers to measurements given by the random variable X is determined by the behavior of $F(x)$. This function is used extensively in the material that follows. Sometimes we want to compute certain distribution functions, exactly or approximately. Sometimes the distribution function is given, and from it we want to deduce properties of the system, or properties of extensions of the system. When this happens we find that the concepts of sample space, event space, and probability measure, which are of paramount importance in building the theory of probability, often fade into the background and become almost unseen, and functions such as the distribution function become the most important entities. However, we must always keep the background in mind.

EXAMPLE 3.5

A simple distribution function is given by Example 2.17 using the random variable defined in Example 3.2. From those we have $F(x) = 0$ for all $x < 0$, $F(x) = .081$ for all $0 \leq x < 1$, and $F(x) = 1$ for all $x \geq 1$. So we have Figure 3.5.

EXAMPLE 3.6

Consider the situation of Example 3.4. Table 3.2 gives $F(x) = \Pr[X \leq x]$ for $x = -1, 0, 1, 2, 3, 4$. Since X must be an integer, the value of $F(x)$ for x not an integer must be the same as the value for the *nearest* integer next below x. Thus, the graph of $F(x)$ will be flat between integers, with jumps at $x = 0, 1, 2, 3$. If x is

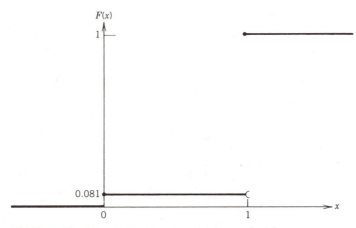

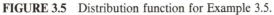

FIGURE 3.5   Distribution function for Example 3.5.

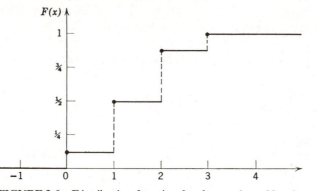

FIGURE 3.6   Distribution function for the number of heads in three coin tosses.

negative, then $X(s) \le x$ is impossible, $[X \le x] = \emptyset$, and $F(x) = 0$. If x exceeds 3, then $X(s) \le x$ is certain, $[X \le x] = S$, and $F(x) = 1$. Thus, the graph of $F(x)$ is as shown in Figure 3.6. Observe that the graph has jumps at $x = 0, 1, 2, 3$, and that at each jump the *upper* value is the correct value for $F(x)$.

Several properties of $F(x)$ follow directly from its definition (3.4).

(a) $0 \le F(x) \le 1,$ for $-\infty < x < \infty$

This follows because $F(x)$ is a probability.

If $x_1 \le x_2$, clearly $[X \le x_1] \subset [X \le x_2]$ since whenever $X(s) \le x_1$ necessarily $X(s) \le x_2$, and it follows that any sample point s in $[X \le x_1]$ must also be in $[X \le x_2]$. Thus, $\Pr[X \le x_1] \le \Pr[X \le x_2]$ and we have the following property.

(b)   If $x_1 \le x_2$ then $F(x_1) \le F(x_2)$

This implies that if x increases, then $F(x)$ must also increase (or, at least, not decrease). $F(x)$ is said to be a *nondecreasing* function of x. We observe this phenomenon in Figure 3.6.

In addition

(c) $\lim_{x \to +\infty} F(x) = F(\infty) = 1,$ $\lim_{x \to -\infty} F(x) = F(-\infty) = 0$

Intuitively these formulas are clear since it would seem that, as $x \to +\infty$, $F(x) = \Pr[X \le x] \to \Pr[X < +\infty] = \Pr[S] = 1$ and, as $x \to -\infty$, $F(x) = \Pr[X \le x] \to \Pr[X \le -\infty] = \Pr[\emptyset] = 0$, since $-\infty < X(s) < \infty$ for every $s \in S$ (see Figure 3.6).

A rigorous proof of (c) requires use of axiom (iii) for probability (Section 2.2). Events of the form $[i - 1 < X \le i]$ are mutually exclusive for all integers i, and

$$S = \cdots \cup [-1 < X \le 0] \cup [0 < X \le 1] \cup \cdots \cup [n - 1 < X \le n]$$

$$\cup [n < X \le n + 1] \cup \cdots$$

$$[X \le n] = \cdots \cup [-1 < X \le 0] \cup [0 < X \le 1] \cup \cdots \cup [n - 1 < X \le n]$$

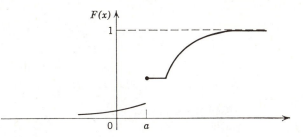

FIGURE 3.7 A possible distribution function.

Thus,

$$
\begin{aligned}
1 = \Pr[S] &= \sum_{i=-\infty}^{+\infty} \Pr[i-1 < X \le i] \\
&= \lim_{n \to +\infty} \sum_{i=-\infty}^{n} \Pr[i-1 < X \le i] \\
&= \lim_{n \to +\infty} \Pr[X \le n] \\
&= \lim_{n \to +\infty} F(n)
\end{aligned}
$$

Thus, $\lim_{n \to +\infty} F(n) = 1$. The case where x is not an integer is accounted for by considering the largest integer below x. The proof that $\lim_{x \to -\infty} F(x) = 0$ is similar.

A fourth property of $F(x)$ is more delicate. Suppose that x is made to approach a fixed value x_0 *from the right*, that is, $x \to x_0$ and $x \ge x_0$. We denote this by the symbol $x \to x_0 +$. Then, clearly, any value $X(s)$ of the random variable will be $\le x_0$ if, and only if, it is $\le x$ for each $x \ge x_0$. Using an argument similar to the one used in the preceding paragraph, it follows that $F(x) \to F(x_0)$ as $x \to x_0 +$. Thus, we have the property

(d) $\quad \lim_{x \to x_0 +} F(x) = F(x_0)$

We say that $F(x)$ is *continuous on the right*. A similar property for approach from the left does not hold true in general, since we have seen in Figure 3.6 that $F(x)$ may have *jumps* and, thus, may not be continuous from both sides. If a jump does occur at x_0, then $F(x_0)$ is always the *upper* value.

A possible graph for $F(x)$ is shown in Figure 3.7. It can be shown that any function $F(x)$ satisfying properties (a) to (d) is the distribution function of *some* random variable.

EXAMPLE 3.7

The function

$$
F(x) = \begin{cases} 0, & \text{if} \quad x < 0 \\ x + \tfrac{1}{2}, & \text{if} \ 0 \le x \le \tfrac{1}{2} \\ 1, & \text{if} \quad x > \tfrac{1}{2} \end{cases}
$$

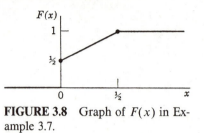

FIGURE 3.8   Graph of $F(x)$ in Example 3.7.

can be graphed as shown in Figure 3.8. We easily see that $F(x)$ has the properties given by (a) through (d). Hence $F(x)$ is the distribution function of some random variable X. Notice that $F(x)$ has a jump at $x = 0$ and that $F(0) = \frac{1}{2}$, the *upper* value. This implies that $\Pr[X = 0] = \frac{1}{2}$, $\Pr[X < 0] = 0$.

EXAMPLE 3.8

In Chapters 7 and 8 we will see that data collected on the times betwen consecutive mine accidents seem to be represented by the following distribution function. (See Figure 3.9.)

$$F(x) = \begin{cases} 0, & \text{if } x \le 0 \\ 1 - e^{-\lambda x}, & \text{if } x > 0 \end{cases}$$

The constant λ, which turns out to be $1/240$, will be computed in Chapter 6.

We readily observe that $F(x)$ has the properties given by (a) through (d). This type of distribution function occurs in many applications. We shall learn later that, under some reasonable assumption concerning telephone calls, the random variable $X =$ the time between two telephone calls at an exchange will have the *negative exponential* distribution function of this example. In fact, this distribution function is extremely important to much of our following work.

EXAMPLE 3.9

The function

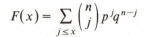

$$F(x) = \sum_{j \le x} \binom{n}{j} p^j q^{n-j}$$

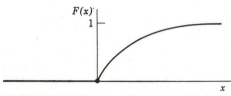

FIGURE 3.9   Graph of $F(x)$ in Example 3.8.

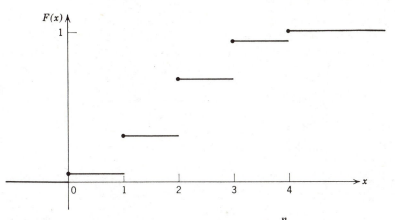

FIGURE 3.10 The distribution function $F(x) = \sum_{j \le x} \binom{n}{j} p^j q^{n-j}$; $p = q = \frac{1}{2}$, $n = 4$.

where $0 \le p \le 1$, $0 \le q \le 1$, $p + q = 1$, and n is a fixed positive integer, can be graphed as shown in Figure 3.10. It is clear from this figure that $F(x)$ has the properties (a) through (d). We shall learn that this distribution function is useful in many applications that involve counting. It is called the *binomial* distribution function.

We notice that in Example 3.9 the points of increase of $F(x)$ occur only at integers. If x is not an integer, $F(x)$ has the same value that it had at the integer nearest x on the left. That is, $F(x)$ proceeds from 0 to 1 in a series of jumps. Except at these jumps, $F(x)$ is constant. In Example 3.7 we observed that $F(x)$ had one jump at 0 but, except for that jump, it proceeded "smoothly" from 0 to 1 as x proceeded from left to right. At $x = \frac{1}{2}$, there was a point at which $F(x)$ was continuous but not differentiable. In Example 3.8 we determined that $F(x)$ was continuous everywhere, but that at the point $x = 0$ it was not differentiable. More complicated examples could be constructed, but the ones above illustrate the following point.

It can be shown, by using deeper analysis, that every distribution function can be decomposed into a weighted sum of three parts in the form

$$F(x) = F_1(x) + F_2(x) + F_3(x)$$

where $F_1(x)$ is a function that changes only by jumps (as in Example 3.9 but the jumps need not occur only at the integers) and $F_2(x)$ is continuous (as for Example 3.8). The function $F_3(x)$ is termed *singular*. Singular functions rarely occur in distribution functions of practical interest, and henceforth we assume that the distribution functions discussed have no singular part, that is, we assume that $F_3(x) \equiv 0$. If further $F_2(x) \equiv 0$, that is, if the function proceeds only by jumps, we say that the random variable X is a *discrete random variable*. If, on the other hand, $F_1(x) \equiv 0$, we say that the random variable X is a *continuous random variable*, although in Section 3.7 we place additional restrictions on X to make formal manipulations easier. Otherwise, the random variable X is said to be of *mixed type*. In most applied work the random variable is either discrete or continuous. We study only these two cases in detail. There are cases of useful random variables that are of mixed type, but in elementary consideration they arise so infrequently that no special discussion is necessary. Usually they can be easily analyzed as special cases.

3.5   Determining Probabilities from the Distribution Function

We observed in Section 3.4 that $F(x)$ gives probabilities of the form $\Pr[X \le x]$. But we also want to compute other probabilities, such as $\Pr[a < X \le b]$ or $\Pr[X > a]$. We accomplish this as follows:

(a)   Since

$$[X \le a] \cup [X > a] = S$$

we have

$$\Pr[X \le a] + \Pr[X > a] = \Pr[S] = 1$$

by using axiom (iii) for probability (Section 2.2). Thus

$$\Pr[X > a] = 1 - \Pr[X \le a] = 1 - F(a)$$

(b)   Since

$$[X \le a] \cup [a < X \le b] = [X \le b]$$

we have

$$\Pr[X \le a] + \Pr[a < X \le b] = \Pr[X \le b]$$

Thus

$$\Pr[a < X \le b] = \Pr[X \le b] - \Pr[X \le a]$$
$$= F(b) - F(a)$$

Therefore, $F(x)$ determines probabilities for all intervals of the form $[a < X \le b]$. We also need to explore probabilities of the form $\Pr[X = a]$. This is done in Sections 3.6 and 3.7, which follow.

EXAMPLE 3.10

Using the distribution function given in Example 3.7 we determine that

$$\Pr\left[X \le \tfrac{1}{4}\right] = F\left(\tfrac{1}{4}\right) = \tfrac{3}{4}$$
$$\Pr\left[0 < X \le \tfrac{1}{4}\right] = F\left(\tfrac{1}{4}\right) - F(0) = \tfrac{1}{4}$$
$$\Pr\left[X > \tfrac{1}{4}\right] = 1 - F\left(\tfrac{1}{4}\right) = \tfrac{1}{4}$$
$$\Pr\left[X \le \tfrac{1}{8}\right] = \tfrac{5}{8}$$

$$\Pr\left[X > \tfrac{1}{8}\right] = 1 - F\left(\tfrac{1}{8}\right) = \tfrac{3}{8}$$

EXAMPLE 3.11

Continuing Example 3.8 we have

(a) $\Pr[X \le 120] = 1 - e^{-120/240} = 1 - e^{-0.5} = 0.39$

(b) $\Pr[240 < X \le 360] = (1 - e^{-1.5}) - (1 - e^{-1}) = 0.14$

(c) $\Pr[X > 480] = e^{-2} = 0.14$

We can interpret these probabilities in the setting of the example.

(a)   The probability that the time between two consecutive accidents is 120 days or less is .39.

(b)   The probability that the time between two consecutive accidents is more than 240 days but less than (or equal to) 360 days is .14.

(c)   The probability that the time between two consecutive accidents is more than 480 days is .14.

3.6   Discrete Random Variables

In Section 3.4 we learned that a discrete random variable can be characterized as one whose distribution function $F(x)$ proceeds in jumps (see Figure 3.10 and the discussion that follows). The only nonzero contributions to $F(X)$ occur at some discrete set of points $x_0, x_1, x_2 \ldots$ (which, in the example, were integers but, in general, need not be). Therefore, probabilities of nonzero value must all be contributed by the sets of the form

$$[X = x_i], \qquad i = 0, 1, 2, \ldots$$

Thus we can define a function that takes nonzero values only at the x_i. We let the value it takes at each x_i be the amount that $F(x)$ jumps at that point. That is, we define a new function

$$p(x) = \Pr[X = x] \tag{3.5}$$

called the *density function* for the discrete random variable X. Clearly $p(x)$ will have these properties:

(a) $p(x) = 0$ *unless x is one of* $x_0, x_1, x_2, \ldots$.

(b) $0 \le p(x_i) \le 1$, for each x_i in the range.

(c) $\sum_i p(x_i) = \sum_i \Pr[X = x_i] = 1$.

Here the sums are to be taken over the entire range of X.

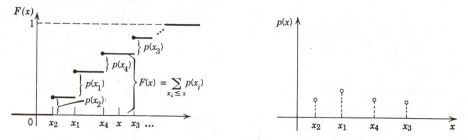

FIGURE 3.11   Relation between the distribution function $F(x)$ and the density function $p(x)$ for a discrete random variable.

The functions $F(x)$ and $p(x)$ are related by the formula

$$F(x) = \Pr[X \le x] = \Pr[X = x_i \text{ for some } x_i \le x]$$

$$= \sum_{x_i \le x} \Pr[X = x_i]$$

$$F(x) = \sum_{x_i \le x} p(x_i) \qquad (3.6)$$

where the summation is extended over all indices i for which $x_i \le x$.

This relation is illustrated in Figure 3.11. From (3.5) and (3.6) we observe that $p(x)$ completely determines probabilities for events of the form

$$[X \le a], [X < a], [X > a], [X \ge a]$$
$$[X = a], [a < X < b], [a \le X < b], [a < X \le b], [a \le X \le b]$$

We also notice, from (3.6), that the probabilities for these events are defined completely either by the distribution function $F(x)$ or by the density function $p(x)$, when X is a discrete random variable. Hence, in practical application, we work with either $F(x)$ or $p(x)$, whichever is the easier one to use.

In cases where the range of X consists of a set of integers, we shall frequently use the notation p_x or p_i, in place of $p(x)$ or $p(i)$, for the density function. Thus, $p_i = \Pr[X = i]$ for i an integer, $0 \le p_i \le 1$, $\sum p_i = 1$, and $F(X) = \sum_{i \le x} p_i$.

A useful analogy in studying the probability distribution induced on the real line by a random variable is the analogy of a *mass distribution*. If we assume that a total mass of 1 unit is spread along the real line in such a way that the *probability* associated with any set of points is just the *mass* of the set, then

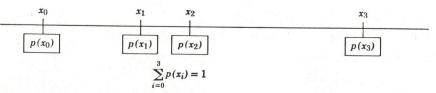

FIGURE 3.12   A physical interpretation of a density function as a mass distribution.

probabilities can be interpreted as masses. With this interpretation the distribution function $F(x)$ is just the mass of the part of the line lying to the left of x (and including x). A *discrete* distribution is one in which the mass distribution consists entirely of *point masses* at the points $x_0, x_1, x_2, \ldots$. The density function $p(x)$, in this case, gives only the mass at the point x and, hence, is zero unless x is one of the points $x_0, x_1, x_2, \ldots$, and $p(x_i)$ is the mass attached to the point x_i (see Figure 3.12).

We find such a physical model frequently useful for suggesting many operations in probability theory that are analogous to well-known formulas in mechanics.

EXAMPLE 3.12

In Example 3.6, in which X represents the number of heads in three tosses of a coin, the distribution function $F(x)$ is given by Figure 3.6. The corresponding density function (see Figure 3.13) is given by

$$p_0 = p(0) = \Pr[X = 0] = \tfrac{1}{8}$$
$$p_1 = p(1) = \Pr[X = 1] = \tfrac{3}{8}$$
$$p_2 = p(2) = \Pr[X = 2] = \tfrac{3}{8}$$
$$p_3 = p(3) = \Pr[X = 3] = \tfrac{1}{8}$$

and

$$p(x) = 0 \qquad \text{for } x \neq 0, 1, 2, 3$$

EXAMPLE 3.13

The preceding example can be generalized as follows. Suppose that an honest coin is tossed n times and let $X =$ the number of heads observed. Then X is a discrete random variable with range $0, 1, 2, \ldots, n$. If i is an integer in this range, $0 \le i \le n$, then

$$p_i = p(i) = \Pr[X = i] = \Pr[i \text{ heads in } n \text{ tosses}]$$

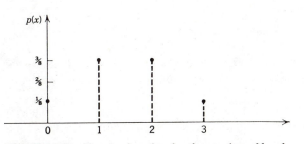

FIGURE 3.13 Density function for the number of heads in three coin tosses.

By formula (2.13) we know that

$$p_i = \binom{n}{i} \frac{1}{2^n} \tag{3.7}$$

for $i = 0, 1, 2, \ldots, n$. If i has any value other than these values, $p_i = 0$.

For each real x,

$$F(x) = \sum_{i \le x} p_i = \frac{1}{2^n} \sum_{i \le x} \binom{n}{i}$$

$F(x)$ will be a step-function with jumps at $x = 0, 1, \ldots, n$, and holding constant between jumps. $F(x) = 0$ for $x < 0$ and $F(x) = 1$ for $x \ge n$. This function is sketched for $n = 4$ in Figure 3.10.

EXAMPLE 3.14

In Example 2.4 the experiment of tossing a coin until the first head appeared was discussed. If the random variable X represents the number of tosses required, then, from the results of that section, we observe that X is a discrete random variable whose range is the set of all positive integers $1, 2, 3, \ldots$ (an infinite but countable sequence) and whose density function is given by

$$p_i = \Pr[X = i] = \frac{1}{2^i}, \qquad \text{for } i = 1, 2, 3, \ldots$$

Clearly this set $\{p_i\}$ does define a density function, since

$$p_i = 0, \qquad \text{unless } i = 1, 2, \ldots$$

$$0 \le p_i \le 1, \qquad \text{for } i = 1, 2, \ldots$$

$$\sum_{i=1}^{\infty} p_i = 1$$

The last point can be determined easily from Formula A.3 of Appendix A. Thus,

$$\sum p_i = \sum \frac{1}{2^i} = \frac{1/2}{1 - 1/2} = 1$$

3.7 Continuous Random Variables

We noted in Section 3.5 that a random variable X is said to be continuous if its distribution function is continuous. This definition, although usually used in the advanced theory of probability, would force us to adopt tools of analysis beyond the scope of this book to study such distributions. Thus, we shall restrict our definition of a continuous random variable. In so doing, we shall not seriously restrict the set of random variables available to us for application, but shall simplify the necessary mathematics, Thus, for our purposes, we use the following.

DEFINITION A random variable X is *continuous* if its distribution function $F(x)$

(a) is continuous,

(b) has a derivative $f(x) = (d/dx)F(x)$ for all values of x, with possibly a finite number of exceptions, and

(c) this derivative is piecewise continuous. $\qquad\qquad$ (3.8)

Under these circumstances, the range of X will consist of one or more intervals, finite or infinite.

In this case we readily observe that probabilities of the form $\Pr[X = x]$ are all zero;

$$\Pr[X = x] = 0 \qquad\qquad (3.9)$$

This follows from the fact that $\Pr[X = x] \le \Pr[x - h < X \le x] = F(x) - F(x - h)$; and this difference approaches 0 as $h \to 0 +$ if $F(x)$ is continuous. (Note that this is an example of an event with probability 0 that is not necessarily impossible or $\varnothing$.) It follows that formula (3.5) is not an appropriate definition for the density function here. It is more useful to try to generalize formula (3.6). The natural generalization of a sum here is an integral. Thus we define the *density function* $f(x)$, in this case to be the derivative given in property (b) of the definition

$$f(x) = \frac{d}{dx}F(x)$$

Then formula (3.6) can be generalized by

$$F(x) = F(x) - F(-\infty)$$
$$= \int_{-\infty}^{x} \frac{d}{dt}F(t)\, dt$$
$$= \int_{-\infty}^{x} f(t)\, dt \qquad\qquad (3.10)$$

The analogy with (3.6) is clear, with the sum being replaced by an integral. We notice from (3.10) that the distribution function of a continuous random variable is determined once its density function is given, and conversely.

By using the result of Section 3.5(b) we have

$$\Pr[a < X \le b] = F(b) - F(a) = \int_{a}^{b} f(t)\, dt \qquad\qquad (3.11)$$

We recall that $\Pr[X = b] = 0$, by (3.9), and thus the left side of (3.11) can be written as $\Pr[a \le X \le b]$ or $\Pr[a < X < b]$ with equal validity.

The function $f(x) = F'(x)$ has these properties:

(i) $f(x) = 0$ if x is not in the range of X (since outside of this range $F(x)$ is constant and its derivative is 0).

(ii) $f(x) \ge 0$ for all x (since $F(x)$ is a nondecreasing function).

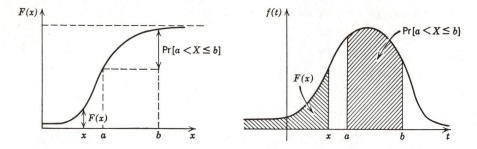

FIGURE 3.14   The determination of the probabilities for the event $[a < X \leq b]$ from the distribution function $F(x)$ and the density function $f(t)$ for a continuous random variable.

(iii) $\displaystyle\int_{-\infty}^{\infty} f(x)\, dx = 1$ [by letting $x \to +\infty$ in (3.10)].

(iv) $f(x)$ is piecewise continuous.

Any function that has these properties is the density function for *some* continuous random variable.

The connection between the functions f and F is illustrated in Figure 3.14. *Probabilities* are represented by *areas* under the f curve. The total area under the f curve is 1.

An alternate interpretation again is to consider the probability distribution as a one-dimensional *mass distribution* of total mass 1 distributed along the x-axis. The probability associated with any interval is then equal to its mass; $F(x)$ is the mass of the interval from $-\infty$ to x; $f(x)$ is the mass density (mass per unit length) at the point x.

Since any actual measuring device can have only a finite number of possible readings, no real-life experiment and measurement can possibly yield a continuous random variable. Nonetheless, even for practical applications, it is convenient to have such random variables in our mathematical model. They are frequently much easier to handle analytically, and we shall observe that often a simple continuous distribution can be used to approximate a complicated discrete distribution (see Figure 3.15).

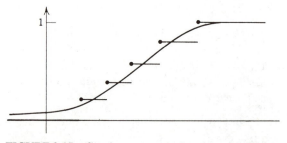

FIGURE 3.15   Continuous approximation to a discrete distribution function.

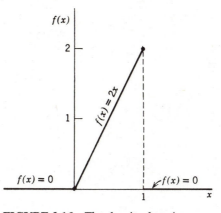

FIGURE 3.16   The density function
$$f(x) = \begin{cases} 0, & \text{if } x \leq 0 \\ 2x, & \text{if } 0 < x \leq 1 \\ 0, & \text{if } x > 1 \end{cases}$$

EXAMPLE 3.15

The function

$$f(x) = \begin{cases} 2x, & \text{if } 0 \leq x \leq 1 \\ 0, & \text{otherwise} \end{cases}$$

is a density function satisfying (i) to (iv) above, since $f(x) \geq 0$, for all x, $\int_{-\infty}^{\infty} f(x)\,dx = \int_{-\infty}^{0} 0\,dx + 2\int_{0}^{1} x\,dx + \int_{1}^{\infty} 0\,dx = 1$, and $f(x)$ is piecewise continuous with a discontinuity at $x = 1$.

A graph of $f(x)$ is given in Figure 3.16.

The distribution function corresponding to this density function is, by (3.10),

$$F(x) = \int_{-\infty}^{x} f(x)\,dx = \begin{cases} 0, & \text{if } x < 0 \\ x^2, & \text{if } 0 \leq x \leq 1 \\ 1, & \text{if } x > 1 \end{cases}$$

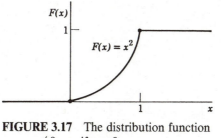

FIGURE 3.17   The distribution function
$$F(x) = \begin{cases} 0, & \text{if } x \leq 0 \\ x^2, & \text{if } 0 < x \leq 1 \\ 1, & \text{if } x > 1 \end{cases}$$

A graph of $F(x)$ is given in Figure 3.17.

From these two functions we can determine probabilities such as

$$\Pr\left[X \le \tfrac{1}{2}\right] = F\left(\tfrac{1}{2}\right) = \int_0^{1/2} 2x\,dx = \tfrac{1}{4}$$

$$\Pr\left[\tfrac{1}{4} < X \le \tfrac{3}{4}\right] = \int_{1/4}^{3/4} 2x\,dx = F\left(\tfrac{3}{4}\right) - F\left(\tfrac{1}{4}\right) = \tfrac{1}{2}$$

Notice that, since $\Pr[X = b] = 0$ for any b, we obtain immediately

$$\Pr\left[\tfrac{1}{4} < X \le \tfrac{3}{4}\right] = \Pr\left[\tfrac{1}{4} \le X \le \tfrac{3}{4}\right] = \Pr\left[\tfrac{1}{4} \le X < \tfrac{3}{4}\right] = \Pr\left[\tfrac{1}{4} < X < \tfrac{3}{4}\right]$$

and in all cases this probability is $\tfrac{1}{2}$.

EXAMPLE 3.16

The *negative exponential* distribution function

$$F(x) = \begin{cases} 1 - e^{-\lambda x}, & \text{if } 0 \le x < \infty \\ 0, & \text{if } x < 0 \end{cases}$$

discussed in Example 3.8 is the distribution function of a continuous random variable. The corresponding density function

$$F(x) = F'(x) = \begin{cases} \lambda e^{-\lambda x}, & \text{if } 0 \le x < \infty \\ 0, & \text{if } x < 0 \end{cases}$$

is graphed in Figure 3.18.

Observe that, in this case,

$$\Pr[X > x] = \int_x^\infty f(t)\,dt = \int_x^\infty \lambda e^{-\lambda t}\,dt$$

$$= -e^{-\lambda t}\Big|_x^\infty = e^{-\lambda x}$$

for $x \ge 0$.

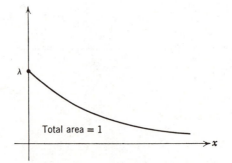

FIGURE 3.18 The negative exponential density function.

EXAMPLE 3.17

The function

$$F(x) = \begin{cases} 1 - e^{-x^b}, & \text{if } x \geq 0 \\ 0, & \text{if } x < 0 \end{cases}$$

where b is a positive constant, is a continuous distribution function. Its derivative

$$f(x) = \begin{cases} bx^{b-1}e^{-x^b}, & \text{if } x > 0 \\ 0, & \text{if } x < 0 \end{cases}$$

exists for all $x \neq 0$ and is at least piecewise continuous. Hence $f(x)$ is the density function, termed a *Weibull* density function, for some continuous random variable. This density function has found widespread application in the area of mechanical reliability theory (see Exercises 16 and 17 for a further discussion).

EXAMPLE 3.18

One of the most important density functions in the field of statistics is the *normal density* function. We shall discuss it in more detail in Chapter 7. We merely note here that the density function is given by the formula

$$f(x) = \frac{1}{\sqrt{2\pi}\,\sigma} e^{-(x-\mu)^2/2\sigma^2}, \qquad -\infty < x < \infty$$

In Chapter 7 we show that $f(x)$ is a proper density function for a continuous random variable. The constants μ, σ^2 represent two parameters that must be specified. In the special case $\mu = 0$, $\sigma = 1$, we speak of the *standard normal* density function. The graph of $f(x)$ is a bell-shaped curve symmetric about $x = \mu$ as shown in Figure 3.19.

Because of its wide applicability in describing real-life phenomena and, perhaps more importantly, because of the Central Limit Theorem (Section 7.10), the normal density function is one of the most useful continuous densities in probability and statistical theory.

The normal probability *distribution* function cannot be expressed simply in closed form. Of course, it is given by

$$F(t) = \int_{-\infty}^{t} \frac{1}{\sqrt{2\pi}\,\sigma} e^{-(x-\mu)^2/2\sigma^2} \, dx$$

which is a nonelementary integral. By the change of variable

$$z = \frac{x - \mu}{\sigma}$$

the computation of $F(t)$ can be transformed to the computation of the distribution function of a *standard* normal distribution ($\mu = 0$, $\sigma = 1$), which is tabulated

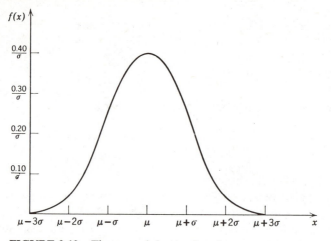

FIGURE 3.19   The normal density function.

in every elementary statistics book. A short table (Table B.1) is given in Appendix B.

A few examples will illustrate the use of this table in determining probabilities of interest.

First, suppose that $\mu = 0$, $\sigma = 1$, so that X itself is a standard normally distributed random variable. Then

(a)
$$\Pr[\,X \le 0\,] = F(0) = .5000$$

Since X is a continuous random variable, we also have

$$\Pr[\,X < 0\,] = .5000$$

(b)
$$\Pr[\,X > 1\,] = 1 - \Pr[\,X \le 1\,]$$

$$= 1 - F(1)$$

$$= 1 - .8413$$

$$= .1587$$

(Notice that $\Pr[\,X \ge 1\,]$ will also be .1587.)

(c)
$$\Pr[\,1 < X \le 2\,] = \Pr[\,X \le 2\,] - \Pr[\,X \le 1\,]$$

$$= F(2) - F(1) = .9772 - .8413$$

$$= .1359$$

Since this density function is symmetric about 0, we have

$$F(-x) = 1 - F(x)$$

Consequently,

(d) $\Pr[X \le -1] = F(-1) = 1 - F(1) = .1587$

(e) $\Pr[X > -1] = 1 - \Pr[X \le -1]$

$\qquad = \Pr[X \le 1]$

$\qquad = .8413$

(f) $\Pr[X > -\frac{1}{2}] = 1 - \Pr[X \le -\frac{1}{2}]$

$\qquad = \Pr[X \le \frac{1}{2}]$

$\qquad = .6915$

(g) $\Pr[-1 < X < -\frac{1}{2}] = \Pr[X \le -\frac{1}{2}] - \Pr[X \le -1]$

$\qquad = .1498$

(h) $\Pr[-\frac{1}{2} < X \le \frac{1}{3}] = F\left(\frac{1}{3}\right) - F(-\frac{1}{2})$

$\qquad = F(\frac{1}{3}) - [1 - F(\frac{1}{2})]$

$\qquad = F(\frac{1}{3}) - 1 + F(\frac{1}{2})$

$\qquad = .6306 - 1 + .6915$

$\qquad = .3221$

Suppose next that $\mu = 60$, $\sigma = 3$, so that X is normal but not standard normal. The substitution $z = (x - 60)/3$ will transform the integrals needed to compute probabilities for X into integrals that can be evaluated using the standard normal table. For example,

(i) $\Pr[X \le 60] = \displaystyle\int_{-\infty}^{60} f(x)\, dx$

$\qquad = \dfrac{1}{3\sqrt{2\pi}} \displaystyle\int_{-\infty}^{60} e^{-(x-60)^2/18}\, dx$

$\qquad = \dfrac{1}{\sqrt{2\pi}} \displaystyle\int_{-\infty}^{0} e^{-z^2/2}\, dz$

if the substitution $z = (x - 60)/3$ is made. Thus we determine from the standard normal table that $\Pr[X \le 60] = .5000$.

(j) $\Pr[57 \le X \le 64] = \displaystyle\int_{57}^{64} f(x)\, dx$

$\qquad = \dfrac{1}{3\sqrt{2\pi}} \displaystyle\int_{57}^{64} e^{-(x-60)^2/18}\, dx$

$\qquad = \dfrac{1}{\sqrt{2\pi}} \displaystyle\int_{-1}^{4/3} e^{-z^2/2}\, dz$

$\qquad = F(4/3) - F(-1)$

$\qquad = .9088 - (1 - .8413)$

$\qquad = .5701$

using the substitution $z = (x - 60)/3$ and the standard normal table. Here F refers to the standard normal distribution function given in the table.

Many other special probability distribution and density functions arise in applications. A more thorough discussion of the most important of these special distributions will be deferred until Chapter 7.

Exercises 3

1. Using the results of Examples 2.17, 3.2, and 3.5, describe in words the following events and compute their probabilities.

   (a) $\{s: X(s) \le x\}$ when $x < 0$.   (c) $\{s: X(s) \le x\}$ when $0 < x < 1$.
   (b) $\{s: X(s) \le x\}$ when $x \le 0$.   (d) $\{s: X(s) \le x\}$ when $x \ge 1$.

2. The function

$$F(x) = \begin{cases} 1 - e^{-\lambda x}, & x > 0 \\ 0, & x \le 0 \end{cases}$$

   is a distribution function for each $\lambda > 0$. Using it, determine

   (a) $\Pr[X > 1]$.   (c) $\Pr[2 < X < 3]$.
   (b) $\Pr[2 \le X \le 3]$.   (d) $\Pr[X > y]$ for any $y > 0$.

3. The function

$$F(x) = \begin{cases} C \ln x, & 1 \le x \le 2 \\ 0, & x < 1 \\ 1, & x > 2 \end{cases}$$

   is a distribution function for some C. Determine that C and show $F(x)$ is a distribution function.

4. In Example 6.17 we will study some data on the time between flights of a particular aphid studied by Broadbent. We will see that $X =$ time between two consecutive flights of the aphid has the distribution function approximately given by

$$F(x) = \begin{cases} 0, & x \le 0 \\ 1 - e^{-x/15}, & x > 0 \end{cases}$$

   Determine the following probabilities numerically and give a physical interpretation of them in view of the experiment.

   (a) $\Pr[X \le 30]$.   (e) $\Pr[X < 45]$.
   (b) $\Pr[X \le 3]$.   (f) $\Pr[X > 45]$.
   (c) $\Pr[3 < X \le 30]$.   (g) $\Pr[30 < X \le 45]$.
   (d) $\Pr[X < 30]$.

5. For what value of C is the following a distribution function?

$$F(x) = \begin{cases} 0, & x < 0 \\ \sin x, & 0 \le x \le C \\ 1, & x > C \end{cases}$$

6. Later we will give data on an experiment conducted by Mosteller and Wallace (1963)[1] on the use of language by Alexander Hamilton. Their study attempted to unravel a longstanding question of who wrote the 12 disputed *Federalist Papers*. We will show that if $X =$ the number of times Hamilton used the word "an" in a given text, then a good distribution function for that random variable is

$$F(x) = \begin{cases} 0, & \text{if } x < 0 \\ \sum_{i \le x} \dfrac{(1.24)^i e^{-1.24}}{i!}, & \text{if } x \ge 0 \end{cases}$$

Using this result, compute the following probabilities.

(a) $\Pr[X \le 0]$.   (e) $\Pr[0 \le X \le 1]$.
(b) $\Pr[0 < X < 1]$.   (f) $\Pr[X > 4]$.
(c) $\Pr[0 < X \le 1]$. (g) $\Pr[X \ge 4]$.
(d) $\Pr[0 \le X < 1]$. (h) $\Pr[X = 0]$.

7. An investigation of work times of a hospital admittance desk found that the admittance nurse performs two jobs depending on the type of patient she is admitting. Either she merely types out a history of the patient, or the patient needs immediate assistance and she also conducts him to his room. Measurements taken on the time to do her job lead to the conjecture that her work time has the following distribution function.

$$F(x) = \begin{cases} 0, & x < 0 \\ p(1 - e^{-2p\lambda x}) + q(1 - e^{-2q\lambda x}), & x \ge 0 \end{cases}$$

where p, q, and λ are positive constants with $p + q = 1$. Let $p = 1/3$, $q = 2/3$, $\lambda = 1/60$. Let $X =$ the time required by this nurse to admit a patient. Determine

(a) $\Pr[X \le 60]$.   (c) $\Pr[X > 20]$.
(b) $\Pr[X \le 5]$.   (d) $\Pr[80 < X \le 150]$.

8. Recall Examples 2.17, 3.2, and 3.5. Suppose a type-0 call has just occurred. Determine the density function for the distribution function $F(x)$ given in Example 3.5.

9. Suppose that

$$S = \{ s \colon 5 \le s \le 6 \}$$

[1]F. Mosteller and D. L. Wallace (1963), "Inference in an Authorship Problem," *Journal of the American Statistical Association*, **58**, pp. 275–309.

and that probabilities are assigned to subintervals $c < s < d$ of the interval from 5 to 6 so that intervals (c, d) of equal length are equally likely. Let $X(s) = s$. Determine

(a) $\Pr[x \leq 5\frac{1}{4}]$.

(b) $\Pr[X \leq 5\frac{3}{4}]$.

(c) $\Pr[5\frac{1}{8} < X \leq 5\frac{2}{3}]$.

(d) $F(x)$.

(e) $f(x)$.

(f) Sketch $F(x)$ and $f(x)$.

(*Note*: The probability density function here is termed a *uniform* density function.)

10. An honest coin is tossed until the first head appears. Define a sample space for this experiment. Let Y be the random variable that assigns to each sample point the number of tails until the first head appears. Determine and sketch the distribution function $F(y)$.

11. A study at a bank led to the conjecture that the number of customer arrivals in a 15-minute period is a discrete random variable X with density function

$$p_i = \begin{cases} \lambda^i e^{-\lambda}/i!, & i = 0, 1, 2, \ldots \\ 0, & \text{otherwise} \end{cases}$$

Show that this is a proper density function. This density function is called the *Poisson density function*. We shall encounter it again in later chapters.

12. Using the Poisson density function in Exercise 11 determine the following probabilities in terms of the constant λ

(a) $\Pr[X \geq 1]$.

(b) $\Pr[X < 2]$.

(c) $\Pr[2 \leq X \leq 4]$.

Interpret these probabilities in terms of bank arrivals.

13. Consider a sequence of independent experiments each having only two possible outcomes, "success" or "failure," so that $\Pr["\text{success}"] = p$ and $\Pr["\text{failure}"] = q$, $p + q = 1$, on each trial. Define the random variable X to be the number of trials required until the first "success" occurs. If $p > 0$, show that X has a range of all positive integers $1, 2, 3, \ldots$ and density function

$$p_i = pq^{i-1} = (1 - q)q^{i-1}$$

(Notice that Exercise 10 is a special case of this with $Y = X - 1$ and $p = q = \frac{1}{2}$.)

14. The lifetime (in hours) of a certain piece of equipment is a continuous random variable X having range $0 < x < \infty$ and density function

$$f(x) = \begin{cases} \dfrac{xe^{-x/10}}{100}, & x \geq 0 \\ 0, & \text{otherwise} \end{cases}$$

(a) Show that this is a density function.

(b) Compute the probability that the lifetime exceeds 20 hours, $\Pr[X > 20]$.

15. A continuous random variable X is said to be *triangularly distributed* on the interval $[a, b]$ if it has density function

$$f(x) = \begin{cases} 0, & \text{if } x < a \\ C(x - a), & a \leq x \leq \dfrac{a+b}{2} \\ C(b - x), & \dfrac{a+b}{2} \leq x \leq b \\ 0, & x > b \end{cases}$$

Sketch $f(x)$. For what value of C is $f(x)$ a density function? For the particular case $a = 1$, $b = 3$, determine

(a) $\Pr[X \geq 2]$.   (c) $\Pr[X \geq \frac{7}{3}]$.

(b) $\Pr[\frac{5}{3} \leq X \leq \frac{7}{3}]$.   (d) $F(x)$ and sketch it.

16. In the study of the reliability of electrical and mechanical parts, the lifetime X of a given part can be considered as a random variable. If X has the density function

$$f(x) = \frac{b}{\theta - x_0} \left(\frac{x - x_0}{\theta - x_0} \right)^{b-1} \exp\left[-\left(\frac{x - x_0}{\theta - x_0} \right)^b \right], \qquad x_0 \leq x < \infty$$

then X is said to have a general *Weibull* distribution. By using the substitution $u = [(x - x_0)/(\theta - x_0)]^b$, show that $f(x)$ is a density function for $b > 0$, $0 < x_0 < \theta$. Find $F(x)$ for such a random variable. Show that $\ln\{\ln[1/(1 - F(x))]\}$ is a linear function of $\ln(x - x_0)$ with slope b. One calls b the *Weibull slope*. Show that $\Pr[X \leq \theta]$ does not depend on b or x_0. θ is called the *characteristic life* of a part.

17. If X has a Weibull distribution, as in Exercise 16, determine

(a) $\Pr[X \leq a]$, if $a > x_0$.

(b) $\Pr[X \geq a]$, if $a > x_0$.

(c) $\Pr[a \leq X \leq c]$, if $c > a > x_0$.

18. Using the table of the normal distribution function given in Appendix B, determine these probabilities for a random variable having a standard normal distribution.

(a) $\Pr[X \leq 3]$.   (f) $\Pr[-2 < X < 2]$.

(b) $\Pr[X \leq -2]$.   (g) $\Pr[\frac{1}{2} < X < \frac{3}{4}]$.

(c) $\Pr[X \geq 2]$.   (h) $\Pr[-\frac{1}{2} < X < \frac{3}{4}]$.

(d) $\Pr[X \geq 3]$.   (i) $\Pr[-\frac{3}{4} < X < -\frac{1}{2}]$.

(e) $\Pr[2 < X < 3]$.   (j) $\Pr[-3 < X < 3]$.

19. Sketch the standard normal density function and shade the areas under it corresponding to each of the probabilities in Exercise 18.

20. Suppose X is a normally distributed random variable with parameters μ and σ. Using the standard normal table, find

(a) $\Pr[X < \mu]$.

(b) $\Pr[\mu - \sigma < X < \mu + \sigma]$.

(c) $\Pr[\mu - 2\sigma < X < \mu + 2\sigma]$.

(d) $\Pr[\mu - 1.96\sigma < X < \mu + 1.96\sigma]$.

(e) $\Pr[\mu - 1.645\sigma < X < \mu + 1.645\sigma]$.

(f) $\Pr[|X - \mu| > 3\sigma]$.

21. Cramer (1954)[1] studied the mean daily temperature in Stockholm in June for the years 1841 to 1940. He found that these temperatures were normally distributed with $\mu = 14.28°C$ and $\sigma = 1.57°C$. Using these results and the table in Appendix B compute the following probabilities for $X =$ mean daily temperature on a day in June in Stockholm.

(a) $\Pr[X \le 10]$.

(b) $\Pr[13.50 < X \le 14.70]$.

(c) $\Pr[X > 20.20]$.

(d) $\Pr[18.30 < X \le 19.80]$.

(e) $\Pr[X \le 20.70]$.

(f) $\Pr[X > 14.28]$.

(g) $\Pr[14.28 < X \le 19.0]$.

(h) $\Pr[9.50 < X \le 14.28]$.

(i) $\Pr[9.50 < X \le 11.72]$.

(j) $\Pr[13.78 < X \le 15.00]$.

22. Determine the density function for the random variable in Exercise 6.

23. Determine the density function for the distribution function in Exercise 7. Sketch that density function. On the same graph, sketch $f(x) = \lambda e^{-\lambda x}$ for the same λ to see how these two density functions differ.

24. Parsons (1975)[2] gives the following data (Table 3.3) for passengers arriving at a check-in point for domestic flights that leave between 4:00 A.M. and 10:00 A.M. If we take these percentages as our estimate of probabilities and if we let E_i be the event "the customer checks in between i and $i - 10$ minutes early," $i = 10, 20, \ldots 90$, and $E_0 = $"the passenger is late," what is the probability of the events E_{50} and E_0? What is the probability of the event $E_{50} \cup E_{40}$? Compute the following probabilities.

(a) $\Pr[E_{90} \cup E_{10}]$.

(b) $\Pr[E_{90} \cup E_{80} \cup E_{70}]$.

(c) $\Pr\left[\bigcup_{i=10}^{90} E_i\right]$.

(d) $\Pr\left[\left(\bigcup_{i=10}^{50} E_i\right) \cup (E_{60})\right]$.

(e) $\Pr\left[\left(\bigcup_{i=10}^{50} E_i\right) \cup \left(\bigcup_{i=60}^{90} E_i\right)\right]$.

[1] H. Cramer (1954), *Mathematical Methods of Statistics*, (2nd edition), Princeton University Press, Princeton, New Jersey.

[2] Ralph M. Parsons Company (1975), *The Apron and Terminal Building Planning manual*, Report No. FAA-RD-75-191, prepared for the U.S. Department of Transportation, Federal Aviation Administration, Systems Research and Development Service, Washington, D.C.

TABLE 3.3
Percentage of Passengers Arriving at Indicated Minutes before Flight Time

90 −80	80 −70	70 −60	60 −50	50 −40	40 −30	30 −20	20 −10	10 −0	LATE
1%	2%	5%	9%	14%	20%	20%	17%	10%	2%

25. In his study of power generating equipment, DiMarco (1972)[1] found 1061 cases of double system failure in 10 years of data on three turbine-boiler systems. Table 3.4 gives the number of times a failure in system i ($i = 1, 2, 3$) was followed by a failure in system j. If $[X = i]$ = the first failure occurred in system $i = 1, 2, 3$ and $[Y = j]$ = the second failure was in system $j = 1, 2, 3$, compute a relative frequency estimate of the $\Pr[X = i]$ for $i = 1, 2, 3$ and $\Pr[Y = j]$ for $j = 1, 2, 3$.

TABLE 3.4
Sequential Boiler Failures

i	j:	1	2	3
1		412	54	165
2		54	19	25
4		165	24	143

26. A random variable X having range $0 < x < \infty$ and density function of the form

$$f(x) = \begin{cases} Cx^{\alpha-1}e^{-\beta x}, & 0 < x < \infty \\ 0, & \text{otherwise} \end{cases}$$

for some positive constants α, β, and C is said to have a *gamma density*.

(a) Whenever $\alpha > 0$, $\beta > 0$, show that $f(x)$ is a density function provided that the constant C has the value

$$C = \frac{\beta^{\alpha}}{\Gamma(\alpha)}$$

where $\Gamma(\alpha)$ is defined in Appendix A.10.

(b) Sketch $f(x)$ in the cases $\alpha = 2, 4$, $\beta = 1$. Notice that the case $\alpha = 1$ gives the negative exponential distribution.

The results of Exercises 27–30 and 31–33 are formalized and extended in Sections 7.3, 7.5, and 7.10.

27. Using the results of Examples 3.5 and 3.2, look at 10 consecutive telephone calls. We ask for the density function for the random variable X = the number of calls out of the 10 that are of type-0. [*Hint:* See Example 2.12,

[1]A. E. DiMarco (1972), *The Intermediate-Term Security of a Power Generating System*, Ph.D. dissertation, Department of Industrial and Operations Engineering, University of Michigan, Ann Arbor, Michigan. Reprinted with permission of the author.

formula (2.13), to help in the computation here.] Sketch this density for $X = 0, 1, 2, \ldots, 6$.

28. In Exercise 27 above suppose we look at 20 consecutive calls rather than 10. Let X be the number of type-0 calls among these 20.

29. (continuation) In Exercise 27 above suppose we observe 50 consecutive calls rather than 10. Let X be the number of type-0 calls among these 50. What is the density function of X? Sketch it for $x = 0, 1, 2, \ldots, 6$.

30. (continuation) What happens to the shape of the density function for X when Exercises 27, 28, and 29 are compared? What assumptions have you made about these calls in using formula (2.13)?

31. Sketch the density function for the $F(x)$ described in Exercise 6.

32. Repeat Exercise 31 with 1.24 replaced by 3.47.

33. Repeat Exercise 31 with 1.24 replaced by 6.

4 · Sets of Random Variables and Random Sequences

4.1 Introduction

In Chapter 3 we were concerned with the properties of a single random variable defined on a given sample space. In many applications it is important to study two or more random variables defined on the same sample space. Thus, if we consider the experiment of selecting a person from a set S of people, and if X represents the height of the person selected and Y represents the person's weight, then X and Y are both random variables defined on the *same* sample space S. Space module designs might be interested in relations between these two quantities; that is, they would be interested in the *pair* (X, Y) of random variables.

In this chapter we consider the problem of two random variables, their associated distribution and density functions, and some properties, such as independence of the random variables. These concepts are extended easily in later sections to the case of many random variables defined on the same sample space. These ideas then lead naturally to the concept of a *random sequence* or a *discrete parameter random process*. We give some preliminary discussion to such processes. These concepts are extended and elaborated on in Chapters 9 to 13.

4.2 Pairs of Random Variables

Let X and Y be two random variables defined on some given sample space S. (For example, $X =$ height and $Y =$ weight of a randomly selected individual.) Each random variable is a function whose domain is S and whose range is some set of real numbers, as described in Section 3.2 (see Figure 4.1).

Suppose now that we consider X and Y to have values on real axes that are coordinate axes of the ordinary two-dimensional plane. In this manner, each sample point s defines a pair of real numbers, $x = X(s)$ and $y = Y(s)$, where (x, y) can be treated as the coordinates of some point in the plane. Thus, the pair (X, Y) of random variables can be considered as a function that to each point s in the sample space assigns a point (x, y) in the plane (see Figure 4.2). We shall find this geometric interpretation valuable.

By using this idea, we observe that we can construct correspondences between certain *events* (subsets of S) and certain *regions* in the (x, y) plane. For example, for fixed x_1 and y_1, the event $A = [X = x_1 \text{ and } Y = y_1]$ is the set of all sample points s having X-value x_1 and Y-value y_1, $\{s : X(s) = x_1 \text{ and } Y(s) = y_1\}$. The corresponding region in the (x, y)-plane, therefore, will consist only of the single point (x_1, y_1). [See Figure 4.3(a).]

73

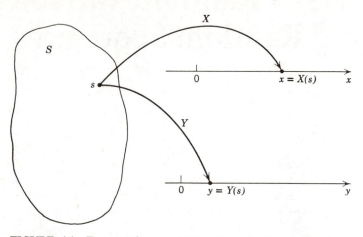

FIGURE 4.1 Two random variables X and Y considered as real-valued functions on the sample space.

The event $B = [X \leq x_1] = \{s : X(s) \leq x_1\}$ will correspond to the region consisting of all points (x, y) whose x-coordinate is less than or equal to x_1, $\{(x, y) : x \leq x_1\}$; that is, the region to the *left* of or on the line $x = x_1$. [See Figure 4.3(b).] Similarly, the event $C = [Y \leq y_1] = \{s : Y(s) \leq y_1\}$ will correspond to the region *below* or on the line $y = y_1$. [See Figure 4.3(c).]

If we consider the event consisting of all sample points, s, for which *both* $X(s) \leq x_1$ *and* $Y(s) \leq y_1$,

$$[X \leq x_1 \text{ and } Y \leq y_1] = [X \leq x_1] \cap [Y \leq y_1] = B \cap C \qquad (4.1)$$

then this will correspond to the *intersection* of the two shaded regions in Figures 4.3(b) and 4.3(c). [See Figure 4.3(d).]

Events of these forms, and the corresponding regions in the (x, y)-plane, will be the basic building blocks needed to analyze the probability structure of pairs of random variables.

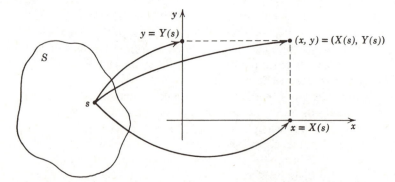

FIGURE 4.2 (X, Y) as a function from S to the plane.

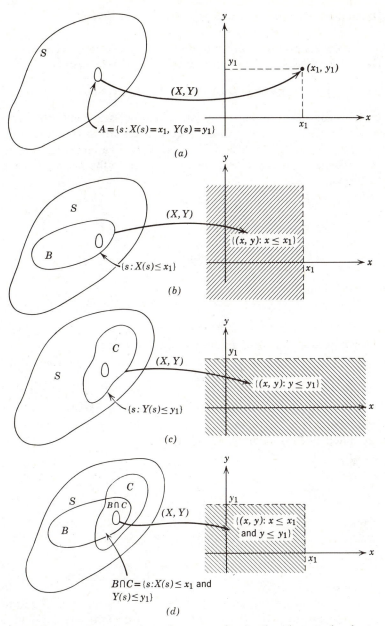

FIGURE 4.3   Correspondence between points in the plane and subsets of S.

4.3   Distribution Functions and Independent Random Variables

All of the events discussed in Section 4.2, being subsets of the sample space, will have probabilities. Thus, we can talk about probabilities of the form

$$\Pr[\,X \le x_1\,], \qquad \Pr[\,Y \le y_1\,], \qquad \Pr[\,X \le x_1 \text{ and } Y \le y_1\,]$$

The first two, considered as functions of x_1 and y_1, are simply the *distribution functions* of X and Y, as defined in Chapter 3. The third probability will depend on *both* x_1 *and* y_1 and, hence, is a function of two variables termed the *joint distribution function of X and Y*. For simplicity we drop the subscripts from the x and y and define

$$F_X(x) = \Pr[\,X \le x\,] = \Pr[\,A_1\,] \tag{4.2}$$

$$F_Y(y) = \Pr[\,Y \le y\,] = \Pr[\,A_2\,] \tag{4.3}$$

$$F(x, y) = \Pr[\,X \le x \text{ and } Y \le y\,] = \Pr[\,A_1 \cap A_2\,] \tag{4.4}$$

where

$$A_1 = \{\,s : X(s) \le x\,\}, \qquad A_2 = \{\,s : Y(s) \le y\,\} \tag{4.5}$$

$F(x, y)$ is called the *joint distribution function* of X and Y. If we must specify exactly which random variables are involved, we will write $F_{X,Y}(x, y)$ in place of $F(x, y)$. Thus, to each region in the (x, y)-plane of the form shaded in Figure 4.3(d), $F(x, y)$ assigns a number, a *probability*.

Recall that two events, A and B, were termed *independent* if $\Pr[\,A \cap B\,] = \Pr[\,A\,]\Pr[\,B\,]$. Extending these ideas, we define the random variables X and Y to be *independent* if the events A_1 and A_2 are independent *for every x and y*. Thus X and Y are independent random variables if

$$\Pr[\,A_1 \cap A_2\,] = \Pr[\,A_1\,]\Pr[\,A_2\,] \tag{4.6}$$

or

$$\Pr[\,X \le x \text{ and } Y \le y\,] = \Pr[\,X \le x\,]\Pr[\,Y \le y\,] \tag{4.7}$$

or, in terms of the distribution functions,

$$F(x, y) = F_X(x)F_Y(y) \tag{4.8}$$

for every value of x and y.

Thus the condition for independence is that $F(x, y)$ factor in the form (4.8). Actually, it is easy to show that if $F(x, y)$ factors into any product of two nonnegative functions, one of x only and the other of y only,

$$F(x, y) = u(x)v(y) \tag{4.9}$$

then, for some nonzero constant c, $cu(x) = F_X(x)$ and $(1/c)v(y) = F_Y(y)$, so that $F(x, y) = cu(x) \cdot (1/c)v(y) = F_X(x)F_Y(y)$ and we can conclude that X and Y are independent.

EXAMPLE 4.1

A Geiger counter is being bombarded by subatomic particles. Let X and Y represent the time intervals between the first and second and between the second and third clicks of the counter. Thus, X and Y are random variables with ranges $0 < x < \infty, 0 < y < \infty$. It can be shown that they have joint distribution function

$$F(x, y) = \begin{cases} 1 - e^{-ax} - e^{-ay} + e^{-a(x+y)}, & 0 < x < \infty, 0 < y < \infty \\ 0, & \text{otherwise} \end{cases}$$

where a is some positive constant. Thus, the probability that the first time interval does not exceed 2 sec and that the second time interval does not exceed 1 sec is

$$\Pr[X \le 2 \text{ and } Y \le 1] = 1 - e^{-2a} - e^{-a} + e^{-3a}$$

Here we observe that

$$F(x, y) = \begin{cases} (1 - e^{-ax})(1 - e^{-ay}), & 0 \le x < \infty, 0 \le y < \infty \\ 0, & \text{otherwise} \end{cases}$$

and it follows that $F(x, y)$ factors into the form (4.9) where

$$u(x) = \begin{cases} 1 - e^{-ax}, & 0 \le x < \infty; \\ 0, & \text{otherwise}; \end{cases} v(y) = \begin{cases} 1 - e^{-ay}, & 0 \le y < \infty \\ 0, & \text{otherwise} \end{cases}$$

Hence, X and Y are independent in this case. Here $u(x) = F_X(x)$ and $v(y) = F_Y(y)$.

4.4 Properties of Joint Distribution Functions

Since $F(x, y)$ is a function of two variables, its graphical representation would require three dimensions [one for x, one for y, and one for $F(x, y)$]. We notice the following properties that correspond to the properties of the distribution function of a single random variable discussed in Section 3.4.

(a) $0 \le F(x, y) \le 1$, for $-\infty < x < \infty, -\infty < y < \infty$. This follows since $F(x, y)$ is a probability.

(b) If $x_1 \le x_2$ and $y_1 \le y_2$, then $F(x_1, y_1) \le F(x_2, y_2)$. This follows since the region consisting of all points (x, y) satisfying $x \le x_1$ and $y \le y_1$ will lie *inside* the region $x \le x_2$ and $y \le y_2$ (see Figure 4.4). It follows that $[X \le x_1$ and $Y \le y_1] \subset [X \le x_2$ and $Y \le y_2]$ and, hence, that $F(x_1, y_1) \le F(x_2, y_2)$.

$$\lim_{\substack{x \to \infty \\ y \to \infty}} F(x, y) = F(\infty, \infty) = 1$$

$$\lim_{x \to -\infty} F(x, y) = F(-\infty, y) = 0$$

$$\lim_{y \to -\infty} F(x, y) = F(x, -\infty) = 0$$

These properties seem reasonable since the event $[X \le x$ and $Y \le y]$ approaches a certain event if $x \to \infty$ and $y \to \infty$ and approaches an impossible

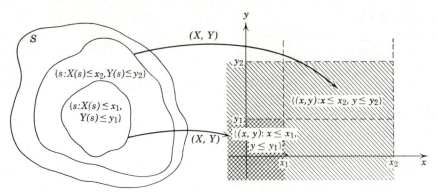

FIGURE 4.4   Graph of sets $[X \le x_1; \; Y \le y_1] \subset [X \le x_2; \; Y \le y_2]$ for $x_1 \le x_2$ and $y_1 \le y_2$.

event if $x \to -\infty$ or if $y \to -\infty$. A rigorous proof would follow the lines of the proof of property (c) of Section 3.4.

(d)
$$\lim_{x \to a+} F(x, y) = F(a, y)$$

$$\lim_{y \to b+} F(x, y) = F(x, b)$$

Again, this seems reasonable and a proof would be similar to the proof of property (d) of Section 3.4.

We notice that the function $F(x, y)$ in Example 4.1 satisfies properties (a) to (d). However, consider the following example.

EXAMPLE 4.2

Let

$$F(x, y) = \begin{cases} 1 - e^{-(x+y)}, & 0 \le x < \infty, 0 \le y < \infty \\ 0, & \text{otherwise} \end{cases}$$

We easily check that this function satisfies properties (a) to (d). In spite of this fact, $F(x, y)$ is *not* a joint distribution function. To determine this assume that there are random variables X and Y having $F(x, y)$ as joint distribution function. Consider the events

$$A = [X \le 1 \text{ and } Y \le 2]$$

$$B = [X \le 2 \text{ and } Y \le 1]$$

$$A \cap B = [X \le 1 \text{ and } Y \le 1]$$

and

$$C = [X \le 2 \text{ and } Y \le 2]$$

Clearly $A \subset C, B \subset C$, and, hence, $A \cup B \subset C$. Thus, $\Pr[A \cup B] \le \Pr[C]$ or $\Pr[C]$

$- \Pr[A \cup B] \geq 0$. But

$$\Pr[C] - \Pr[A \cup B] = \Pr[C] - \{\Pr[A] + \Pr[B] - \Pr[A \cap B]\}$$

using property (c) of Section 2.3. If these probabilities are all expressed in terms of $F(x, y)$, we have

$$\Pr[C] - \Pr[A \cup B] = F(2,2) - \{F(1,2) + F(2,1) - F(1,1)\}$$

$$= (1 - e^{-4}) - (1 - e^{-3}) - (1 - e^{-3}) + (1 - e^{-2})$$

$$= -(e^{-2} - e^{-1})^2 < 0$$

contradicting the fact that a probability must be ≥ 0. Hence, $F(x, y)$ is *not* a joint distribution function.

Example 4.2 shows that properties (a) to (d) are not sufficient to guarantee that $F(x, y)$ is a distribution function. An additional property is necessary.

(e) Whenever $a \leq b$ and $c \leq d$, then

$$F(a,c) - F(a,d) - F(b,c) + F(b,d) \geq 0$$

The left side of (e) represents $\Pr[a < X \leq b \text{ and } c < Y \leq d]$ (see Section 4.5) and hence (e) follows since a probability must be nonnegative.

The calculation in Example 4.2 shows that (e) fails in case $a = c = 1$, $b = d = 2$ and, hence, that the function of this example cannot be a distribution function.

If we let $y \to \infty$, then the event $[X \leq x \text{ and } Y \leq y]$ approaches the event $[X \leq x \text{ and } Y < \infty]$, which is the same as the event $[X \leq x]$ since the condition $Y < \infty$ is *always* satisfied (every real number is less than ∞). It follows that, as $y \to \infty$,

$$F(x, y) = \Pr[X \leq x \text{ and } Y \leq y]$$

$$\to \Pr[X \leq x] = F_X(x)$$

or

$$F(x, \infty) = F_X(x) \tag{4.10}$$

This argument is not quite rigorous, but it can be justified in a manner similar to the one for property (c).

Similarly,

$$F(\infty, y) = F_Y(y) \tag{4.11}$$

Formulas (4.10) and (4.11) show how the *individual* distribution functions of X and Y can be found from their *joint* distribution function. When the individual distribution functions are obtained in this way, they are sometimes termed *marginal* distribution functions.

The preceding discussion applies equally to discrete as well as to continuous random variables. Consider the following discrete example.

EXAMPLE 4.3

An honest coin is tossed three times and the following random variables are defined.

X = the number of heads on the first two tosses

$$(X = 0, 1, \text{ or } 2)$$

Y = the number of heads on the third toss

$$(Y = 0 \text{ or } 1)$$

The sample space S here consists of eight sample points

$$S = \left\{ \begin{matrix} s_1 & s_2 & s_3 & s_4 & s_5 & s_6 & s_7 & s_8 \\ HHH, & HHT, & HTH, & THH, & THT, & TTH, & HTT, & TTT \end{matrix} \right\}$$

All are equally likely, each having probability $\frac{1}{8}$.

The corresponding values of X and Y are given in Table 4.1.

We can then determine the following events.

$$[X \le 2 \text{ and } Y \le 1] = \{s_1, s_2, s_3, s_4, s_5, s_6, s_7, s_8\} = S$$

$$[X \le 2 \text{ and } Y \le 0] = \{s_2, s_5, s_7, s_8\}$$

$$[X \le 1 \text{ and } Y \le 1] = \{s_3, s_4, s_5, s_6, s_7, s_8\}$$

$$[X \le 1 \text{ and } Y \le 0] = \{s_5, s_7, s_8\}$$

$$[X \le 0 \text{ and } Y \le 1] = \{s_6, s_8\}$$

$$[X \le 0 \text{ and } Y \le 0] = \{s_8\}$$

Because the above events are all combinations of mutually exclusive, equally likely elementary events of the form $\{s_i\}$, we can construct Table 4.2 for the joint distribution function $F(x, y)$.

This table gives the value of $F(x, y)$ only for certain values of x and y, namely, $x = 0, 1, 2$ and $y = 0, 1$. However, the distribution function $F(x, y)$ is

TABLE 4.1
Two Random Variables for the
Same Coin-Tossing Experiment

S	$X(s)$	$Y(s)$
$s_1 = HHH$	2	1
$s_2 = HHT$	2	0
$s_3 = HTH$	1	1
$s_4 = THH$	1	1
$s_5 = THT$	1	0
$s_6 = TTH$	0	1
$s_7 = HTT$	1	0
$s_8 = TTT$	0	0

TABLE 4.2
The Joint Distribution Function
$F(x, y)$ **of** X, Y

x \ y	0	1
0	$\frac{1}{8}$	$\frac{1}{4}$
1	$\frac{3}{8}$	$\frac{3}{4}$
2	$\frac{1}{2}$	1

defined for *all* values of x and y. For values not in this table we need only use the fact that X and Y must take only these integer values. Thus,

$$F(\tfrac{3}{2}, \tfrac{1}{2}) = \Pr[X \le \tfrac{3}{2} \text{ and } Y \le \tfrac{1}{2}]$$
$$= \Pr[X \le 1 \text{ and } Y \le 0]$$
$$= F(1, 0) = \tfrac{3}{8}$$
$$F(7, 0) = \Pr[X \le 7 \text{ and } Y \le 0]$$
$$= \Pr[X \le 2 \text{ and } Y \le 0]$$
$$= F(2, 0) = \tfrac{1}{2}$$

The marginal distribution functions $F_X(x)$ and $F_Y(y)$ can be obtained from (4.10) and (4.11) and are given in Table 4.3 for these integer values of x and y. Again, these functions are extended to other values of X and Y by the fact that X and Y can take only these integer values.

The graphs of $F_X(x)$ and $F_Y(y)$ are, therefore, the step-functions sketched in Figure 4.5.

Notice that

$$F(0, 0) = \tfrac{1}{4} \cdot \tfrac{1}{2} = F_X(0) F_Y(0)$$
$$F(0, 1) = \tfrac{1}{4} \cdot 1 = F_X(0) F_Y(1)$$
$$F(1, 0) = \tfrac{3}{4} \cdot \tfrac{1}{2} = F_X(1) F_Y(0)$$
$$F(1, 1) = \tfrac{3}{4} \cdot 1 = F_X(1) F_Y(1)$$
$$F(2, 0) = 1 \cdot \tfrac{1}{2} = F_X(2) F_Y(0)$$
$$F(2, 1) = 1 \cdot 1 = F_X(2) F_Y(1)$$

TABLE 4.3
Distribution Functions for X **and** Y

x	$F_X(x) = \Pr[X \le x]$	y	$F_Y(y) = \Pr[Y \le y]$
0	$\frac{1}{4}$	0	$\frac{1}{2}$
1	$\frac{3}{4}$	1	1
2	1		

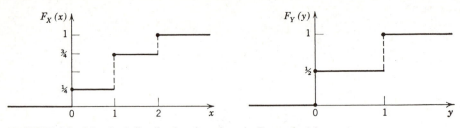

FIGURE 4.5   Marginal distribution functions in Example 4.3.

and, in general,

$$F(x, y) = F_X(x) \cdot F_Y(y)$$

Thus, we observe that these random variables X and Y are independent. Their joint distribution is given in Table 4.2 and their marginal distributions are given in Table 4.3 and are graphed in Figure 4.5. Notice that $F(x, y)$ does indeed satisfy properties (a) to (e).

4.5   Determining Probabilities from a Joint Distribution Function

If X and Y are given random variables, we frequently want to find the probabilities of certain events defined in terms of X and Y, for which we know the joint distribution function. Thus, any event of the form $[X \le x_1$ and $Y \le y_1]$ has probability $F(x_1, y_1)$. For more complicated events it is useful to sketch the region in the (x, y)-plane, in which the values of X and Y must lie, to define the required event. [For example, the region corresponding to the event $[X \le x_1$ and $Y \le y_1]$ is shaded in Figure 4.3(d).]

An event of the form $[a < X \le b$ and $Y \le y_1]$, for fixed $a < b$ and y_1, would correspond to the region R sketched in Figure 4.6.

In Figure 4.6 the event $[X \le b$ and $Y \le y_1]$ corresponds to the region shaded ($\backslash\backslash\backslash\backslash$), and in the event $[X \le a$ and $Y \le y_1]$ corresponds to the region shaded ($/////$). These two events will have probabilities $F(b, y_1)$ and $F(a, y_1)$. Clearly

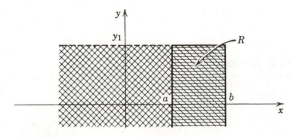

FIGURE 4.6   Region R defining the event $[a < X \le b$ and $Y \le y_1]$.

the region R corresponding to the event $[a < X \le b$ and $Y \le y_1]$ will be the difference between these two regions [shaded ($\equiv$) in the figure]. Hence

$$\Pr[a < X \le b \text{ and } Y \le y_1] = F(b, y_1) - F(a, y_1)$$

Analogously, we observe that

$$\Pr[X \le x_1 \text{ and } a < Y \le b] = F(x_1, b) - F(x_1, a)$$

Next, we wish to determine the probability of a more general event of the form $R = [a < X \le b$ and $c < Y \le d]$, for fixed $a < b, c < d$. Consider the events

$$A = [X \le b \text{ and } Y \le d] \supset R$$
$$B_1 = [X \le a \text{ and } Y \le d]$$
$$B_2 = [X \le b \text{ and } Y \le c]$$

We notice that any point of A that does not lie in R must lie in B_1 or B_2 (Figure 4.7). Thus,

$$R \cup (B_1 \cup B_2) = A$$

where, clearly, R and $B_1 \cup B_2$ are mutually exclusive.

Hence,

$$\Pr[R] + \Pr[B_1 \cup B_2] = \Pr[A]$$
$$\Pr[R] = \Pr[A] - \Pr[B_1 \cup B_2]$$
$$= \Pr[A] - (\Pr[B_1] + \Pr[B_2] - \Pr[B_1 \cap B_2])$$
$$= \Pr[A] - \Pr[B_1] - \Pr[B_2] + \Pr[B_1 \cap B_2]$$

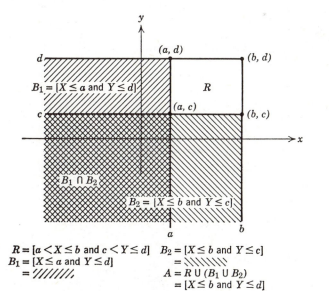

$R = [a < X \le b \text{ and } c < Y \le d]$ $B_2 = [X \le b \text{ and } Y \le c]$
$B_1 = [X \le a \text{ and } Y \le d]$ $= \backslash\backslash\backslash\backslash\backslash$
$\quad = /\!/\!/\!/\!/\!/$ $A = R \cup (B_1 \cup B_2)$
$\qquad\qquad\qquad\qquad\qquad\quad = [X \le b \text{ and } Y \le d]$

FIGURE 4.7 Regions corresponding to certain joint events.

But

$$\Pr[A] = F(b,d), \qquad \Pr[B_1] = F(a,d), \qquad \Pr[B_2] = F(b,c)$$

and

$$\Pr[B_1 \cap B_2] = \Pr[X \le a \text{ and } Y \le c] = F(a,c)$$

Thus,

$$\Pr[a < X \le b \text{ and } c < Y \le d] = F(b,d) - F(a,d) - F(b,c) + F(a,c)$$

$$(4.12)$$

[Compare (4.12) with Section 4.4, condition (e).]

Intuitively, to determine the probability of the event R we must deduct from the probability of the larger event A the probabilities of B_1 and B_2. But (Figure 4.7) this will deduct the probability of the intersection $B_1 \cap B_2$ *twice*; hence, it must be added back in once to obtain the required probability.

In case X and Y are independent, $F(x, y) = F_X(x) F_Y(y)$, formula (4.12) can be rewritten as follows.

$$\begin{aligned} \Pr[a < X \le b \text{ and } c < Y \le d] &= F_X(b)F_Y(d) - F_X(a)F_Y(d) \\ &\quad - F_X(b)F_Y(c) + F_X(a)F_Y(c) \\ &= [F_X(b) - F_X(a)][F_Y(d) - F_Y(c)] \\ &= \Pr[a < X \le b]\Pr[c < Y \le d] \end{aligned} \qquad (4.13)$$

In other words, independence of X and Y implies that every event of the form $[a < X \le b]$ is independent of every event of the form $[c < Y \le d]$.

EXAMPLE 4.4

Consider the distribution function in Example 4.1.

$$F(x, y) = \begin{cases} 1 - e^{-ax} - e^{-ay} + e^{-a(x+y)}, & 0 \le x < \infty \quad 0 \le y < \infty \\ 0, & \text{otherwise} \end{cases}$$

and, for simplicity, assume that $a = 1$. Thus,

$$F(1,1) = 1 - e^{-1} - e^{-1} + e^{-2} = .399$$

$$F(1,2) = 1 - e^{-1} - e^{-2} + e^{-3} = .447$$

$$F(2,1) = 1 - e^{-2} - e^{-1} + e^{-3} = .447$$

$$F(2,2) = 1 - e^{-2} - e^{-2} + e^{-4} = .748$$

and, by (4.12),

$$\begin{aligned} \Pr[1 < X \le 2 \text{ and } 1 < Y \le 2] &= F(2,2) - F(1,2) - F(2,1) + F(1,1) \\ &= .748 - .447 - .447 + .399 \\ &= .253 \end{aligned}$$

4.6 Density Functions: Discrete Case

A pair of random variables (X, Y) is said to have a *discrete* joint distribution if X and Y individually are *discrete* random variables in the sense discussed in Section 3.6. If X and Y individually are *continuous* random variables, as discussed in Section 3.7, then the pair (X, Y) is said to have a *continuous* joint distribution. The *mixed* case occurs when one of the random variables is discrete and the other continuous. When this happens, we usually can easily modify our discussion to take care of it (see Example 4.13). The discrete random variables that will be most important in our later discussions all will have ranges consisting of a set of *integers*. Thus, let us now assume that X and Y are *integer-valued* random variables. Then define

$$p(i, j) = \Pr[X = i \text{ and } Y = j] \tag{4.14}$$

If we interpret probability as a two-dimensional mass distribution, $p(i, j)$ represents the point mass attached to the point with coordinates (i, j). If the pair (i, j) is *not* a point in the joint range of (X, Y), then $p(i, j) = 0$.

Clearly $p(i, j)$ represents a function that assigns a number to each *pair* of integers (i, j). This function is termed the *joint density function* of X and Y. It has these properties:

(a) $0 \le p(i, j) \le 1$ for each i, j

(b) $\sum_{i, j} p(i, j) = \sum_i \sum_j p(i, j) = 1$

(c) $F(x, y) = \sum_{i \le x} \sum_{j \le y} p(i, j)$

If

$$p_i = \Pr[X = i], \qquad q_j = \Pr[Y = j]$$

then

$$p_i = \sum_j \Pr[X = i \text{ and } Y = j] = \sum_j p(i, j)$$

with a similar formula for q_j. Thus, we have

$$p_i = \sum_j p(i, j), \qquad q_j = \sum_i p(i, j) \tag{4.15}$$

The functions p_i and q_j are sometimes referred to as the *marginal density functions* of X and Y.

If X and Y are *independent* random variables, formula (4.13) can be applied as follows:

$$\begin{aligned}
p(i, j) &= \Pr[X = i \text{ and } Y = j] \\
&= \Pr[i - 1 < X \le i \text{ and } j - 1 < Y \le j] \\
&= \Pr[i - 1 < X \le i]\Pr[j - 1 < Y \le j] \\
&= \Pr[X = i]\Pr[Y = j] \\
&= p_i q_j
\end{aligned}$$

Thus, if X and Y are independent,

$$p(i, j) = p_i q_j \qquad (4.16)$$

and we can show that, conversely, the identity (4.16) implies that X and Y are independent.

EXAMPLE 4.5

Consider the coin-tossing experiment in Example 4.3. A coin is tossed three times. $X =$ the number of heads on the first two tosses and $Y =$ the number of heads on the third toss. The values of $p(i, j)$ are easily computed from Table 4.1 and are given in Table 4.4.

Identity (4.16) can be checked, in this case, from Table 4.4. Since X and Y are independent here, (4.16) must hold true. The distribution function $F(x, y)$ can also be determined from property (c) above and can be compared with the values in Table 4.2. Thus,

$$F(1,1) = \sum_{i \le 1} \sum_{j \le 1} p(i, j)$$

$$= p(0,0) + p(0,1) + p(1,0) + p(1,1)$$

$$= \tfrac{3}{4} = \left(\tfrac{3}{4}\right) \cdot (1) = F_X(1) F_Y(1)$$

4.7 Density Functions: Continuous Case

As in the single-variable case, formula (4.14) cannot be extended directly to define a density function in the continuous case since normally in this case, $\Pr[X = x \text{ and } Y = y] \equiv 0$. Instead we seek to generalize property (c) of Section 4.6, which relates the distribution and density functions. Since the natural continuous analogue of summation is integration, we define the *joint density function*, $f(x, y)$, of the continuous random variables X, Y as a function that, when

TABLE 4.4
Joint Density $p(i, j)$ for $X =$ Number of Heads on Two Coin Tosses and $Y =$ Number of Heads on Third Toss

i \ j	0	1	p_i
0	$\frac{1}{8}$	$\frac{1}{8}$	$\frac{1}{4}$
1	$\frac{1}{4}$	$\frac{1}{4}$	$\frac{1}{2}$
2	$\frac{1}{8}$	$\frac{1}{8}$	$\frac{1}{4}$
q_j	$\frac{1}{2}$	$\frac{1}{2}$	

integrated, yields $F(x, y)$; that is, we define $f(x, y)$ by

$$F(x, y) = \int_{-\infty}^{x} \int_{-\infty}^{y} f(s, t) \, dt \, ds \qquad (4.17)$$

To guarantee the existence and uniqueness of $f(x, y)$ we confine our attention here to pairs of continuous random variables, (X, Y), whose distribution function, $F(x, y)$, is continuous for all (x, y) and has a continuous mixed partial derivative $\partial^2 F / \partial x \, \partial y$ for all but possibly a finite number of values of x and of y. In this case, we can differentiate (4.17) twice, and can express $f(x, y)$ in terms of $F(x, y)$ as

$$f(x, y) = \frac{\partial^2 F(x, y)}{\partial x \, \partial y} \qquad (4.18)$$

Conversely, if (4.18) holds true, and $f(x, y)$ is sufficiently smooth, we can integrate and show that (4.17) must hold true.

The joint density function $f(x, y)$ satisfies (4.17) and has the following properties.

(a) $f(x, y) \geq 0$ for all x, y.

(b) $\int_{-\infty}^{\infty} \int_{-\infty}^{\infty} f(x, y) \, dy \, dx = 1$.

(c) $f(x, y)$ is continuous for all except possibly finitely many values of x or of y.

Property (b) follows from (4.17) by letting $x \to \infty$, $y \to \infty$, and property (c) holds true by our assumption concerning $\partial^2 F / \partial x \, \partial y$. Property (a) is true because of properties (b) and (e) of Section 4.4, although a rigorous proof of (a) would involve expressing the partial derivative of $F(x, y)$ as a limit of a difference quotient.

If R is any region of the two-dimensional plane, then the probability that (X, Y) lies in this region is obtained by integrating $f(x, y)$ over the region. In particular,

$$\Pr[a < X \leq b \text{ and } c < Y \leq d] = \int_{a}^{b} \int_{c}^{d} f(x, y) \, dy \, dx \qquad (4.19)$$

Since $\Pr[X = a] = 0 = \Pr[Y = c]$ [see (3.7)], it follows that the integral in (4.19) also represents $\Pr[a \leq X \leq b \text{ and } c \leq Y \leq d]$ and $\Pr[a < X < b \text{ and } c < Y < d]$. Thus, if $f(x, y)$ is represented geometrically by a surface, the volumes under this surface give probabilities (see Figure 4.8).

If we interpret $f(x, y)$ as a mass density, then integration over a plane region amounts to determining the total mass in that region.

By (4.10),

$$F_X(x) = F(x, \infty) = \int_{-\infty}^{x} \int_{-\infty}^{\infty} f(s, t) \, dt \, ds$$

and hence

$$f_X(x) = F_X'(x) = \int_{-\infty}^{\infty} f(x, t) \, dt$$

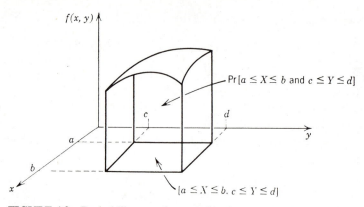

FIGURE 4.8   Probability as volume under the joint density surface.

or

$$f_X(x) = \int_{-\infty}^{\infty} f(x, y)\, dy \tag{4.20}$$

where $f_X(x)$ is the density function of the random variable X. Similarly,

$$f_Y(y) = \int_{-\infty}^{\infty} f(x, y)\, dx \tag{4.21}$$

When obtained in this way, $f_X(x)$ and $f_Y(y)$ are sometimes termed the *marginal density functions* of X and Y.

If X and Y are independent, the differentiation of formula (4.8) yields

$$f(x, y) = \frac{\partial^2}{\partial x\, \partial y}[F_X(x)F_Y(y)] = F_X'(x)F_Y'(y)$$

$$f(x, y) = f_X(x)f_Y(y) \tag{4.22}$$

Thus, if X and Y are independent, their joint density function factors into the product of the individual density functions. Conversely, if (4.22) holds true, then integration yields (4.8) and we would conclude that X and Y are independent.

Notice that if $f(x, y)$ is known, then $f_X(x)$ and $f_Y(y)$ can be determined by (4.20) and (4.21), but that, conversely, knowledge of $f_X(x)$ and $f_Y(y)$ does *not* in general determine $f(x, y)$. In the *independent* case, however, $f_X(x)$ and $f_Y(y)$ *do* determine $f(x, y)$, by (4.22).

EXAMPLE 4.6

Let X and Y be continuous random variables with ranges $0 < x < \infty$, $0 < y < \infty$ and joint density function

$$f(x, y) = \begin{cases} e^{-(x+y)}, & 0 \le x < \infty, 0 \le y < \infty \\ 0, & \text{otherwise} \end{cases}$$

In this case

$$f_X(x) = \int_{-\infty}^{\infty} f(x, y)\, dy$$

$$= \int_{0}^{\infty} e^{-(x+y)}\, dy, \qquad \text{since } f(x, y) = 0 \quad \text{for } y < 0$$

$$= e^{-x} \int_{0}^{\infty} e^{-y}\, dy$$

$$= \begin{cases} e^{-x}, & \text{for } 0 \le x < \infty \\ 0, & \text{for } -\infty < x < 0 \end{cases}$$

Similarly,

$$f_Y(y) = \begin{cases} e^{-y}, & 0 < y < \infty \\ 0, & \text{otherwise} \end{cases}$$

Notice that $f(x, y) = f_X(x) f_Y(y)$, so that X and Y are independent here. To compute a joint probability, we must integrate $f(x, y)$. For example,

$$\Pr[0 < X < 1 \text{ and } 1 < Y < 2] = \int_{0}^{1} \int_{1}^{2} f(x, y)\, dy\, dx$$

$$= \int_{0}^{1} \int_{1}^{2} e^{-(x+y)}\, dy\, dx$$

$$= \left(\int_{0}^{1} e^{-x}\, dx \right) \left(\int_{1}^{2} e^{-y}\, dy \right)$$

$$= (1 - e^{-1})(e^{-1} - e^{-2}) = .148$$

EXAMPLE 4.7

Let X and Y be continuous random variables with ranges $0 < x < \infty$, $0 < y < \infty$ and joint density function

$$f(x, y) = \begin{cases} xe^{-x(y+1)}, & 0 \le x < \infty, 0 \le y < \infty \\ 0, & \text{otherwise} \end{cases}$$

$$f_X(x) = \int_{0}^{\infty} xe^{-x(y+1)}\, dy = xe^{-x} \int_{0}^{\infty} e^{-xy}\, dy$$

$$= e^{-x}, \qquad 0 \le x < \infty$$

$$f_Y(y) = \int_{0}^{\infty} xe^{-x(y+1)}\, dx$$

$$= \frac{1}{(y+1)^2}, \qquad 0 \le y < \infty$$

We observe that $f(x, y) \ne f_X(x) f_Y(y)$, so that X and Y are *not* independent here. [In both integrations formula (A.10) in the Appendix was used.]

EXAMPLE 4.8   The Bivariate Normal Distribution

The natural generalization to bivariate distributions of the normal distribution discussed in Example 3.18 is the *bivariate normal distribution*. Two random variables, X and Y, are said to have this distribution if they have ranges $-\infty < x < \infty$, $-\infty < y < \infty$ and joint density function of the form

$$f(x, y) = \frac{1}{2\pi\sigma_X\sigma_Y\sqrt{1-\rho^2}} \exp\left\{\frac{-1}{2(1-\rho^2)}\left[\left(\frac{x-\mu_X}{\sigma_X}\right)^2\right.\right.$$

$$\left.\left. -2\rho\left(\frac{x-\mu_X}{\sigma_X}\right)\left(\frac{y-\mu_Y}{\sigma_Y}\right) + \left(\frac{y-\mu_Y}{\sigma_Y}\right)^2\right]\right\} \quad (4.23)$$

The quantities μ_X, μ_Y, σ_X, σ_Y, and ρ are constants with $\sigma_X > 0$, $\sigma_Y > 0$, and $-1 < \rho < 1$. Calculus computations show (i) that this is indeed a density function, (ii) that the marginal densities $f_X(x)$ and $f_Y(y)$ are normal, and (iii) that X and Y are independent if, and only if, $\rho = 0$.

The bivariate normal occurs rather often in sampling-type problems. An interesting area of application is concerned with gunfire. It is supposed that, because of barrel whip, aiming errors, windage, projectile defects, and many other things, a gun aimed to hit a point (x', y') will, in general, hit a different point, and the probability distribution for the point of impact can be modeled as a pair of random variables, (X, Y), where X is the horizontal deviation of the hit point and Y is the range deviation of the hit point, measured from some given origin. The probability density function for the impact point is given by (4.23) with $\mu_X = x'$ and $\mu_Y = y'$. Depending on the type of fire and the characteristics of the weapon, ρ may be practically 0 or it may be large.

EXAMPLE 4.9

The function

$$f(x, y) = \begin{cases} 2e^{-x-y}, & 0 \le x \le y, 0 \le y < \infty \\ 0, & \text{otherwise} \end{cases}$$

satisfies all of the properties of a joint density function. Furthermore,

$$f(x, y) = (2e^{-x})(e^{-y})$$

within the joint range. But

$$f_X(x) = 2\int_x^\infty e^{-x-y}\,dy$$

$$= 2e^{-x}\int_x^\infty e^{-y}\,dy$$

$$= 2e^{-2x}$$

$$f_Y(y) = 2\int_0^y e^{-x-y}\,dx$$

$$= 2e^{-y}[1 - e^{-y}]$$

Clearly

$$f(x, y) \neq f_X(x) f_Y(y)$$

and, thus, X and Y are not independent random variables. This illustrates the principle that for X and Y to be independent, $f(x, y)$ must factor into a function of x multiplied by a function of y for *all x and y, not just within the joint range of X and Y.*

4.8 Conditional Distributions

If two random variables are independent, all of the properties of their joint distribution are determined once their marginal density or distribution functions are known. In practical applications this provides considerable simplification, since these marginal distributions can be estimated separately (as, for example, in chapter 8) and from them the joint probability distribution and density functions can be determined using (4.8) and (4.16) or (4.22). In the case where the random variables are not independent the joint probabilities can no longer be determined in this simple fashion. Then the question is: What must we know to determine the joint probability density and distribution function of two random variables that are not independent?

We recall that if A and B are two events not necessarily independent, the joint probability of A and B is

$$\Pr[A \cap B] = \Pr[A|B]\Pr[B]$$

$$= \Pr[B|A]\Pr[A]$$

where $\Pr[A|B]$ ($\Pr[B|A]$) is the condition probability of A given that B has occurred (of B given that A has occurred). See Section 2.6.

In particular, if A and B are events of the form $A = [Y \leq y]$, $B = [X \leq x]$, we write

$$\Pr[Y \leq y | X \leq x] = \Pr[A|B]$$

If we hold x fixed and consider this probability as a function of y, it is termed the *conditional distribution function of Y given that $X \leq x$*:

$$G(y|x) = \Pr[Y \leq y | X \leq x]$$

$G(y|x)$ is well defined as $F(x, y)/F_X(x)$ as long as $F_X(x) \neq 0$. Otherwise $G(y|x)$ is not defined.

If X and Y are *discrete* random variables, then for any pair (i, j) in the joint range of (X, Y) let us define the following conditional probability.

$$p_{ij} = \Pr[Y = j | X = i]$$

$$= \frac{\Pr[X = i \text{ and } Y = j]}{\Pr[X = i]} = \frac{p(i, j)}{p_i}, \qquad p_i \neq 0 \qquad (4.24)$$

If p_{ij} is considered as a function of j, for fixed i, it is termed the *conditional density function* of Y given that $X = i$. For fixed i, clearly p_{ij} is a discrete density function since, using (4.15),

$$\sum_{j} p_{ij} = \sum_{j} \frac{p(i, j)}{p_i} = \left(\frac{\displaystyle\sum_{j} p(i, j)}{p_i} \right)$$

$$= \frac{p_i}{p_i} = 1$$

The notation p_{ij} does not display which random variable is the Y-variable whose conditional probabilities are involved and which is the X-variable on which the condition is placed. When using this notation, therefore, we must make the distinction clear from the discussion.

In case X and Y are continuous random variables, in analogy with (4.24) we define the quantity

$$f(y|x) = \frac{f(x, y)}{f_X(x)} \tag{4.25}$$

to be the *conditional density function of Y given that $X = x$* (provided $f_X(x) > 0$). For fixed x, $f(y|x)$ clearly is a density function since, using (4.20),

$$\int_{-\infty}^{\infty} f(y|x)\, dy = \int_{-\infty}^{\infty} \frac{f(x, y)\, dy}{f_X(x)} = \frac{f_X(x)}{f_X(x)} = 1$$

Also, since $f(x, y) \geq 0$ and $f_X(x) > 0$ we know that $f(y|x) \geq 0$.

To interpret $f(y|x)$ as the density function of a certain conditional distribution a limiting argument is necessary: For given x, y, and h, with $h > 0$ and

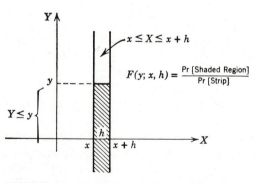

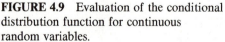

FIGURE 4.9   Evaluation of the conditional distribution function for continuous random variables.

$f_x(x) \neq 0$, define

$$F(y; x, h) = \Pr[Y \leq y | x \leq X \leq x + h]$$
$$= \frac{\Pr[x \leq X \leq x + h \text{ and } Y \leq y]}{\Pr[x \leq X \leq x + h]}$$
$$= \frac{\int_x^{x+h} \int_{-\infty}^y f(s, t) \, dt \, ds}{\int_x^{x+h} f_X(s) \, ds} \qquad (4.26)$$

(See Figure 4.9.)

Then define the *conditional distribution function* of Y, given $X = x$, by

$$F(y|x) = \lim_{h \to 0+} F(y; x, h)$$

Notice that this is *not* the same as $G(y|x)$ above.

So we have defined two conditional distributions $G(y|x)$ and $F(y|x)$ where

$$G(y|x) = \Pr[Y \leq y | X \leq x] = F(x, y)/F_x(x)$$

and we would like to interpret $F(y|x)$ as

$$F(y|x) = \Pr[Y \leq y | X = x]$$

We must show that this interpretation makes sense, because the conditional probability $\Pr[Y \leq y | X = x]$ normally will not exist according to the definition of conditional probability, since $\Pr[X = x] = 0$ in the continuous case.

To evaluate the limit defining $F(y|x)$ we use l'Hôpital's rule, which, we recall, states that in computing the limit of a quotient that reduces to $0/0$ (as here) one should compute instead the limit of the quotient of derivatives. We easily check that the hypotheses necessary to apply this rule are satisfied here. Thus,

$$F(y|x) = \lim_{h \to 0+} F(y; x, h) = \lim_{h \to 0+} \frac{\int_x^{x+h} \int_{-\infty}^y f(s, t) \, dt \, ds}{\int_x^{x+h} f_X(s) \, ds}$$

$$= \lim_{h \to 0+} \frac{\int_{-\infty}^y f(x + h, t) \, dt^{\,1}}{f_X(x + h)}$$

$$= \frac{\int_{-\infty}^y \lim_{h \to 0+} f(x + h, t) \, dt}{\lim_{h \to 0+} f_X(x + h)}$$

$$= \frac{\int_{-\infty}^y f(x, t) \, dt}{f_X(x)}$$

$$= \int_{-\infty}^y \frac{f(x, t)}{f_X(x)} \, dt = \int_{-\infty}^y f(t|x) \, dt \qquad (4.27)$$

where the interchange of limit and integral is also easily justified.

---

[1] Using the fact that $(d/dh) \int_x^{x+h} \varphi(s) \, ds = \varphi(x + h)$ by the Fundamental Theorem of Calculus.

Comparison of (4.27) with (3.10) shows that the conditional density function, $f(y|x)$, bears the same relation to the conditional distribution function, $F(y|x)$, as the ordinary density function bears to the ordinary distribution function. That is, $dF(y|x)/dy = f(y|x)$. (Notice that $dF(y|x)/dx$ is not a probability.)

Thus, we can define $\Pr[Y \leq y | X = x]$ by

$$F(y|x) = \int_{-\infty}^{y} f(t|x)\, dt = \Pr[Y \leq y | X = x]$$

If (4.24) and (4.25) are written in the form

$$p(i, j) = p_i p_{ij}, \qquad f(x, y) = f_X(x) f(y|x) \qquad (4.28)$$

clearly one natural way of determining the joint density function of two given random variables is to find the density function of one of those variables and the conditional density of the other, given the first, and multiply.

Notice also that if X and Y are *independent* $[p(i, j) = p_i q_j$ or $f(x, y) = f_X(x) f_Y(y)]$, then

$$p_{ij} = q_j \qquad \text{or} \qquad f(y|x) = f_Y(y) \qquad (4.29)$$

where q_j represents the discrete density of Y. In the independent case, the conditional density of Y is the same as the unconditional density of Y.

EXAMPLE 4.10

Consider Example 4.7 with $f(x, y) = xe^{-x(y+1)}$ for $0 \leq x < \infty$, $0 \leq y < \infty$. In this case

$$f(y|x) = \frac{xe^{-x(y+1)}}{e^{-x}} = xe^{-xy}$$

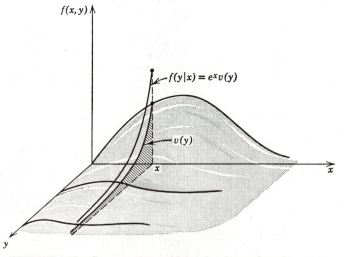

FIGURE 4.10 Cross sections of joint density surface $f(x, y) = xe^{-x(y+1)}$, with relation to $f(y|x)$.

for $0 \leq x < \infty$, $0 \leq y < \infty$. If x has a value for which $f_X(x) \neq 0$ while $f(x, y) = 0$, that is, if $x > 0$ but $y \leq 0$, then $f(y|x) = 0$. If $x \leq 0$, $f(y|x)$ is undefined.

This situation is sketched in Figure 4.10. Observe that, for a fixed value of x, the graph of $f(y|x)$ is related to the curve on the $f(x, y)$ surface of points with fixed x-coordinate. If we denote this curve by $v = v(y) = e^{-x(y+1)}$ (x fixed), then $f(y|x) = e^x v(y)$. The factor e^x is required to adjust the curve so that the total area under it is 1, a necessary requirement for any probability density function.

Notice also that we can compute the opposite conditional density function

$$f(x|y) = \frac{f(x, y)}{f_Y(y)} = \frac{xe^{-x(y+1)}}{1/(y+1)^2} = x(y+1)^2 e^{-x(y+1)}$$

for $0 \leq x < \infty$, $0 \leq y < \infty$. Notice that

$$F(x|y) = 1 - e^{-x(y+1)} - x(y+1)e^{-x(y+1)}, \qquad 0 \leq x < \infty$$

and

$$G(x|y) = \frac{x(y+1)e^{-x(y+1)}}{y}, \qquad 0 \leq x < \infty$$

Clearly

$$F(x|y) \neq G(x|y)$$

EXAMPLE 4.11

In Example 4.6 observe that

$$f(y|x) = \frac{e^{-(x+y)}}{e^{-x}} = e^{-y} = f_Y(y)$$

as must be the case since X and Y are independent as shown in that example.

EXAMPLE 4.12

The following kind of argument has been used to model the length of time that occurs between successive demands for airplane spare parts. Suppose that the random variable X represents the time between successive calls for spare parts and

$$f(x|\lambda) = \lambda e^{-\lambda x}, \qquad 0 \leq x < \infty$$

The value λ may be a value of a random variable Λ. Suppose Λ has density function given by

$$f_\Lambda(\lambda) = \begin{cases} \lambda e^{-\lambda}, & 0 \leq \lambda < \infty \\ 0, & \text{otherwise} \end{cases}$$

That is, we suppose that the time between demands depends on the value taken by another random variable Λ. Given that $\Lambda = \lambda$, the time between demands is a

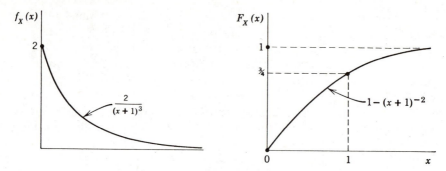

FIGURE 4.11   The marginal density and distribution function for X in Example 4.12.

negative exponentially distributed random variable. The random variable Λ has the given marginal density function. Physically Λ might be interpreted as a characteristic of the particular item in the airplane that has broken down.

We want to know the marginal density function of X, the interdemand time. We notice immediately, using (4.28), that

$$f(x,\lambda) = \lambda^2 e^{-\lambda(x+1)}, \qquad 0 \le x < \infty, 0 \le \lambda < \infty$$

We easily check that $f(x,\lambda)$ is a proper joint density function. From it, we obtain

$$f_X(x) = \int_0^\infty \lambda^2 e^{-\lambda(x+1)}\, d\lambda$$

$$= \frac{2}{(x+1)^3}, \qquad 0 \le x < \infty$$

$$F_X(x) = \int_0^x f_X(s)\, ds$$

$$= \begin{cases} 1 - (1+x)^{-2}, & 0 \le x < \infty \\ 0, & \text{otherwise} \end{cases}$$

$f_X(x)$ and $F_X(x)$ are shown in Figure 4.11. We can then determine probabilities for various levels of demand as follows.

$$\Pr[X \le 1] = F_X(1) = 1 - (1+1)^{-2} = \tfrac{3}{4}$$

$$\Pr[X > 2] = 1 - F_X(2) = 1 - \left[1 - (1+2)^{-2}\right] = \tfrac{1}{9}$$

$$\Pr[1 \le X \le 3] = F_X(3) - F_X(1) = \left[1 - (1+3)^{-2}\right] - \left[1 - (1+1)^{-2}\right]$$

$$= \tfrac{1}{4} - \tfrac{1}{16} = \tfrac{3}{16}$$

Notice that these probabilities are quite different from those obtained from $f(x|\lambda)$ for a given λ (for example, for a given item).

EXAMPLE 4.13

In the previous examples we required that both random variables be either continuous or discrete. All of our results carry over to the mixed case in which one of the variables is discrete and the other is continuous. To illustrate these ideas, we consider the following problem, which is of importance in the area of waiting-line theory (see Chapter 13).

Suppose that people arrive singly and at random at a theater ticket booth, line up for service, are served, and depart. The number M_t of individuals arriving in a time interval of fixed length t is one random variable of interest here. This is a discrete random variable with range $m = 0, 1, 2, \ldots$. Let us assume that it has a density function of the form

$$\Pr[M_t = m] = \frac{e^{-\lambda t}(\lambda t)^m}{m!} \tag{4.30}$$

for $m = 0, 1, 2, \ldots$, where λ is a given positive constant. The justification and realism of such an assumption about M_t will be discussed later. Another random variable of interest is the time T that it takes the ticket seller to service a single customer. This is a continuous random variable that we assume has probability density function

$$f_T(t) = \begin{cases} \lambda e^{-\lambda t}, & 0 < t < \infty \\ 0, & \text{otherwise} \end{cases} \tag{4.31}$$

Again, the reason for this assumption will be discussed later. Still another random variable of interest is the number N of arrivals during the time T that one customer is being served. We wish to find the probability density function of this random variable N.

Let us start by considering the pair (T, N). Formula (4.30) tells us what the probability density of N would be if T had a fixed value, $T = t$. In other words, (4.30) can be used to give the *conditional* density of N given that $T = t$:

$$f(n|t) = \Pr[N = n | T = t] = \frac{e^{-\lambda t}(\lambda t)^n}{n!} \tag{4.32}$$

for $n = 0, 1, 2, \ldots$. The *joint* density function of (T, N) is then obtained by multiplying (4.31) and (4.32), using (4.28):

$$f(t, n) = f_T(t) f(n|t)$$
$$= \frac{\lambda e^{-\lambda t} e^{-\lambda t}(\lambda t)^n}{n!}$$
$$= \frac{\lambda e^{-2\lambda t}(\lambda t)^n}{n!}$$

For $n = 0, 1, 2, \ldots$ and $0 < t < \infty$. Notice that all these manipulations carry through in spite of the fact that T is continuous and N is discrete. To find the density function of N alone we apply (4.21), integrating $f(t, n)$, since T is continuous:

$$p_n = \Pr[N = n] = \int_0^\infty f(t, n)\, dt$$
$$= \int_0^\infty \frac{\lambda e^{-2\lambda t}(\lambda t)^n}{n!}\, dt$$
$$= \frac{1}{n!} \int_0^\infty e^{-2\lambda t}(\lambda t)^n\, d(\lambda t)$$

This integral is evaluated in formula A.10, Appendix A, substituting $\lambda t = x$. We obtain, for $n = 0, 1, 2, \ldots,$

$$p_n = \frac{1}{n!} \frac{n!}{2^{n+1}}$$

$$= \left(\frac{1}{2}\right)^{n+1} \tag{4.33}$$

Clearly this example has many analogues, such as orders coming in to a factory, particles being counted in a Geiger counter, demands coming to a computer facility, breakdown and repair of electronic equipment, et cetera. It is interesting to note that here the resulting density, p_n, does not involve the constant λ. Our discussion is also easily generalized to cover the case when the value of λ in the density of M_t differs from that in the density of T (see Exercise 17).

In Example 2.4 a distribution almost identical to the distribution of (4.33) occurred when we discussed the experiment of tossing a coin until the first head appeared. Coin-tossing, card-playing, et cetera, at first might appear to have little connection with the real world. Yet in this problem, which is considerably more complicated and "real" than a simple coin-tossing experiment, we obtain the same density function. Hence, all of the analysis for the coin-tossing experiment is valid for the random variable N. In fact, we could *simulate* the arrival process during a service time by the simple expedient of flipping a coin. These relations abound in scientific and engineering studies and are studied under categories such as *similitude* or *simulation*. When those relations can be found they provide the scientist with a powerful tool for studying the behavior of things that are difficult to observe (for example, N) by watching the behavior of a much more mundane and easily observed phenomena (coin-tossing). Thus we would be mistaken if we concluded that seemingly unrealistic ideas such as coin-tossing and card-playing are devoid of important, useful applications.

4.9 Many Random Variables

In the previous sections we have restricted our attention to two random variables. These concepts extend immediately to more than two random variables defined on the same sample space. In this section we indicate some of the extensions.

Let $X_1, X_2, \ldots, X_n$ be any collection of n random variables defined on the same sample space. Then their *joint distribution function* is defined as

$$F(x_1, x_2 \ldots x_n) = \Pr[X_1 \le x_1 \text{ and } X_2 \le x_2 \text{ and} \ldots \text{and } X_n \le x_n]$$

For brevity of notation we write

$$F(x_1, x_2 \ldots x_n) = \Pr[X_1 \le x_1; X_2 \le x_2; \ldots; X_n \le x_n] \tag{4.34}$$

Again, we can consider the discrete or continuous cases. If the random variables are discrete (with integer ranges), then their joint density function is

$$p(i_1, i_2, \ldots, i_n) = \Pr[X_1 = i_1; X_2 = i_2; \ldots; X_n = i_n] \qquad (4.35)$$

The marginal densities of one or more of the random variables are obtained by summing (4.35) appropriately. For example,

$$p_{X_n}(i_n) = \Pr[X_n = i_n] = \sum_{i_1} \sum_{i_2} \cdots \sum_{i_{n-1}} p(i_1, i_2, \ldots, i_n) \qquad (4.36)$$

$$p_{X_1, \ldots, X_{n-1}}(i_1, \ldots, i_{n-1}) = \Pr[X_1 = i_1; \ldots; X_{n-1} = i_{n-1}]$$

$$= \sum_{i_n} p(i_1, i_2, \ldots, i_n) \qquad (4.37)$$

Conditional densities are defined similarly:

$$p(i_n | i_1, \ldots, i_{n-1}) = \Pr[X_n = i_n | X_1 = i_1; \ldots; X_{n-1} = i_{n-1}]$$

$$= \frac{p(i_1, i_2, \ldots, i_n)}{p_{X_1, \ldots, X_{n-1}}(i_1, \ldots, i_{n-1})} \qquad (4.38)$$

or

$$p(i_1, i_2, \ldots, i_n) = p_{X_1, \ldots, X_{n-1}}(i_1, \ldots, i_{n-1}) p(i_n | i_1, \ldots, i_{n-1}) \qquad (4.39)$$

In the continuous case, assuming that $F(x_1, x_2, \ldots, x_n)$ has a continuous mixed partial derivative, we define the *joint density* function as

$$f(x_1, x_2, \ldots, x_n) = \frac{\partial^n F(x_1, x_2, \ldots, x_n)}{\partial x_1 \, \partial x_2 \ldots \partial x_n} \qquad (4.40)$$

Integration of this function gives probabilities, such as

$$\Pr[a_1 \le X_1 \le b_1; \ldots; a_n \le X_n \le b_n] = \int_{a_n}^{b_n} \cdots \int_{a_1}^{b_1} f(x_1, x_2, \ldots, x_n) \, dx_1 \ldots dx_n \qquad (4.41)$$

Again, marginal densities are obtained by integration. For example,

$$f_{X_n}(x_n) = \int_{-\infty}^{\infty} \cdots \int_{-\infty}^{\infty} f(x_1, x_2, \ldots, x_n) \, dx_1 \ldots dx_{n-1} \qquad (4.42)$$

$$f_{X_1, \ldots, X_{n-1}}(x_1, \ldots, x_{n-1}) = \int_{-\infty}^{\infty} f(x_1, x_2, \ldots, x_n) \, dx_n \qquad (4.43)$$

Conditional densities can also be defined:

$$f(x_n | x_1, \ldots, x_{n-1}) = \frac{f(x_1, x_2, \ldots, x_n)}{f_{X_1, \ldots, X_{n-1}}(x_1, \ldots, x_{n-1})} \qquad (4.44)$$

or

$$f(x_1, x_2, \ldots, x_n) = f_{X_1, \ldots, X_{n-1}}(x_1, \ldots, x_{n-1}) f(x_n | x_1, \ldots, x_{n-1}) \qquad (4.45)$$

The random variables $X_1, X_2, \ldots, X_n$ are said to be *mutually independent* if

$$p(i_1, i_2, \ldots, i_n) = p_{i_1}^{(1)} p_{i_2}^{(2)} \cdots p_{i_n}^{(n)} \tag{4.46}$$

where $p_i^{(j)}$ is the probability density function of X_j, $p_i^{(j)} = p_{X_j}(i) = \Pr[X_j = i]$. In the continuous case *mutual independence* means

$$f(x_1, x_2, \ldots, x_n) = f_{X_1}(x_1) f_{X_2}(x_2) \cdots f_{X_n}(x_n) \tag{4.47}$$

If the variables are mutually independent, then

$$p(i_n | i_1, \ldots, i_{n-1}) = p_{i_n}^{(n)} \tag{4.48}$$

or

$$f(x_n | x_1, \ldots, x_{n-1}) = f_{X_n}(x_n) \tag{4.49}$$

EXAMPLE 4.14

Suppose two coins are tossed, and let X_1 represent the number of heads on the first coin tossed (0 or 1), let X_2 represent the number of heads on the second coin tossed (0 or 1), and let X_3 represent the number of matches (0 or 1). By the result of Example 2.5 we observe that the *pairs* (X_1, X_2), (X_1, X_3), and (X_2, X_3) are each independent, but that the set (X_1, X_2, X_3) is not mutually independent.

4.10   Discrete Parameter Random Processes: Independent Case

Now let $\{X_1, X_2, \ldots, X_n, \ldots\}$ be a sequence of random variables in the sense that X_1 comes first in the sequence, X_2 next, et cetera. Then $\{X_1, X_2, \ldots, X_n, \ldots\}$ is called a *random sequence* or a *discrete parameter random process*.

The simplest possible discrete parameter random process, $\{X_1, X_2, \ldots\}$, is a sequence of mutually independent random variables (every finite subset is mutually independent). In this case

$$p(i_n | i_1, \ldots, i_{n-1}) = p_{i_n}^{(n)}$$

and

$$p(i_1, i_2, \ldots, i_n) = p_{i_1}^{(1)} p_{i_2}^{(2)} \cdots p_{i_n}^{(n)} \tag{4.50}$$

Here only the individual probability density functions $p_i^{(1)}, p_i^{(2)}, \ldots$ of $X_1, X_2, \ldots$ must be determined to compute the joint probabilities.

If, in addition, all the random variables have the same distribution, $p_i^{(j)} = p_i$ for each j, the sequence is said to be a sequence of independent, identically distributed random variables. In this case (4.50) becomes

$$p(i_1, i_2, \ldots, i_n) = p_{i_1} p_{i_2} \cdots p_{i_n} \tag{4.51}$$

Simple examples of these processes are easy to construct. Given a sequence of tosses of an honest coin as the random experiment, the sequence $\{X_1, X_2, \ldots\}$, where X_n is taken to be 1 or 0 depending on whether the nth toss is a head or a tail, will be a sequence of independent, identically distributed random variables. Each X_n would have values 0 and 1 and common probability density function $p_0 = \frac{1}{2}$, $p_1 = \frac{1}{2}$. A slightly more useful model is the following.

EXAMPLE 4.15

Consider a sequence $\{X_1, X_2, \ldots\}$ of independent random variables, each of which can take only the values $+1$ and -1, with probabilities p and q, respectively ($p + q = 1$). Thus $p_1^{(n)} = p$, $p_{-1}^{(n)} = q$, for each n. The joint probability density function is

$$p(i_1, i_2, \ldots, i_n) = p^k q^{n-k}$$

where $i_j = \pm 1$ for each j, and k represents the number of $+1$'s in the set $(i_1, i_2, \ldots, i_n)$.

This simple model has many important applications to fields such as variations in inventory size, gambling wins and losses, Brownian motion, switching circuits ($+1 =$ closed switch, $-1 =$ open switch), binary communication codes, and many others.

In many studies (see Exercise 3.25, for example) one collects data on some phenomena (power generating equipment in that exercise) by noting when some event of interest occurs (i.e., when an equipment outage occurs). Then one can define T_n as the time at which the nth event occurs (the time of the nth failure). Each T_n could take values in the set of nonnegative real numbers. Then $T_1, T_2, \ldots, T_n, \ldots$ would be a discrete parameter random process.

When comparing Example 4.15 with this process one notes that in the former each X_n takes values in some discrete set $\{-1, 1\}$ while in the latter each random variable, T_n, takes values in a continuous set (the set of nonnegative real numbers). One says that the discrete parameter random process in Example 4.15 had a discrete (or countable) *state space*. In this example the state space is uncountable or continuous. We will exploit these concepts in later chapters.

4.11 Testing for Independence

We have discussed the definition of independent random variables, but one must ask how one determines whether random variables are independent or not based on data available from physical experiments. The simple answer to such a question is that we will never be absolutely sure, based only on presented data, that the random variables that generate that data are independent. The best we can hope for is to say that it is highly likely (or unlikely), that is, that the data exhibit behavior we would expect of independent (or dependent) random variables. Therefore, within the limits of our data we will proceed along the path indicated until something (perhaps more data) shows us to be wrong.

In this way we can establish a useful procedure involving the probability of obtaining data that exhibits the behavior expected of independent random variables. Such procedures are called tests of hypotheses in statistical analysis. One test of independence studied in courses in statistics is called the chi-squared test (χ^2 test).

Suppose that we have a set of n pieces of data that can be partitioned in two ways according to the values taken by two discrete random variables, say, X and Y. There are c possible values that X can take and there are r possible values that Y can take. That is, suppose we can arrange the data as shown in Table 4.5, called a *contingency table*. Here the n_{ij} represent the number of pieces of data that are in both category x_i and y_j (e.g., n_{13} represents the number of pieces of data for which $X = x_1$ and $Y = y_3$). Now

$$\frac{n_{11} + n_{12} + n_{13} + \cdots + n_{1c}}{n} = \sum_{j=1}^{c} \frac{n_{1j}}{n}$$

is a relative frequency estimate of the probability that X takes the value x_1. (See Section 2.10.) Call this probability *estimate $\hat{p}_1$*. Similarly,

$$\sum_{j=1}^{c} \frac{n_{ij}}{n} = \hat{p}_i, \qquad \text{for each } i = 1, 2, \ldots, r$$

$\hat{p}_i$ is an *estimate* of $p_i = \Pr[X = x_i]$. On the other hand,

$$\hat{q}_j = \frac{n_{1j} + n_{2j} + n_{3j} + \cdots + n_{rj}}{n} = \sum_{i=1}^{r} \frac{n_{ij}}{n}$$

is a relative frequency estimate of the probability $q_j = \Pr[Y = y_j]$. The ratio n_{ij}/n is an estimate of the probability $p(i, j) = \Pr[X = i \text{ and } Y = j]$, that is, the entries in the table, when divided by n, are estimates of the *joint* density of X and Y.

If X and Y are independent then $p(i, j) = p_i q_j$, and we would expect that n_{ij}/n and $\hat{p}_i \hat{q}_j$ would be very nearly the same number since both estimate the same thing. Random variation in our data might make the two values somewhat different, of course. As it turns out, it is easier to work with the whole numbers n_{ij} instead of fractions. So we ask: Are the terms n_{ij} sufficiently close to $n\hat{p}_i \hat{q}_j$ to assume that X and Y are independent?

TABLE 4.5
Two-Way Classification of Data

x \ Y	Y_1	Y_2	Y_3	$\cdots$	Y_c
x_1	n_{11}	n_{12}	n_{13}	$\cdots$	n_{1c}
x_2	n_{21}	n_{22}	n_{23}	$\cdots$	n_{2c}
x_3	n_{31}	n_{32}	n_{33}	$\cdots$	n_{3c}
$\vdots$	$\cdots$	$\cdots$	$\cdots$	$\cdots$	$\cdots$
x_r	n_{r1}	n_{r2}	n_{r3}	$\cdots$	n_{rc}

Now the question is: How close is close enough? One useful way to answer this question is to proceed as follows. Form the ratio

$$R_{ij} = (n_{ij} - n\hat{p}_i\hat{q}_j)^2 / (n\hat{p}_i\hat{q}_j)$$

Intuitively, R_{ij} is a measure of the relative distance between what the data should be if the events $[X = x_i]$ and $[Y = y_j]$ were independent and what the data are. Now compute

$$\chi^2 = \sum_{i=1}^{r} \sum_{j=1}^{c} R_{ij}$$

which measures the total distance between what the data should be if X and Y were independent and what the data actually are.

If χ^2 is a very large value it seems reasonable to assume that X and Y are not independent. The data results are just too far from what one would expect from independence. On the other hand, if χ^2 is small then there is no good reason based on this data to assume that X and Y are dependent. Therefore, we proceed as though these random variables are independent.

The key question then boils down to: How big does χ^2 have to be before we decide one way or the other? The answer depends on how sure one wants to be of giving the correct answer. There is no mathematical answer to that dilemma. One simply must decide the answer based on the purpose of the experiment remembering that since we are basing our decision on only a few of many outcomes of our experiment we can never be *absolutely* sure of any answer we give. In the sciences it is common to say that one would like to be 95% sure of giving the correct answer, or 99%. Then, one can proceed as follows:

In Table B.3 in Appendix B we give those values that χ^2 must achieve in order to say with 95% or 99% confidence that the random variables are independent. This means that the chance is only 5% or 1% that the computed χ^2 value will exceed the value given in the table if the random variables are truly independent.

Now the values in Table B.3 depend not only on which 95% or 99% we choose but also on a parameter called the "degrees of freedom" (d.f.). For this χ^2 test this parameter is always $(c - 1)(r - 1)$, the number of columns less one multiplied by the number of rows less one, in Table 4.5.

There is one potential problem with this procedure. If, for some i and j, $n\hat{p}_i\hat{q}_j$ is very small then that one term can make our computed χ^2 value very large (since it appears in the denominator of R_{ij}), even if the data otherwise strongly suggest that X and Y are independent. To circumvent this possibility, if $n\hat{p}_i\hat{q}_j$ is small one simply combines the row or column containing that value with the next higher or lower row or column. How small must $n\hat{p}_i\hat{q}_j$ be to force us to do this? Some researchers use a rule of thumb called *the rule of 5*. If $n\hat{p}_i\hat{q}_j < 5$ then some row or column is combined with adjacent rows or columns until every cell has $n\hat{p}_i\hat{q}_j \geq 5$. This combining, of course, changes the degrees of freedom. These must be calculated based on the final table used.

TABLE 4.6
Types of Consecutive Telephone Calls

TYPE OF CALL n	TYPE OF CALL $n+1$		Total
	0	1	
0	42 (33.6)	373 (381.4)	415
1	373 (381.4)	4352 (4342.4)	4725
Total	415	4725	5140

EXAMPLE 4.16

Return to Examples 2.17 and 3.2. We ask if a call of type 0 (or 1) has just occurred is the next call's type independent of this one or not? Table 4.6, from Hall (1969), is based on 5140 telephone calls. The terms in parentheses are the estimated counts ($n\hat{p}_i\hat{q}_j$) assuming independence. Using this table we compute $\chi^2 = 2.48$.

From Table B.3 in Appendix B we find, using $(c-1)(r-1) = 1$ degrees of freedom, the 95% and 99% of values of 3.841 and 6.635, respectively. Our computed χ^2 is less than either of these values so we can proceed with the study of those telephone calls as if the type of the nth call and the type of the $(n+1)$st call were independent. (See Exercises 3.27 to 3.29.)

TABLE 4.7
Particle Count of Westgren's Data[a]

i \ j	0	1	2	3	4	5 or 6	Total
0	210	126	35	7	0	1	379
	(92.8)	(137.4)	(85.2)	(41.0)	(15.5)	(7.1)	379
1	134	281	117	29	1	1	563
	(137.9)	(204.1)	(126.7)	(60.8)	(23.0)	(10.5)	563
2	27	138	108	63	16	3	355
	(86.9)	(128.7)	(79.9)	(38.4)	(14.5)	(6.6)	355
3	10	20	76	38	24	6	174
	(42.6)	(63.1)	(39.1)	(18.8)	(7.1)	(3.3)	174
4 or 5	2	2	16	32	23	20	95
	(22.8)	(33.7)	(21.0)	(10.8)	(3.8)	(1.7)	93
Total	383	567	352	169	64	31	1564

SOURCE: S. Chandrasekhar (1943), "Stochastic Problems in Physics and Astronomy," *Reviews of Modern Physics*, **15**, pp. 2–89. Reprinted with permission of the author and the American Institute of Physics.
[a]Actual, with estimated in parentheses assuming independence.

EXAMPLE 4.17

In his test of the Smoluchowski theory, Westgren collected data (see Example 2.15) for two different times and counted the number of gold particles in his area of liquid each time. Table 4.7 gives his data: $X = i$ if there are i particles the first time counted and $Y = j$ if there are j particles the second time counted, $i = 0, 1, 2, \ldots, 5$, $j = 0, 1, 2, \ldots, 6$. We ask if the counts at these two times are independent of each other (i.e., are X and Y independent random variables).

Some of these tabular values are small but it will soon appear that they are not as much of a problem as they might otherwise be. There really is no need to compute χ^2 here. From Table B.3 in Appendix B the value of χ^2 at the 99% level is only 37.566. R_{11} alone is about 148, so we are sure that our computed χ^2 will exceed the tabulated value. Thus, it is quite clear that the particle counts at one time and those at another, later, time are not independent of each other.

Exercises 4

1. Two random variables X and Y have ranges $X = 0, 1$ and $Y = 0, 1, 2$. For points (x, y) in the joint range, the joint distribution function $F(x, y)$ is given by the following table, with the function extended to points outside the range in the usual way.

y \ x	0	1	2
0	$\frac{1}{12}$	$\frac{1}{3}$	$\frac{2}{3}$
1	$\frac{1}{6}$	$\frac{7}{12}$	1

Compute

(a) $\Pr[0 < X < 2 \text{ and } 0 < Y < 2]$

(b) $F_X(x)$ and $F_Y(y)$

(c) $\Pr[X = 1, Y = 1]$

(d) $\Pr[X = 0, Y < 1\frac{1}{2}]$

2. Two random variables X and Y have ranges $0 < X < 1$ and $0 < Y < 1$ and joint distribution function

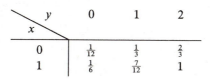

$$F(x, y) = \begin{cases} 0, & \text{for } x \le 0 \text{ or } y \le 0 \\ xy, & \text{for } 0 < x < 1 \text{ and } 0 < y < 1 \\ x, & \text{for } 0 < x < 1 \text{ and } y \ge 1 \\ y, & \text{for } 0 < y < 1 \text{ and } x \ge 1 \\ 1, & \text{for } x \ge 1 \text{ and } y \ge 1 \end{cases}$$

(a) Sketch the joint range of X and Y.

(b) Check that $F(x, y)$ satisfies properties (a) to (e) of Section 4.4.

(c) Compute $F_X(x)$ and $F_Y(y)$.

(d) Compute

(i) $\Pr[X > \frac{1}{2}, Y > \frac{1}{2}]$

(ii) $\Pr[X > \frac{1}{2}, Y < 2]$

(iii) $\Pr[X = \frac{1}{2}, Y = \frac{1}{2}]$

3. Two parts of the same machine have lifetimes X and Y, where these are independent random variables each having the negative exponential distribution

$$F(x) = \begin{cases} 1 - e^{-\lambda x}, & \text{for } x \geq 0 \\ 0, & \text{for } x < 0 \end{cases}$$

where λ is a positive constant.

(a) Compute the joint distribution function $F(x, y)$.

(b) Compute $\Pr[X < \lambda \text{ and } Y > 2\lambda]$.

4. Let X and Y be two discrete random variables with joint density function given by

x \ y	1	2	3
1	$\frac{1}{12}$	$\frac{1}{6}$	$\frac{1}{12}$
2	$\frac{1}{6}$	$\frac{1}{4}$	$\frac{1}{12}$
3	$\frac{1}{12}$	$\frac{1}{12}$	0

Compute the probability of the following events.

(a) X is less than $2\frac{1}{2}$.

(b) X is even.

(c) XY is even.

(d) Y is odd, given that X is odd.

5. A company manufactures items, each of which can have zero, one, or two defects with probabilities .7, .2, and .1. If an item has two defects, however, it is caught by the inspectors and is replaced by a perfect item before delivery. Let X represent the original number of defects in an item produced and Y the number of defects in the corresponding delivered item. Give the joint density function of X and Y and the density function of Y alone, in tabular form. If a delivered item is observed to have zero defects, what is the probability that it is a replacement?

6. In a long-range radio communication system, errors in transmission result from either human error in sending or receiving or natural interference (static). Suppose that a series of 20 symbols are sent. Each has probability .10 of being in error because of natural interference and probability .05 of being in error because of human factors only. Assume independence of the error possibilities of different symbols.

(a) What is the probability of 2 natural interference errors and 1 human error in 20 symbols?

(b) If 2 out of 20 symbols are found to be in error, what is the probability that both were the result of human error?

7. Let X and Y be random variables having joint density function $f(x, y) = e^{-(x+y)}$ for $x \geq 0$, $y \geq 0$ and $f(x, y) = 0$ otherwise. Compute

   (a) $\Pr[X > 1, Y > 1]$.　　(e) $\Pr[X = Y]$.
   (b) $\Pr[X > 1 \text{ or } Y > 1]$.　　(f) $\Pr[X > 1 | Y > 1]$.
   (c) $\Pr[X > 1]$.　　(g) $\Pr[X > Y | Y > 1]$.
   (d) $\Pr[X + Y \leq 4]$.

8. Two random variables X and Y have joint density function $f(x, y) = cxy$ for $0 \leq x \leq y \leq 1$ and $f(x, y) = 0$ otherwise.

   (a) Sketch the joint range of X and Y.

   (b) Find the constant c.

   (c) Compute $f_X(x)$ and $f_Y(y)$.

   (d) Are X and Y independent? Explain.

9. A gun is aimed at a certain point (taken as the origin of a coordinate system). Because of random factors the actual hit-point can be any point (X, Y) in a circle of radius R about the origin. Assume that the joint density function of X and Y is constant in this circle.
   $$f(x, y) = C \quad \text{for } x^2 + y^2 \leq R^2 \quad \text{and} \quad f(x, y) = 0 \quad \text{otherwise}$$

   (a) Compute C.

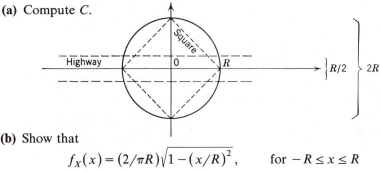

   (b) Show that
   $$f_X(x) = (2/\pi R)\sqrt{1 - (x/R)^2}, \qquad \text{for } -R \leq x \leq R$$
   $$f_X(x) = 0, \qquad\qquad\qquad\qquad \text{otherwise}$$

   (c) If the target area is a square with diagonal $2R$ centered at the origin, what is the probability of hitting the target?

   (d) If the target is a highway of width $R/2$ centered along the x-axis, what is the probability of hitting the highway?

   (e) Are X and Y independent? Explain.

10. Let X, Y, and Z be random variables with joint density function $f(x, y, z) = 12x^2yz$ for $0 \leq x \leq 1, 0 \leq y \leq 1, 0 \leq z \leq 1$ and $f(x, y, z) = 0$ otherwise. Compute
    $$\Pr[X < Y < Z]$$

11. In Exercise 5 compute (in tabular form) the conditional density
    $$p_{ij} = \Pr[Y = j | X = i] \qquad \text{for } i = 0, 1, 2, \ j = 0, 1$$

12. Let X and Y be continuous random variables with joint density function,
    $f(x, y) = e^{-y}$ for $0 \leq x \leq y$ and $f(x, y) = 0$ otherwise.

(a) Compute the individual density functions of X and Y.

(b) Are X and Y independent? Explain.

(c) Determine the conditional densities $f(y|x)$ and $f(x|y)$.

(d) Compute $\Pr[X \le 1 | Y = 2]$.

13. If X and Y are continuous random variables with joint density function that is given by the following, compute the constant c, the individual density and distribution functions of X and Y, $f(y|x)$ and $f(x|y)$.

(a) $f(x, y) = cxy$ for $0 < x < 1,\ 0 < y < 1$.

(b) $f(x, y) = cx(1 - y)$ for $0 < x < 1,\ 0 < y < 1$.

(c) $f(x, y) = cxe^{-y}$ for $0 < x < y$.

(d) $f(x, y) = c(1 - x^2 - y^2)$ for $x^2 + y^2 \le 1$.

14. In Example 4.10 make a sketch similar to Figure 4.10 but indicate the surface cross sections for fixed y and their relation to $f(x|y)$.

15. If X represents the time interval between successive automobiles that pass a given intersection, then the event $[X > a]$ means that the next car has not arrived a time units after the first car has passed. If we observe that this event has occurred, then $Y = X - a$ represents the remaining time that we must wait until the next car passes. Suppose X has density function $f(x) = e^{-x}$ for $x \ge 0$ and $f(x) = 0$ otherwise.

(a) Compute the conditional distribution function of Y
$$G(y) = \Pr[Y \le y | X > a] = \Pr[X - a \le y | X > a]$$

(b) Compute the conditional density $g(y) = G'(y)$.

(c) Compare $g(y)$ to $f(y)$ and comment concerning the relation between the conditional distribution of Y and the distribution of the original X.

16. In Example 4.13 compute the conditional density function $f(t|n)$ and interpret its meaning.

17. In Example 4.13 assume that M_t has discrete density $e^{-\lambda t}(\lambda t)^m / m!$ for $m = 0, 1, 2, \ldots$ and T has density $\mu e^{-\mu t}$ for $t \ge 0$, where λ and μ are not necessarily the same.

(a) Compute the joint density of T and N.

(b) Show that N has density $p_n = \mu \lambda^n / (\lambda + \mu)^{n+1}$ for $n = 0, 1, 2, \ldots$.

(c) Show that $\Pr[N \ge 1] < \frac{1}{2}$ if and only if $\lambda < \mu$.

18. Prove that if $x_1 \le x_2$ and $y_1 \le y_2$ then
$$\Pr[X \le x_1 \text{ and } Y \le y_1] \le \Pr[X \le x_2 \text{ and } Y \le y_2]$$

19. In Example 4.10 determine $F(y|x)$ and $G(y|x)$.

20. In Example 4.6 determine $F(y|x)$ and $G(y|x)$. Then determine $F(x|y)$ and $G(x|y)$. Thus, show that if X and Y are independent in this case, $F(y|x) = G(y|x) = F_Y(y)$ and $F(x|y) = G(x|y) = F_X(x)$.

21. Show that in general if X and Y are independent, $F(x|y) = G(x|y) = F_X(x)$ and $F(y|x) = G(y|x) = F_Y(y)$.

22. Using the data in Exercise 3.25 compute a relative frequency estimate of the probabilities $\Pr[X = i \text{ and } Y = j]$ for $i, j = 1, 2, 3$.

23. (continuation) Does the χ^2 test support the assumption that the random variables X and Y are independent?

24. (continuation) Compute a relative frequency estimate of

    (a) $\Pr[Y = j | X = i] = p_{ij}$.

    (b) Using Bayes' formula (Section 2.7) compute an estimate of $\Pr[X = i | Y = j]$. Compare this with a direct estimate of this quantity as in part (a). Give a physical interpretation based on DiMarco's setting of these conditional probabilities.

25. Consider the following data on automobile trips in the area of Washington, D.C., for 1970. The measurements are in hundreds of trips per day.[1]

			Destination		
Origin	A.C. (1)	F.C. (2)	P.W.C. (3)	Other (4)	Total
(1) Arlington Co.	34.6	5.4	0.2	45.9	86.1
(2) Fairfax Co.	29.0	62.8	1.1	65.2	158.1
(3) Prince William Co.	3.1	6.9	18.2	10.0	38.2
Total	66.7	75.1	19.5	121.1	282.4

Let $[X = i]$ be the event "the trip origin is area i," $i = 1, 2, 3$, and let $[Y = j]$ be the event "the trip destination is area j," $j = 1, 2, 3, 4$. Assume that one can assign probabilities to the events using a relative frequency definition based on the numbers in the above table. Then

(a) Test if X and Y are independent.

(b) Compute $\Pr[X = i]$ for $i = 1, 2, 3$.

(c) Compute $\Pr[Y = j]$ for $j = 1, 2, 3, 4$.

(d) Compute $\Pr[Y = j | X = i]$ for $i = 1, 2, 3$ and $j = 1, 2, 3, 4$.

(e) Compute $\Pr[X = i | Y = j]$ for $i = 1, 2, 3$ and $j = 1, 2, 3, 4$ using Bayes' theorem.

(f) Using the context of this example explain the difference between $\Pr[Y = j | X = i]$ and $\Pr[X = i | Y = j]$.

---

[1]**American Institute of Planners**, (1970) U.S. Government Printing Office, Washington, D.C.

5 · Functions of Random Variables

5.1   Introduction

In previous chapters we have considered the basic properties of sample spaces, event spaces, probability functions, random variables, and distribution functions for one or more random variables and for discrete or continuous random variables. In some applications we need to go further. For example, in the study of system lifetimes we often are confronted with a system of many components. If any one of the components fails then the system fails. To determine the lifetime probabilities of the system we might be able to determine the lifetime probabilities of each component and then define the system lifetime as the *minimum* or *shortest* of the lifetimes of all the components. In this way, of course, we are asserting that the lifetime of the system is a function of the lifetimes of the components and that function is the "minimum" function. In this chapter we will examine a few basic concepts of functions of random variables.

5.2   Functions of One Random Variable

Suppose X is a random variable, that is (see Section 3.2), X is a function whose domain is the sample space and whose range is a set of real numbers; for each sample point s, $X(s)$ is a number. One common technique for constructing new random variables out of old ones is to use an ordinary function Q that maps real numbers into real numbers and whose domain includes the range of X. A new random variable Y is constructed by combining the function Q with the function X. For each sample point s,

$$Y(s) = y$$

whenever

$$y = Q(x) \qquad \text{where } x = X(s)$$

i.e.,

$$Y(s) = Q(X(s))$$

(See Figure 5.1.) The natural notation for the new random variable Y is

$$Y = Q(X)$$

110

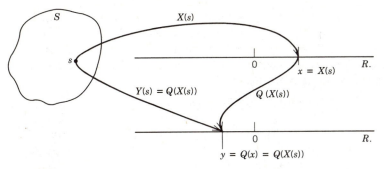

FIGURE 5.1   The mapping $Y = Q(X)$ or $Y(s) = Q(X(s))$.

EXAMPLE 5.1

Suppose Q is a simple linear function, such as

$$Q(x) = 2x + 3$$

Then

$$Y = Q(X) \qquad \text{if } Y = 2X + 3$$

For each sample point s, if $X(s) = x$, then $Y(s) = 2x + 3$. If $X(s) = 4$, then $Y(s) = 11$.

If X has a distribution function F_X, then $Y = Q(X)$ will have a distribution function F_Y. The ease with which F_Y can be determined in terms of F_X will depend on the complexity of the function Q. In general

$$F_Y(y) = \Pr[Y \le y] = \Pr[Q(X) \le y]$$

In the continuous case this probability can be expressed in terms of the density function f_X of X,

$$F_Y(y) = \int_{D_y} f_X(x)\, dx$$

where the integration is extended over the set D_y of x values for which $Q(x) \le y$, $D_x = \{x \colon Q(x) \le y\}$. This may be a simple or a complicated set, so the integral may be simple or complicated to evaluate.

5.3   Linear Functions

Expanding on Example 5.1, let us assume Q is a general linear function, $Q(x) = ax + b$. Then the corresponding random variable is

$$Y = aX + b$$

This can be interpreted as saying that Y differs from X only in origin and scale of measurement. Thus, if we have a random experiment in which the value of X is a temperature in degrees Celsius, then $Y = \frac{9}{5}X + 32$ would represent the same temperature in degrees Fahrenheit. If X represents a dimensional measurement in inches, $Y = 25.4X$ will represent the same measurement in millimeters.

The distribution and density functions of Y can easily be determined in terms of the corresponding functions of X.

$$F_Y(y) = \Pr[Y \le y] = \Pr[aX + b \le y]$$

$$= \Pr\left[X \le \frac{y-b}{a}\right], \qquad \text{provided } a \text{ is positive}$$

$$= F_X\left(\frac{y-b}{a}\right) \tag{5.1}$$

i.e., $F_Y(y) = F_X(x)$, where $x = (y-b)/a$. The density, in the continuous case, will be

$$f_Y(y) = \frac{dF_Y(y)}{dy}$$

$$= \frac{dF_X(x)}{dx} \cdot \frac{dx}{dy} \qquad \text{where } x = \frac{y-b}{a}$$

$$= f_X(x) \cdot \frac{1}{a} = \frac{1}{a} f_X\left(\frac{y-b}{a}\right) \tag{5.2}$$

In the discrete case we have

$$P_Y(y) = p_X\left(\frac{y-b}{a}\right) \tag{5.3}$$

These formulas are valid if a is positive. Some obvious modification is needed if a is negative.

EXAMPLE 5.2

In Example 3.18 we introduced the normal density function and showed that if X is a normally distributed random variable whose density function parameters are μ, σ then by the change of variable $y = (x - \mu)/\sigma$ the calculation of probabilities could be reduced to calculations with a standard normally distributed random variable ($\mu = 0$, $\sigma = 1$). We can now establish the reason behind that result. Recall

$$f_X(x) = \frac{1}{\sqrt{2\pi}\sigma} e^{-(x-\mu)^2/(2\sigma^2)}, \qquad -\infty < x < \infty$$

Then the linear transformation

$$Y = \frac{X - \mu}{\sigma}$$

gives, using (5.2) with $a = 1/\sigma$ and $b = -\mu/\sigma$,

$$f_Y(y) = \frac{1}{\sqrt{2\pi}} e^{-y^2/2}$$

which is the standard normal density given in Example 3.18. Thus, as we used this result in that example, the linear transformation $Y = (X - \mu)/\sigma$ allows us to compute probabilities for any normally distributed random variable using only the table for the standard normal tables as shown in Chapter 3.

5.4   Power Functions

Another useful function is the power function, $Q(x) = x^n$. For simplicity, let us confine our attention to the *square* function, $Q(x) = x^2$. The extension to higher powers is immediate.

Let

$$Y = X^2$$

The crucial difference between this and the linear function is that normally *two* values of X will correspond to the same value of Y, i.e., $X = \pm 2$ both yield $Y = 4$. Regardless of the range of X, the range of Y must be positive.

$$\begin{aligned}
F_Y(y) &= \Pr[Y \le y] = \Pr[X^2 \le y] \\
&= \Pr[|X| \le \sqrt{y}], \quad \text{for } y > 0 \\
&= \Pr[-\sqrt{y} \le X \le \sqrt{y}] \\
&= F_X(\sqrt{y}) - F_X(-\sqrt{y})
\end{aligned} \tag{5.4}$$

$F_Y(y) = 0$ if $y \le 0$. If X is continuous, (5.4) then can be used to compute the density function of Y.

EXAMPLE 5.3

Suppose X is standard normal, $f_X(x) = (1/\sqrt{2\pi})e^{-x^2/2}$, and $Y = X^2$. Then by (5.4)

$$\begin{aligned}
F_Y(y) &= F_X(\sqrt{y}) - F_X(-\sqrt{y}) \\
&= 2F_X(\sqrt{y}) - 1, \quad \text{using the symmetry } F(-x) = 1 - F(x) \\
f_Y(y) &= \frac{d}{dy} F_Y(y) \\
&= 2F_X'(\sqrt{y}) \cdot \frac{1}{2\sqrt{y}} \\
&= \frac{1}{\sqrt{y}} f_X(\sqrt{y}) \\
&= \frac{1}{\sqrt{2\pi y}} e^{-y/2}, \quad \text{for } y > 0
\end{aligned}$$

with $f_Y(y) = 0$ for $y \le 0$. This particular density appears in statistics under the name of a *chi-squared* (χ^2) *density with one degree of freedom*.

5.5 Functions of Several Random Variables

Suppose that X and Y are random variables and Q now is a function of *two* variables. We then can define a new random variable $U = Q(X, Y)$ by the obvious definition that $U(s) = u$ whenever $X(s) = x$, $Y(s) = y$, and $Q(x, y) = u$.

For example, if X and Y represent the lifetimes of two randomly chosen pieces of electronic equipment, we may be interested in the average lifetime, $(X + Y)/2$, or the time at which the first failure occurs, $\min(X, Y)$, or the time at which the second failure occurs, $\max(X, Y)$. Each of these quantities is a random variable whose value depends on knowning the values of both X *and* Y. Each is a function of X and Y.

In an obvious fashion our definition can be extended to functions of more than two random variables.

If $U = Q(X, Y)$, then the distribution function of U, $F_U(u) = \Pr[U \le u] = \Pr[Q(X, Y) \le u]$, can be obtained by integrating or summing the joint density function of X and Y over the two-dimensional set D_u of all pairs of (x, y)-values for which $Q(x, y) \le u$.

$$D_u\{(x, y): Q(x, y) \le u\}$$

$$F_U(u) = \int_{D_u} \int f(x, y) \, dx \, dy \tag{5.5}$$

Again the complexity of the integration will depend on the complexity of the function Q and of the domain D_u.

EXAMPLE 5.4

Suppose $U = \max(X, Y)$. Then

$$U \le u \qquad \text{if, and only if, } X \le u \quad \text{and} \quad Y \le u$$

Thus $D_u = \{(x, y): x \le u \text{ and } y \le u\}$ and

$$F_U(u) = \Pr[U \le u] = \Pr[X \le u \text{ and } Y \le u]$$
$$= F(u, u)$$

In this case it is easy to represent the distribution of the single random variable U in terms of the joint distribution function $F(x, y)$ of the two random variables X and Y.

5.6 Sums of Random Variables

If X and Y are random variables, one of the simplest and most important functions to consider is the sum function

$$U = X + Y$$

If X and Y are continuous with joint density function $F(x, y)$, then

$$F_U(u) = \Pr[U \le u] = \Pr[X + Y \le u] = \int_{D_u} \int f(x, y) \, dy \, dx \tag{5.6}$$

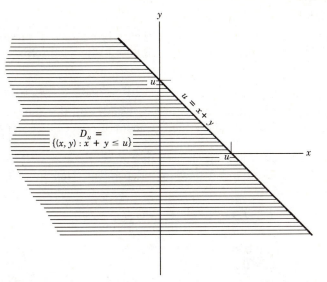

FIGURE 5.2 · The domain D_u for the summing function $Q(X, Y) = X + Y$.

where

$$D_u = \{(x, y): x + y \le u\}$$

The situation is displayed in Figure 5.2. Thus,

$$F_U(u) = \int_{-\infty}^{\infty} \int_{-\infty}^{u-x} f(x, y) \, dy \, dx$$

Now substitute $y = t - x$ (x fixed) in the inner integral and interchange the order of integration:

$$F_U(u) = \int_{-\infty}^{\infty} \int_{-\infty}^{u} f(x, t - x) \, dt \, dx$$

$$= \int_{-\infty}^{u} \int_{-\infty}^{\infty} f(x, t - x) \, dx \, dt \qquad (5.7)$$

This formula can be simplified if X and Y are assumed to be *independent*, $f(x, y) = f_X(x) f_Y(y)$. In this case (5.7) becomes

$$F_U(u) = \int_{-\infty}^{u} \left[\int_{-\infty}^{\infty} f_X(x) f_Y(t - x) \, dx \right] dt$$

Differentiating with respect to u and using the Fundamental Theorem of Calculus $(d/du \int_a^u g(t) \, dt = g(u))$, we obtain

$$f_U(u) = \frac{d}{du} F_U(u) = \int_{-\infty}^{\infty} f_X(x) f_Y(u - x) \, dx \qquad (5.8)$$

This formula, which gives the density function of the sum, U, of two independent random variables, is termed a *convolution* integral. Such integrals appear in many places in probability theory and in other areas of mathematics.

EXAMPLE 5.5

The time between consecutive mine accidents is a random variable having a negative exponential density (Examples 3.8 and 3.16) of the form

$$\lambda e^{-\lambda x}, \qquad x > 0$$

with $\lambda = 1/240$.

Assume that a mine accident has just occurred. Call this time 0. The density function of the time U until the second accident after time 0 occurs will be given by $U = X + Y$.

X and Y will have the same density functions, $f_X(x) = \lambda e^{-\lambda x}$ and $f_Y(y) = \lambda e^{-\lambda y}$, and we will assume that they are independent. Thus by (5.8) U will have the density function

$$f_U(u) = \int_{-\infty}^{\infty} f_X(x) f_Y(u - x) \, dx$$

But $f_X(x) = 0$ if $x < 0$ and $f_Y(u - x) = 0$ if $u - x < 0$, i.e., if $x > u$, so we need only consider the range $0 \le x \le u$. Thus,

$$f_U(u) = \int_0^u f_X(x) f_Y(u - x) \, dx$$

$$= \int_0^u \lambda e^{-\lambda x} \cdot \lambda e^{-\lambda(u - x)} \, dx$$

$$= \lambda^2 e^{-\lambda u} \int_0^u dx$$

$$= \lambda^2 e^{-\lambda u} u, \qquad \text{for } u > 0$$

$f_U(u) = 0$ for $u \le 0$. This gives the density function for the time until the second accident occurs.

Exercises 5

1. In formula (5.4) the distribution function of $F_Y(y)$ was given for the power function

$$Y = X^2$$

Now suppose that X is a nonnegative continuous random variable (i.e., $F_X(x) = 0$ for $x < 0$).

   (i) In terms of $F_X(x)$ and $f_X(x)$ determine (for $Y = X^2$)

   (a) $F_Y(y)$.   (b) $f_Y(y)$.

   (ii) Let

$$Y = X^3$$

   Determine

   (a) $F_Y(y)$.   (b) $f_Y(y)$.

(iii) Let

$$Y = e^X$$

Determine

(a) $F_Y(y)$.   (b) $f_Y(y)$.

2. (continuation) The key ingredient in Exercise 1 is that each function is an increasing function with a derivative for every $x > 0$. Now suppose

$$Y = Q(X)$$

where Q is any such increasing function and X is a nonnegative continuous random variable. Show

(a) $F_Y(y) = \Pr[Y \le y] = \Pr[Q(X) \le y] = \Pr[X \le Q^{-1}(y)] = F(Q^{-1}(y))$.
(b) $f_Y(y) = f(Q^{-1}(y)) \, dQ^{-1}(y)/dy$.
(Q^{-1} represents the *inverse* function of the function Q.)

3. (continuation) Compute $f_Y(y)$ if

$$Y = Q(X)$$

and Q is a *decreasing* function and X is nonnegative and continuous.

4. (continuation) Use your formulas in Exercises 2 and 3 to compute $f_Y(y)$ where

(a) $Q(X) = e^{-X}$.
(b) $Q(X) = 1/X$.
(c) $Q(X) = 2X$.
(d) $Q(X) = -aX$, $a > 0$.

(In each case X is assumed nonnegative and continuous.)

5. A very important function Q that occurs in computer simulation studies is $Q(x) = F(x)$, where F is the distribution function of X itself. If X is a continuous random variable, F is increasing with density function f. Let $Y = F(X)$ and show that

(a) $F_Y(y) = \begin{cases} 0, & y < 0 \\ y, & 0 \le y < 1 \\ 1, & y \ge 1 \end{cases}$

(b) $f_Y(y) = \begin{cases} 0, & y < 0 \\ 1, & 0 \le y < 1 \\ 0, & y > 1 \end{cases}$

A random variable, Y, with this special density function (or distribution function) is called a *uniform* $[0,1]$ random variable (see Section 7.6).

6. Let X be a random variable with the *beta density function*

$$f(x) = Cx^{n-1}(1-x)^{m-1}, \qquad 0 < x < 1, \; n, m > 0$$

Let $Y = 1/X$ and determine the density function $f_Y(y)$ of Y. What is the range of the random variable Y? $f_Y(y)$ is related to the density function of an F-distribution, widely used in statistical analysis.

7. Suppose $Y = |X|$, the absolute value function. In terms of F_X and f_X determine

   (a) $F_Y(y)$.   (b) $f_Y(y)$.

8. In Exercise 7 suppose X is a standard normal random variable as in Example 3.18. Determine $F_Y(y)$ and $f_Y(y)$ if $Y = |X|$. A problem such as this occurs in certain AC to DC conversion circuits in electrical theory.

9. Suppose X is a standard normally distributed random variable and that
$$Y = Q(X) = e^X$$
   Compute

   (a) $F_Y(y)$,   (b) $f_Y(y)$.

   Here Y is said to be a *log-normally* distributed random variable. It has use in various statistical applications.

10. Let $Y = X^n$, for n a positive integer. In terms of $f_X(x)$ and $F_X(x)$, determine

    (a) $F_Y(y)$.   (b) $f_Y(y)$.

    (*Hint*: Treat the cases n odd and even separately. n odd is easy from Exercise 2.)

11. If $U = \max(X, Y)$, where X and Y have joint density function,
$$f(x, y) = xe^{-x(y+1)}, \qquad 0 < x < \infty, 0 < y < \infty$$
    as in Example 4.7. Determine $F_U(u)$ and $f_U(u)$. Problems in which one determines the distribution of a maximum occur in "parallel systems" where the system fails when the last component fails.

12. Let X, Y be two random variables and let
$$U = \min(X, Y)$$
    In terms of $f(x, y)$ determine

    (a) $F_U(u)$.   (b) $f_U(u)$.

    Problems in which this result is useful occur when one has a "series system" where the system fails as soon as the first component in the system fails.

13. Let X and Y be two independent random variables with density functions
$$f_X(x) = \begin{cases} \lambda e^{-\lambda x}, & x > 0 \\ 0, & x \le 0 \end{cases}$$
$$f_Y(y) = \begin{cases} \lambda^2 y e^{-\lambda y}, & y > 0 \\ 0, & y \le 0 \end{cases}$$
    $\lambda > 0$. If $U = X + Y$ determine $f_U(u)$.

14. Let $X =$ the number of spots turned up on the throw of one die. Let $Y =$ the number of spots turned up on the throw of a second die. Let $U = X + Y$, the sum of the spots turned up on a throw of the two dice. Determine $p_i = \Pr[U = i]$ for U values $i = 2, 3, \ldots, 12$.

15. Let $X =$ the number of tosses of an honest penny required to obtain the first head. Let Y be the number of additional tosses of the same coin required to get the next head *after* the first head appears. X and Y are independent and identically distributed. $U = X + Y$ is the number of tosses needed to get the second head. Compute p_i for U where $i = 2, 3, \ldots$. (See Exercise 3.10.)

16. Let X be a uniformly distributed random variable on $[0, 1]$. (See Exercise 5.) Consider the transformation

$$Y = -\frac{1}{a}\ln(1 - X)$$

Show that Y is an exponentially distributed random variable with parameter a. This exercise illustrates one of the many important uses of the result of Exercise 5.

17. Let X be a uniformly distributed random variable on $[0, 1]$. (See Exercise 5.) Create a transformation relating X and Y wherein Y has density function

$$f(y) = \begin{cases} 2y, & 0 < y < 1 \\ 0, & \text{otherwise} \end{cases}$$

18. Let X be a uniformly distributed random variable on $[0, 1]$. Create a transformation relating X and Y wherein Y has density function

$$\Pr[Y = 1] = p$$
$$\Pr[Y = 0] = q$$

for $p + q = 1$ and $0 < p < 1$.

19. Let X be a uniformly distributed random variable on $[0, 1]$. (See Exercise 5.) Create a transformation relating X and Y wherein Y is uniformly distributed on $[a, b]$, where a, b can be positive, negative, or zero and $a < b$.

6 · Expectation

6.1 Introduction

If X is a random variable we have observed that to describe completely the probability behavior of X we must specify an entire *function*, the distribution function $F(x)$ or the density function $f(x)$ or $p(x_i)$. Frequently, we want a single *number*, instead of a function, that gives us as much information about the random variable as a single number can, namely, where the "center" of the distribution of X lies. Many possible ways of constructing this number exist. The *median*, which is defined as the number t for which $\Pr[X \le t] = \Pr[X \ge t]$ (if such a number exists), and the *mode*, which is defined as the number t for which $f(t)$ or $p(x_i)$ is a maximum (again provided such a number exists uniquely) are two values sometimes used for this purpose. In this chapter we will discuss properties of some of the numbers often used to describe the probability structure of a random variable. We also will show how data can be used to estimate these numbers.

6.2 Expectation

The most important and useful number for the location of the "center" of the distribution of X is the *expectation* or *mean*, usually denoted by $E[X]$. (See Figure 6.1.)

If X is a random variable, then the expectation, $E[X]$, of X is defined by

$$E[X] = \begin{cases} \sum_i x_i p(x_i), & \text{in the discrete case} \\ \int_{-\infty}^{\infty} x f(x)\, dx, & \text{in the continuous case} \end{cases} \tag{6.1}$$

$E[X]$ is defined by (6.1) provided that the sum or integral involved converges *absolutely*:

$$\left(\sum_i |x_i| p(x_i) < \infty \text{ or } \int_{-\infty}^{\infty} |x| f(x)\, dx < \infty \right)$$

Otherwise, we say that $E[X]$ does not exist.

If X is a discrete random variable with range a set of integers and density function $\{ p_i \}$ then (6.1) takes the form

$$E[X] = \sum_i i p_i \tag{6.2}$$

120

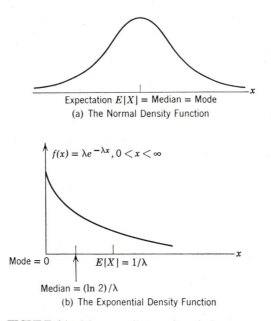

(a) The Normal Density Function

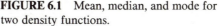

(b) The Exponential Density Function

FIGURE 6.1 Mean, median, and mode for two density functions.

It is important to realize that $E[X]$ is a fixed number, a *constant*. If we interpret the probability distribution of X as a mass distribution of total mass 1, then $E[X]$ represents the *center of gravity* of that mass distribution. Alternately, $E[X]$ can be interpreted as the *weighted average* of the values of X, using the density function, $f(x)$ or $p(x_i)$, as the weight function.

EXAMPLE 6.1

Consider an experiment that has as two possible outcomes "success" and "failure," with probabilities p and q, $p + q = 1$, for example, Examples 2.17 and 3.2. Thus, X has range $\{0,1\}$ and density function $p_0 = q$, $p_1 = p$. Such a random variable is said to have a *Bernoulli* distribution. In this case

$$E[X] = \sum i p_i = 0 p_0 + 1 p_1 = 0q + 1p = p$$

In particular, if $p = .081$ as in Example 2.17, then $E[X] = .081$.

This example shows that $E[X]$ should not be interpreted as a value that we would "expect" X to take. If X can take only the values 0 or 1, then we could never "expect" X to take the value .081.

EXAMPLE 6.2

Suppose that X has the negative exponential density function discussed in Example 3.16 [See Figure 6.1(b)].

$$f(x) = \begin{cases} \lambda e^{-\lambda x}, & \text{for } 0 \le x < \infty \\ 0, & \text{otherwise} \end{cases}$$

Then (using integration by parts, or see Appendix A.10),

$$E[X] = \int_{-\infty}^{\infty} xf(x)\, dx = \int_{0}^{\infty} x(\lambda e^{-\lambda x})\, dx$$

$$= \frac{1}{\lambda} \int_{0}^{\infty} u e^{-u}\, du, \qquad \text{where } u = \lambda x$$

$$= \frac{1}{\lambda}$$

So in Example 3.8 the expected time between two consecutive mine accidents is $1/\lambda = 240$ days.

EXAMPLE 6.3

Let X be a continuous random variable having range $-\infty < x < \infty$ and density function

$$f(x) = \frac{1}{\pi(1 + x^2)}, \qquad -\infty < x < \infty$$

The graph of $f(x)$ is the symmetric, bell-shaped curve sketched in Figure 6.2. To show that $f(x)$ is a density function we must show that $\int_{-\infty}^{\infty} f(x)\, dx = 1$. This can be done directly:

$$\int_{-\infty}^{\infty} \frac{dx}{\pi(1 + x^2)} = \lim_{\substack{a \to -\infty \\ b \to +\infty}} \frac{1}{\pi} \text{Arctan } x \Big|_{a}^{b} = \frac{1}{\pi}\left[\frac{\pi}{2} - \left(\frac{-\pi}{2}\right)\right] = 1$$

A random variable having this density function is said to have a *Cauchy distribution*. For such a random variable definition (6.1) gives

$$E[X] = \frac{1}{\pi} \int_{-\infty}^{\infty} \frac{x}{1 + x^2}\, dx$$

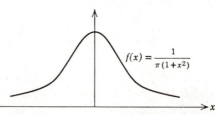

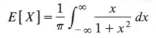
$$f(x) = \frac{1}{\pi(1 + x^2)}$$

FIGURE 6.2 The Cauchy density function.

Now

$$\frac{1}{\pi}\int_{-\infty}^{\infty}\left|\frac{x}{1+x^2}\right|dx = \frac{2}{\pi}\int_0^{\infty}\frac{x}{1+x^2}\,dx = \lim_{t\to\infty}\frac{2}{\pi}\ln(1+t^2) = \infty$$

Thus, the integral does not converge absolutely and $E[X]$ does not exist in this case.

6.3 Expectation of a Function of a Random Variable

Assume that X is a random variable and that Q is a real-valued function (say, a continuous function as in Chapter 5). We wish to discuss the expectation $E[Q(X)]$ of the random variable $Q(X)$.

Since E denotes a weighted average, it seems reasonable that the expectation of the random variable $Q(X)$ should be given by

$$E[Q(X)] = \begin{cases} \sum_i Q(x_i)p(x_i), & \text{in the discrete case} \\ \int_{-\infty}^{\infty} Q(x)f(x)\,dx, & \text{in the continuous case} \end{cases} \tag{6.3}$$

where $f(x)$ or $p(x_i)$ is the density function of X.

Equation (6.3) is true but it does not follow directly from the definition (6.1) of expectation. It is, in fact, a theorem that requires proof.

To obtain $E[Q(X)]$ by using the definition (6.1) directly, we proceed as follows: Denote the random variable $Q(X)$ by Y, $Y = Q(X)$. Let $f_Y(y)$ be the density function of this new random variable Y (see Chapter 5). Then, by (6.1),

$$E[Q(X)] = E[Y] = \begin{cases} \sum_j y_j p_Y(y_j), & \text{discrete case} \\ \int_{-\infty}^{\infty} y f_Y(y)\,d_y, & \text{continuous case} \end{cases} \tag{6.4}$$

The proof that (6.3) and (6.4) always give the same value for $E[Q(X)]$ is beyond the scope of this book and is omitted. Instead we illustrate the identity of (6.3) and (6.4) by some examples.

EXAMPLE 6.4

Suppose that you play a game in which you either win \$1 with probability $\frac{1}{4}$, lose \$1 with probability $\frac{1}{4}$, or tie with probability $\frac{1}{2}$. Let $X =$ your winnings. Then X is a random variable with range $\{-1, 0, 1\}$ and density function $p_{-1} = \frac{1}{4}$, $p_0 = \frac{1}{2}$, $p_1 = \frac{1}{4}$, and $p_i = 0$ otherwise. (See Figure 6.3.)

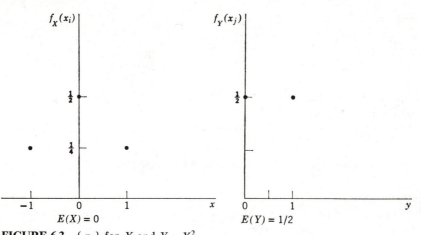

FIGURE 6.3 $\{p_i\}$ for X and $Y = X^2$.

Let $Y = X^2$. Then Y has range $\{0,1\}$ and represents the amount of money that changes hands in the game. By (6.3),

$$E[Y] = E[X^2] = \sum_i x_i^2 p(x_i)$$

$$= (-1)^2 p_{-1} + 0^2 p_0 + 1^2 p_1$$

$$= \tfrac{1}{4} + 0 + \tfrac{1}{4} = \tfrac{1}{2}$$

On the other hand, the alternative formula (6.4) can be applied directly if we note that Y has range $\{0,1\}$ and density function q_j where $q_0 = \Pr[Y = 0] = \Pr[X = 0] = \tfrac{1}{2}$, $q_1 = \Pr[Y = 1] = \Pr[X^2 = 1] = \Pr[X = \pm 1] = \tfrac{1}{4} + \tfrac{1}{4} = \tfrac{1}{2}$, and $q_j = 0$ otherwise (see Figure 6.3). Thus,

$$E[Y] = E[X^2] = \sum_j y_j q_j$$

$$= 0\left(\tfrac{1}{2}\right) + 1\left(\tfrac{1}{2}\right) = \tfrac{1}{2}$$

We notice that, in this case at least, (6.3) and (6.4) give the same result.

6.4   Linear Properties of Expectation

The simplest type of function Q to which formula (6.3) might be applied is a linear function of the type that is discussed in Section 5.3. Suppose that X is a random variable with density function $f(x)$ and define

$$Y = Q(X) = aX + b$$

where a and b are constants. Two random variables, X and Y, that are related by this type of equation can differ only in origin and scale of measurement.

By (6.3),

$$E[Y] = E[aX + b] = \int_{-\infty}^{\infty} (ax + b) f(x) \, dx$$

$$= a \int_{-\infty}^{\infty} x f(x) \, dx + b \int_{-\infty}^{\infty} f(x) \, dx$$

$$= aE[X] + b$$

in the continuous case. By replacing $\int$ with Σ, the same formula is seen to hold true in the discrete case. Thus,

$$E[aX + b] = aE[X] + b \tag{6.5}$$

In other words, (6.5) says: "*The expectation of a linear function of a single random variable is the linear function obtained by replacing the random variable by its expectation.*"

Notice that, in particular, if $a = 0$, then $E[b] = b$, that is, the expectation of a constant is that constant.

6.5 Moments: Mean and Variance

One simple type of function whose expectation might be of interest is the power function $Q(X) = X^k$, for $k = 1, 2, 3, \ldots$. The expectation of this function is termed the kth *moment of the random variable X* and is denoted by μ'_k,

$$\mu'_k = E[X^k] = \begin{cases} \int_{-\infty}^{\infty} x^k f(x) \, dx, & \text{continuous case} \\ \sum_i x_i^k p(x_i), & \text{discrete case} \end{cases} \tag{6.6}$$

for $k = 1, 2, 3, \ldots$. The first moment, μ'_1, is the ordinary expectation of X. This is used frequently and is termed the *mean* of the random variable, usually denoted simply by μ or μ_X,

$$\mu = \mu_X = \mu'_1 = E[X]$$

Of course the integral or sum in (6.6) may not converge absolutely, hence, μ'_k may not exist for certain values of k.

EXAMPLE 6.5

Let X be a continuous random variable that has a constant density function over the range $0 \le x \le 2$ (a *uniform* distribution)

$$f(x) = \begin{cases} \frac{1}{2}, & \text{for } 0 \le x \le 2 \\ 0, & \text{otherwise} \end{cases}$$

In this case

$$\mu'_k = \int_{-\infty}^{\infty} x^k f(x) \, dx = \int_0^2 x^k \cdot \frac{1}{2} \, dx = \frac{2^k}{k+1}$$

In particular

$$\mu = \mu_1' = 1, \qquad \mu_2' = \tfrac{4}{3}$$

To eliminate from the discussion the effect of the position of the origin in the scale of measurement, it is more useful to work with powers of $X - \mu$, the *deviation from the mean*, rather than with powers of X. We define

$$\mu_k = E\left[(X - \mu)^k\right] \tag{6.7}$$

for $k = 1, 2, \ldots$, to be the kth *moment about the mean* of the random variable X. Clearly,

$$\mu_1 = E[X - \mu] = E[X] - \mu = \mu - \mu = 0, \quad \text{by (6.5)}$$

The quantity

$$\mu_2 = E\left[(X - \mu)^2\right] \tag{6.8}$$

is of great importance in the field of statistics. It is called the *variance* of the distribution and is symbolized by σ^2 or $\text{Var}[X]$.

$$\text{Var}[X] = \sigma^2 = \mu_2 = \begin{cases} \int_{-\infty}^{\infty} (x - \mu)^2 f(x)\, dx, & \text{continuous case} \\ \sum_i (x_i - \mu)^2 p(x_i), & \text{discrete case} \end{cases} \tag{6.9}$$

We notice that σ^2 is the weighted average of the values of $(X - \mu)^2$. Thus, if X has a "diffuse" or "spread-out" distribution and takes values far away from μ with comparatively large probability, then σ^2 will be large. If X has a "concentrated" distribution and takes only values near to μ with large probability, then σ^2 will be small. Thus, σ^2 is a measure of the *spread* of the distribution (Figure 6.4).

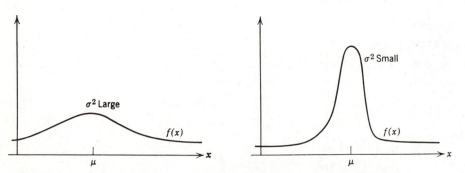

FIGURE 6.4 Illustration of σ^2 as a measure of "spread" of a density function.

The computation of σ^2 frequently can be simplified as follows:

$$\sigma^2 = E\left[(X-\mu)^2\right]$$
$$= E\left[X^2 - 2\mu X + \mu^2\right]$$
$$= E[X^2] - 2\mu E[X] + \mu^2, \qquad \text{by (6.4)}$$
$$= E[X^2] - 2\mu^2 + \mu^2$$

or

$$\sigma^2 = \mu_2' - \mu^2 = E[X^2] - \{E[X]\}^2 \qquad (6.10)$$

EXAMPLE 6.6

Consider a gambling game in which a player wins $1 with probability p or loses $1 with probability q, $p+q=1$. If X represents his winnings, then X has range $\{+1, -1\}$ and density function $p(1)=p$, $p(-1)=q$. X will have mean

$$\mu = E[X] = \sum x_i p(x_i) = 1 \cdot p(1) + (-1) \cdot p(-1) = p - q$$

and second moment

$$\mu_2' = E[X^2] = 1^2 \cdot p(1) + (-1)^2 \cdot p(-1) = p + q = 1$$

The variance σ^2 can be calculated directly using (6.9):

$$\sigma^2 = E\left[(X-\mu)^2\right] = [1-(p-q)]^2 p(1) + [-1-(p-q)]^2 p(-1)$$
$$= [(p+q)-(p-q)]^2 p + [-(p+q)-(p-q)]^2 q$$
$$= 4q^2 p + 4p^2 q = 4pq(p+q) = 4pq$$

The same result can be obtained using (6.10):

$$\sigma^2 = \mu_2' - \mu^2 = 1 - (p-q)^2$$
$$= 1 - p^2 + 2pq - q^2$$
$$= 1 - (p^2 + 2pq + q^2) + 4pq$$
$$= 1 - (p+q)^2 + 4pq$$
$$= 1 - 1 + 4pq = 4pq$$

EXAMPLE 6.7

Consider a random variable X that has the uniform distribution discussed in Example 6.5.

Using (6.10) we obtain

$$\sigma^2 = \mu_2' - \mu^2 = \left(\tfrac{4}{3}\right) - 1^2 = \tfrac{1}{3}$$

The quantity $\sigma = \sqrt{\sigma^2}$ is termed the *standard deviation* of the random variable X. The use of a single letter σ for this quantity rather than for the variance is primarily for dimensional reasons. σ has the same units as X, while σ^2 has the units of X^2. It is important to realize that the expectation $E[X]$ and the various moments of the random variable X are *numbers* determined by the distribution of the random variable, that is, by its density function.

It is possible for two random variables to have the same density function and, hence, the same moments without being, in any sense, the same random variable. They may even be independent random variables. For example, if X is the number appearing on one roll of a die and Y is the number appearing on the second roll of the die, then X and Y clearly have the same ranges, density functions, and moments. However, we cannot say that $X = Y$ always. In fact, this is an unlikely event and has probability only $\tfrac{1}{6}$ of occurring in any given experiment.

An important theorem of analysis, which we shall not prove, states that if X and Y are random variables that have corresponding moments of all orders existing and equal, $E[X^k] = E[Y^k]$ for $k = 0, 1, 2, \ldots$, then X and Y have the same distribution (the same density function).

6.6 The Chebyshev Inequality

The statement in the preceding section that a large σ^2 means a diffuse distribution and a small σ^2 means a concentrated one can be made more precise with the following theorem.

Theorem

Let the random variable X have mean μ and variance σ^2 and let t be any positive number. Then

$$\Pr[|X - \mu| \geq t] \leq \frac{\sigma^2}{t^2} \tag{6.11}$$

This inequality, termed the *Chebyshev inequality*, places a bound on the probability that the random variable will differ from its mean by more than a fixed number t. This bound is directly proportional to σ^2 and inversely proportional to t^2.

Formula 6.11 can be proved as follows (in the continuous case).

PROOF

$$\sigma^2 = E\left[(X-\mu)^2\right] = \int_{-\infty}^{\infty} (x-\mu)^2 f(x)\, dx$$

$$= \int_{-\infty}^{\mu-t} (x-\mu)^2 f(x)\, dx + \int_{\mu-t}^{\mu+t} (x-\mu)^2 f(x)\, dx + \int_{\mu+t}^{\infty} (x-\mu)^2 f(x)\, dx$$

$$\geq \int_{-\infty}^{\mu-t} (x-\mu)^2 f(x)\, dx + \int_{\mu+t}^{\infty} (x-\mu)^2 f(x)\, dx$$

the middle integral being omitted

$$= \int_{|x-\mu| \geq t} (x-\mu)^2 f(x)\, dz \geq \int_{|x-\mu| \geq t} t^2 f(x)\, dx$$

$$= t^2 \int_{|x-\mu| \geq t} f(x)\, dx = t^2 \Pr[|X-\mu| \geq t]$$

Thus,

$$\sigma^2 \geq t^2 \Pr[|X-\mu| \geq t]$$

and Chebyshev's inequality follows.

The proof in the discrete case is similar, but with sums replacing integrals.

The Chebyshev inequality is very crude, and for most reasonable density functions much more precise bounds can be found for the probability of a given deviation from the mean. However, it has the prime advantage of universality. It applies to *all* random variables whose first and second moments exist. Thus, it is very useful in proving general theorems.

6.7   Variance of a Linear Combination

Let X be a random variable having mean μ and variance σ^2 and let Y be a random variable of the form $Y = aX + b$, for some constants a and b. Thus, Y differs from X only in origin and scale of measurement and has mean $\mu_Y = E[Y] = a\mu + b$, by (6.5). Thus,

$$\text{Var}[aX + b] = \text{Var}[Y]$$
$$= E\left[(Y - \mu_Y)^2\right]$$
$$= E\left[(aX + b - a\mu - b)^2\right]$$
$$= E\left[a^2(X - \mu)^2\right]$$
$$= a^2 E\left[(X - \mu)^2\right]$$
$$= a^2 \sigma^2$$
$$\text{Var}[aX + b] = a^2 \text{Var}[X] \tag{6.12}$$

or, equivalently, the standard deviation of $aX + b$ equals $|a|\sigma$. Thus, the additive constant b has no effect on the variance or the standard deviation. This seems reasonable because movement of the origin does not affect the spread of the distribution. A change in scale by the factor a changes the standard deviation by the factor $|a|$.

If X and Y are random variables that differ only in origin and scale of measurement (that is, if X and Y are related by a linear equation of the form $Y = aX + b$), then clearly there is no essential difference in the amount of information that is given by observing the value of X or of Y.

To eliminate any effect of origin or scale, we frequently work with the *standardized random variable* X^* instead of with X. This is defined as follows.

If the random variable X has mean μ and variance σ^2, then

$$X^* = \frac{X - \mu}{\sigma} \tag{6.13}$$

or, equivalently,

$$X = \sigma X^* + \mu$$

Since $X^* = (1/\sigma)X - (\mu/\sigma)$ we observe that X^* always has mean 0 and variance 1 regardless of the values of μ and σ^2, using (6.5) and (6.12).

The use of standardized random variables frequently simplifies discussion. For example, the normal distribution, mentioned in Examples 3.18 and 5.2, depends on two parameters, μ and σ^2, which are, in fact, the mean and the variance (see Section 7.9). Thus, it appears necessary that a table of its distribution function $F(x)$ be constructed for every pair μ, σ^2. The necessity for multiple tables is eliminated, however, by transforming all questions about a random variable X into questions about X^*. Thus, the normal distribution function must only be tabulated for the single case $\mu' = 0$, $\sigma^2 = 1$.

It is left to the reader, as an exercise, to show that if

$$Y = aX + b$$

with $a \geq 0$, then

$$Y^* = X^*$$

[Represent $E[Y]$ and $Var[Y]$ by using (6.5) and (6.12) and find Y^* by using (6.13).] This shows that standardization of random variables does, in fact, eliminate the effect of origin and scale.

6.8   Expectation of a Function of Several Random Variables

The problems involved in taking the expectation of a random variable that is a function of two or more other random variables, for example, $U = Q(X, Y)$, are similar to those with functions of one random variable that were discussed in Section 6.3. Strictly speaking, the density function of U must be known to compute $E[U]$ directly from its definition. However, as in the single-variable case, it can be proved that, instead, one can average $Q(X, Y)$ over all pairs of values in the joint range of X and Y using the joint density function of X and Y as weight function. That is, the following formula holds true.

$$E[Q(X,Y)] = \begin{cases} \displaystyle\int_{-\infty}^{\infty}\int_{-\infty}^{\infty} Q(x,y)f(x,y)\,dx\,dy, & \text{continuous case} \\[2ex] \displaystyle\sum_i \sum_j Q(x_i, y_j)p(x_i, y_j), & \text{discrete case} \end{cases}$$

$$(6.14)$$

where $f(x, y)$ [or $p(x_i, y_j)$] is the joint density function of X and Y. The proof of (6.14) is omitted. The formula generalizes in obvious fashion to functions of more than two random variables.

The simplest types of functions to consider are the sum function and the product function.

For a sum of random variables, (6.14) gives, in the continuous case,

$$\begin{aligned} E[X+Y] &= \int_{-\infty}^{\infty}\int_{-\infty}^{\infty} (x+y)f(x,y)\,dx\,dy \\ &= \int_{-\infty}^{\infty} x \int_{-\infty}^{\infty} f(x,y)\,dy\,dx + \int_{-\infty}^{\infty} y \int_{-\infty}^{\infty} f(x,y)\,dx\,dy \\ &= \int_{-\infty}^{\infty} xf_X(x)\,dx + \int_{-\infty}^{\infty} yf_Y(y)\,dy \\ &= E[X] + E[Y] \end{aligned}$$

$$(6.15)$$

by using formulas (4.20) and (4.21) to obtain $f_X(x)$ and $f_Y(y)$ from $f(x, y)$. The proof of (6.15) in the discrete case is similar. Formula (6.15) can be extended to sums of more than two random variables and, when combined with (6.5), yields the following general *linear* property for expectation.

$$E\left[\sum_{i=1}^{n} a_i X_i\right] = \sum_{i=1}^{n} a_i E[X_i] \tag{6.16}$$

whenever $X_1, \ldots, X_n$ are random variables and $a_1, \ldots, a_n$ are constants.

A simple formula for the expectation of a product (corresponding to (6.15) for a sum) does not hold true in general. However, in the special case when X and Y are *independent*, we have

$$\begin{aligned} E[XY] &= \int_{-\infty}^{\infty}\int_{-\infty}^{\infty} xyf(x,y)\,dx\,dy \\ &= \int_{-\infty}^{\infty}\int_{-\infty}^{\infty} xyf_X(x)f_Y(y)\,dx\,dy \\ &= \int_{-\infty}^{\infty} xf_X(x)\,dx \int_{-\infty}^{\infty} yf_Y(y)\,dy \\ &= E[X]E[Y] \end{aligned}$$

$$(6.17)$$

the proof being similar in the discrete case. Thus, for independent random variables

$$E[XY] = E[X]E[Y] \qquad (6.18)$$

Formula (6.18) generalizes without difficulty to

$$E[\phi_1(X)\phi_2(Y)] = E[\phi_1(x)]E[\phi_2(Y)] \qquad (6.19)$$

when X and Y are independent and ϕ_1 and ϕ_2 are two functions. Both (6.18) and (6.19) also hold true for products of more than two independent factors.

6.9 Variance of a Sum

Suppose X and Y are random variables that have expectations $\mu_X = E[X]$ and $\mu_Y = E[Y]$ and variances $\sigma_{X^2} = E[(X - \mu_X)^2]$ and $\sigma_{Y^2} = E[(Y - \mu_Y)^2]$. Let us attempt to compute the variance of their sum:

$$S = X + Y \qquad (6.20)$$

By (6.15), S will have mean $\mu_S = \mu_X + \mu_Y$. Thus the variance of S is, by using (6.23),

$$
\begin{aligned}
\sigma_S^2 &= E\left[(S - \mu_S)^2\right] \\
&= E\left[(X + Y - \mu_X - \mu_Y)^2\right] \\
&= E\left[\{(X - \mu_X) + (Y - \mu_Y)\}^2\right] \\
&= E\left[(X - \mu_X)^2 + (Y - \mu_Y)^2 + 2(X - \mu_X)(Y - \mu_Y)\right] \\
&= E\left[(X - \mu_X)^2\right] + E\left[(Y - \mu_Y)^2\right] + 2E\left[(X - \mu_X)(Y - \mu_Y)\right]
\end{aligned}
$$

It follows that

$$\sigma_S^2 = \sigma_X^2 + \sigma_Y^2 + 2E\left[(X - \mu_X)(Y - \mu_Y)\right] \qquad (6.21)$$

The last term on the right side of (6.21) is termed the *covariance* of X and Y and, in general, cannot be dispensed with. However, in the special case where X and Y are *independent*, by using (6.19) we find

$$
\begin{aligned}
E\left[(X - \mu_X)(Y - \mu_Y)\right] &= E[X - \mu_X]E[Y - \mu_Y] \\
&= (\mu_X - \mu_X)(\mu_Y - \mu_Y) \\
&= 0
\end{aligned}
$$

Thus, if X and Y are *independent*

$$\sigma_S^2 = \sigma_X^2 + \sigma_Y^2 \qquad (6.22)$$

where $S = X + Y$.

Formula (6.22) readily extends to sums of more than two mutually independent random variables and to linear combinations.

If

$$S = \sum_{i=1}^{n} a_i X_i$$

then

$$\sigma_S^2 = \sum_{i=1}^{n} a_i^2 \sigma_{X_i}^2 \qquad (6.23)$$

if $X_1, X_2, \ldots, X_n$ are independent random variables and $a_1, a_2, \ldots, a_n$ are constants.

EXAMPLE 6.8

The profit of a certain company in any given month is a random variable having mean \$10,000 and standard deviation \$1,000. Assuming that the profits in different months are independent, find the mean and standard deviation of the total profit S in an entire year.

Here $S = \sum_{i=1}^{12} X_i$, where $X_1, X_2, \ldots, X_n$ are independent random variables, each having mean 10,000 and variance $(1,000)^2$. By (6.16) and (6.23), S will have mean and variance

$$\mu_S = \sum_{i=1}^{12} 10,000 = 120,000$$

$$\sigma_S^2 = \sum_{i=1}^{12} (1,000)^2 = 12,000,000$$

Thus, the standard deviation of S is $\sqrt{12,000,000} = \$3,464$. Notice that the standard deviation of S is only about $3\frac{1}{2}$ times the standard deviation of each X, not 12 times as one might possibly guess.

6.10 Correlation Coefficient

In equation (6.21) we defined the quantity

$$E\left[(X - \mu_X)(Y - \mu_Y)\right]$$

as the *covariance* of X and Y. Of some importance, especially in statistical studies involving the bivariate normal density (see Example 4.8), is the quantity

$$\rho = \frac{E\left[(X - \mu_X)(Y - \mu_Y)\right]}{\sigma_X \sigma_Y} \qquad (6.24)$$

called the *correlation coefficient*. This number has the following properties.

(a) $-1 \le \rho \le 1$.
(b) If $Y = aX + b$ then $\rho = +1$ if $a > 0$ and $\rho = -1$ if $a < 0$.

But perhaps its major property comes from the following theorem.

Theorem

(a) If X and Y are independent, $\rho = 0$.

(b) If $\rho = 0$ and if X and Y have a bivariate normal density (Example 4.8), then X and Y are independent.

Thus, bivariate normal random variables are independent if and only if $\rho = 0$. We will leave a proof of this result to the exercises.

The point here is that independent random variables have zero correlation coefficient. However, the converse of this theorem is not true in general. That is, if $\rho = 0$, the theorem only assures us that the random variables are independent in the bivariate normal case.

6.11   The Weak and Strong Laws of Large Numbers

We now have the tools needed to tie together some loose ends. In the process we will see one important application of the Chebyshev inequality and how the axiomatic construction of probability theory that we developed in Section 2.2 ties back to the older relative frequency definition in Section 2.1.

To do all this let $X_1, X_2, \ldots, X_n, \ldots$ be a sequence of mutually independent random variables and let each random variable have range $\{0, 1\}$ with the same density function

$$\Pr[X_j = 0] = q, \qquad \Pr[X_j = 1] = p, \qquad p + q = 1 \qquad (6.25)$$

Then from Example 6.1 and equation (6.9) we have

$$E[X_j] = p; \qquad \mathrm{Var}[X_j] = pq \qquad (6.26)$$

Now form the linear sum of these X_j,

$$S_n = X_1 + X_2 + \cdots + X_n \qquad (6.27)$$

and consider the *average* $S_n/n = (\sum_1^n X_i)/n$. From (6.16) and (6.23) it follows that

$$E\left[\frac{S_n}{n}\right] = p; \qquad \mathrm{Var}\left[\frac{S_n}{n}\right] = pq/n \qquad (6.28)$$

The Chebyshev inequality, with $X = S_n/n$ and $\mu = p$, implies, for each positive value of t,

$$\Pr\left[\left|\frac{S_n}{n} - p\right| \geq t\right] \leq \frac{pq}{t^2 n}$$

Letting $n \to \infty$ we obtain

$$\lim_{n \to \infty} \Pr\left[\left|\frac{S_n}{n} - p\right| \geq t\right] \leq \lim_{n \to \infty} \frac{pq}{t^2 n} = 0$$

i.e.,

$$\lim_{n \to \infty} \Pr\left[\left|\frac{S_n}{n} - p\right| \geq t\right] = 0 \qquad (6.29)$$

Thus, for large values of n, the probability that the ratio S_n/n differs from p by more than any $t > 0$ is small, and as $n \to \infty$ the probability that these two differ by any amount, t, goes to 0.

The result (6.29) is called a *weak law of large numbers*. As we have set up this particular form of the law it tells us some important facts about probabilities. For let us consider an event that we can call a "favorable event." (See Section 2.1.) Let $X_j = 1$ if this "favorable event" occurs the jth time we perform our experiment and $X_j = 0$ otherwise. Then S_n/n is simply the ratio of the (random) number of successes (occurrences of the "favorable event") to the total number of trials. That is, S_n/n is the *relative frequency* of "successes" to "trials." What the weak law of large numbers is telling us is that this relative frequency approaches the probability of "success" for $n \to \infty$ in the sense that the probability that the relative frequency differs from the probability p by any quantity t, however small, is itself small, and in the limit the probability that this relative frequency differs from p by any positive quantity goes to 0.

In this way we demonstrate that the early relative frequency definition of probability is a reasonable method for estimating probabilities defined in the axiomatic structure. Recall (Section 2.1) that within the axiomatic structure of probability there was no concern about how these probabilities were obtained. They could be gotten in many ways. What we have now shown is that a relative frequency method of obtaining these probabilities is a "good" method for getting them in the sense of (6.29). Of course it is not the only way as we observed in Section 2.1, however, it is a widely used way as shown in Section 2.10.

The weak law of large numbers can be extended as follows. The proof still uses the Chebyshev inequality much as we did to get to (6.29), however, we will omit the details.

Theorem   (Weak Law of Large Numbers)

Let $X_1, X_2, \ldots, X_n, \ldots$ be a sequence of mutually independent, identically distributed random variables with $E[X_j] = \mu$, $\text{Var}[X_j] = \sigma^2 < \infty$. Let

$$S_n = X_1 + X_2 + \cdots + X_n$$

Then,

$$\lim_{n \to \infty} \Pr\left[\left|\frac{S_n}{n} - \mu\right| \geq t\right] = 0$$

for each $t > 0$.

There are other versions of these types of theorems that are called *Strong Laws of Large Numbers*. In probability theory a theorem that tells how a sequence of probabilities converges is called a "weak law." A "strong theorem" tells us how the sequence of random variables behaves in the limit. Notice the important distinction here. "Weak laws" are usually much easier to prove than "strong laws." We will state one Strong Law of Large Numbers but we do not have the tools to prove it here.

Theorem (Strong Law of Large Numbers)

Let $X_1, X_2, \ldots, X_n$ be a sequence of mutually independent, identically distributed random variables with $E[X_j] = \mu$, $\text{Var}[X_j] = \sigma^2 < \infty$. Let

$$S_n = X_1 + X_2 + \cdots + X_n$$

Then,

$$\Pr\left[\lim_{n \to \infty} \left| \frac{S_n}{n} - \mu \right| \geq t \right] = 0$$

for each $t > 0$.

Note carefully that the difference between the Weak Law of Large Numbers and the Strong Law of Large Numbers is not just a formula. We did not simply move the $\lim_{n \to \infty}$ operation inside the Pr. These two results are saying very different things. The Strong Law of Large Numbers is saying that the sequence S_n/n is converging to the constant μ. This is what we really want the relative frequency S_n/n to do. It is the strong law that gives the relative frequency definition of probability its greatest credence.

6.12   The Moment-Generating Function

If X is a random variable, then the quantity $e^{X\theta}$ is a function of X for each fixed value of θ. Thus, its expectation $E[e^{X\theta}]$ will be a number depending on θ provided that it exists. Hence, we are led to define the *function*

$$M(\theta) = E[e^{X\theta}] = \begin{cases} \int_{-\infty}^{\infty} e^{x\theta} f(x)\, dx, & \text{continuous case} \\ \sum_i e^{x_i \theta} p(x_i), & \text{discrete case} \end{cases} \tag{6.30}$$

This function is termed the *moment-generating function* of the random variable X. If more than one random variable is involved in the same discussion we use the notation $M_X(\theta)$ in place of $M(\theta)$.

Of course the expectation defining $M(\theta)$ may not exist for *all* real numbers θ. (The integral or sum may not converge absolutely.) However, for most of the random variables that we are concerned with there is an interval of θ-values within which $M(\theta)$ does exist.

EXAMPLE 6.9

Let X have a negative exponential density function, $f(x) = \lambda e^{-\lambda x}$, $x > 0$. Then X has moment-generating function

$$M(\theta) = E[e^{X\theta}] = \int_0^\infty e^{x\theta} \cdot \lambda e^{-\lambda x} \, dx$$

$$= \lambda \int_0^\infty e^{-(\lambda - \theta)x} \, dx$$

$$= \frac{\lambda}{\lambda - \theta} = \frac{1}{1 - (\theta/\lambda)}$$

provided that $\theta < \lambda$.

EXAMPLE 6.10

Let X represent the number of heads in n tosses of an honest coin. Then X has the density function given by (3.7) in Example 3.13:

$$p_i = \binom{n}{i} \frac{1}{2^n}, \qquad i = 0, 1, 2, \ldots, n$$

The moment-generating function will be

$$M(\theta) = E[e^{X\theta}] = \sum_{i=0}^n e^{i\theta} p_i$$

$$= \left[\sum_{i=0}^n e^{i\theta} \binom{n}{i} \right] \frac{1}{2^n} = \frac{(1 + e^\theta)^n}{2^n}$$

using the binomial theorem (A.6 in Appendix A).

Clearly (6.30) represents a method of changing or transforming one function, the density function, into another function, the moment-generating function. We shall find that the moment-generating function of a random variable is often easier to determine than its density function, and that we can use $M(\theta)$ in place of the density function to answer many questions of interest. One reason for this is found in the following theorem, which we quote without proof.

Theorem

(a) If two random variables have the same moment-generating functions, then they must have the same distribution and density functions.

(b) If two moment-generating functions are almost the same, then the same is true of the corresponding distribution functions. More precisely, if a sequence of moment-generating functions converges to a given moment-generating function, $\lim_{n \to \infty} M_n(\theta) = M(\theta)$, then the same is true of the corresponding distribution functions, $\lim_{n \to \infty} F_n(x) = F(x)$.

If we can show that a random variable that is being studied has the same moment-generating function as that of another random variable with a *known* probability distribution, then part (a) of this theorem assures us that the distribution of the original random variable must also be the same.

The most important situation in which the use of the moment-generating function simplifies the discussion exists when independent random variables are added. If

$$S = X + Y$$

where X and Y are independent, then S has moment-generating function

$$M_S(\theta) = E[e^{S\theta}] = E[e^{(X+Y)\theta}]$$
$$= E[e^{X\theta}e^{Y\theta}]$$
$$= E[e^{X\theta}]E[e^{Y\theta}]$$
$$= M_X(\theta)M_Y(\theta)$$

by using (6.19) with $\phi_1(X) = e^{X\theta}$, $\phi_2(Y) = e^{Y\theta}$, and θ being held fixed. Thus, for independent random variables the moment-generating function of a sum is simply the product of the moment-generating functions of the individual terms:

$$M_{X+Y}(\theta) = M_X(\theta)M_Y(\theta) \tag{6.31}$$

A similar formula holds true for sums of more than two independent terms. Use of (6.31) is usually much easier than computing the density function of $S = X + Y$ directly, and the advantage becomes even more marked for sums of more than two terms.

EXAMPLE 6.11

In Example 6.9 the moment-generating function of a random variable that has a negative exponential density function was shown to be

$$\frac{1}{1-(\theta/\lambda)}$$

If U has density function

$$f(u) = \lambda^2 e^{-\lambda u}u, \quad \text{for } u > 0$$

then U has moment-generating function

$$E[e^{U\theta}] = \int_0^\infty e^{u\theta}\lambda^2 e^{-\lambda u}u\,du$$
$$= \lambda^2 \int_0^\infty e^{-(\lambda-\theta)u}u\,du$$
$$= \frac{\lambda^2}{(\lambda-\theta)^2} = \frac{1}{[1-(\theta/\lambda)]^2}, \quad \text{for } \theta < \lambda$$

using integration by parts. If $S = X + Y$, where X and Y are independent negative exponentially distributed random variables, by (6.31) S will have moment-generating function

$$\left[\frac{1}{1-(\theta/\lambda)}\right]\left[\frac{1}{1-(\theta/\lambda)}\right]=\frac{1}{[1-(\theta/\lambda)]^2}$$

which is the same as the moment-generating function of U. Hence, by the preceding theorem, S and U must have the same density function.

This method of finding the density function of $X+Y$ should be compared with the direct calculations in Example 5.5 for the same functions.

The reason for the phrase *moment-generating* function comes from the fact that $e^{X\theta}$ can be expanded in series (A.5 in Appendix A)

$$e^{X\theta}=1+X\theta+\frac{X^2\theta^2}{2!}+\cdots+\frac{X^n\theta^n}{n!}+\cdots$$

and, if the situation is sufficiently regular that the linear property of expectation (6.16) can be applied to this *infinite* series, then the moment-generating function becomes

$$M(\theta)=E[e^{X\theta}]=1+E[X]\theta+E[X^2]\frac{\theta^2}{2!}+\cdots+\frac{E[X^n]\theta^n}{n!}+\cdots$$

$$=1+\mu\theta+\mu_2'\frac{\theta^2}{2!}+\cdots+\mu_n'\frac{\theta^n}{n!}+\cdots \qquad (6.32)$$

Thus the moments μ_n' can all be found by expanding $M(\theta)$ in powers of θ and by noting the coefficients of the terms $\theta^n/n!$. In particular, the mean and second moment, μ and μ_2', can be found by evaluating the first two derivatives at 0, as is usual in computing the coefficients of a Taylor series:

$$\mu=E[X]=M'(0), \qquad \mu_2'=E[X^2]=M''(0) \qquad (6.33)$$

EXAMPLE 6.12

If X has a negative exponential distribution we have found that

$$M(\theta)=\frac{1}{1-(\theta/\lambda)}, \qquad \theta<\lambda$$

This can be expanded in series by using the formula for the sum of a geometric series (at least provided that $-\lambda<\theta<\lambda$)

$$M(\theta)=1+\frac{\theta}{\lambda}+\frac{\theta^2}{\lambda^2}+\cdots+\frac{\theta^n}{\lambda^n}+\cdots$$

$$=1+\frac{\theta}{\lambda}+\frac{2}{\lambda^2}\cdot\frac{\theta^2}{2!}+\cdots+\frac{n!}{\lambda^n}\cdot\frac{\theta^n}{n!}+\cdots$$

It follows from (6.32) that

$$\mu=\frac{1}{\lambda},\ \mu_2'=\frac{2}{\lambda^2},\ldots,\mu^n=\frac{n!}{\lambda^n},\ldots$$

Also

$$\sigma^2 = \mu_2' - \mu^2 = \frac{2}{\lambda^2} - \left(\frac{1}{\lambda}\right)^2 = \frac{1}{\lambda^2}$$

EXAMPLE 6.13

In Example 6.10 we showed that if X is a discrete random variable representing the number of heads in n tosses of an honest coin, then X has moment-generating function $M(\theta) = (1 + e^\theta)^n/2^n$. By (6.33) the mean and second moment of X are given by

$$\mu = M'(0) = \frac{n(1 + e^\theta)^{n-1}e^\theta}{2^n}\Bigg|_{\theta=0} = \frac{n(1+1)^{n-1}}{2^n}$$

$$= \frac{n}{2}$$

$$\mu_2' = M''(0) = \frac{\left[n(n-1)(1+e^\theta)^{n-2}(e^\theta)^2 + n(1+e^\theta)^{n-1}e^\theta\right]}{2^n}\Bigg|_{\theta=0}$$

$$= \frac{\left[n(n-1)(1+1)^{n-2} + n(1+1)^{n-1}\right]}{2^n}$$

$$= \frac{n(n-1) + 2n}{4} = \frac{n^2 + n}{4}$$

Using (6.10) the variance of X is found to be

$$\sigma^2 = \mu_2' - \mu^2 = \frac{n^2 + n}{4} - \left(\frac{n}{2}\right)^2 = \frac{n}{4}$$

The student should try to compute μ and σ^2 directly from the definition of expectation by using the density function $p_i = \binom{n}{i}/2^n$, $i = 0, 1, \ldots, n$, to convince himself that the use of (6.33) offers some advantage in simplicity.

6.13   Probability-Generating Function, z-Transform, and Characteristic Function

If X is an *integer-valued*, discrete random variable with density function p_i, then the moment-generating function can be simplified by the substitution $z = e^\theta$. The resulting function of z is termed the *probability-generating function* of X or, sometimes, the *z-transform* of X. It is defined by

$$\psi(z) = \psi_X(z) = E[z^X] = \sum_i z^i p_i$$

All the formulas of the previous section for moment-generating functions carry over to formulas that involve $\psi(z)$ if we change the variable by using the identity $z = e^\theta$. Thus, for example, (6.33) becomes

$$\mu = \psi'(1), \qquad \mu_2' = \psi''(1) + \psi'(1) \tag{6.34}$$

where one uses the fact that $M(\theta) = \psi(z)$ with $z = e_i^\theta$ and the differentiation formulas

$$M'(\theta) = \frac{d\psi}{dz} \cdot \frac{dz}{d\theta} = \psi'(z)e^\theta = \psi'(z)z$$

$$M''(\theta) = \frac{d}{d\theta}M'(\theta) = \frac{d}{dz}M'(\theta) \cdot \frac{dz}{d\theta} = \left[\psi''(z)z + \psi'(z)\right] \cdot z$$

as well as the fact that $z = 1$ when $\theta = 0$.

For instance, if X is the random variable that is discussed in Examples 6.10 and 6.13 X has probability-generating function

$$\psi(z) = \sum_{i=0}^{n} \frac{z^i\binom{n}{i}}{2^n} = \frac{(1+z)^n}{2^n}$$

by using the binomial theorem (A.6 in Appendix A). Thus,

$$\mu = \psi'(1) = \left. \frac{n(1+z)^{n-1}}{2} \right|_{z=1} = \frac{n}{2}$$

$$\mu_2' = \psi''(1) + \psi'(1) = \left. \frac{n(n-1)(1+z)^{n-2}}{2^n} \right|_{z=1} + \frac{n}{2}$$

$$= \frac{n(n-1)}{4} + \frac{n}{2} = \frac{n^2+n}{4}$$

which is an even simpler computation than the one in Example 6.13.

We have ignored the possibility that the moment-generating function or probability-generating function might not exist. (The defining sum or integral might not converge absolutely.) There is no great loss in generality here since most of what we have done could also be done by using still another generating function that always *does* exist. This is the *characteristic function* or *Fourier transform* of the random variable X. To define it we must introduce complex numbers. The *characteristic function* is defined by

$$\phi(t) = E[e^{itX}] = E[\cos tX + i\sin tX]$$

for real t. (Here $i = \sqrt{-1}$.) Since

$$|e^{itX}| = \sqrt{\cos^2 tX + \sin^2 tX} = 1$$

the function whose expectation is being taken is bounded and, hence, the defining integral or sum must converge absolutely,

$$\left|E[e^{itX}]\right| = \left|\int_{-\infty}^{\infty} e^{itx}f(x)\,dx\right| \le \int_{-\infty}^{\infty} |e^{itx}|f(x)\,dx$$

$$= \int_{-\infty}^{\infty} 1 \cdot f(x)\,dx = 1$$

These functions play a crucial role in the extensions of the theory that we have discussed. We do not develop them now because the tools that we have described are sufficient for our needs.

6.14   Estimating Moments

For many important classes of random variables it turns out that the density function of a random variable is completely determined as soon as one or two of its moments are known. Thus, we seek methods of *estimating* certain moments based on a set of *observations* of values of the random variable, i.e., based on a set of *data*.

Suppose we are given a data set consisting of n observations $x_1, x_2, \ldots, x_n$ of the random variable X. This is a set of n numbers obtained from observing the value of X in n independent experiments. Such a set of numbers is termed a *sample*. We wish to study methods of *estimating* the mean $\mu = E[X]$ and the variance $\sigma^2 = \text{Var}[X]$ using this set of numbers.

(a) **Estimating Expectations from Data**   Since by its definition (6.1) the mean $\mu = E[X]$ is a density average it seem reasonable to estimate μ by the corresponding *sample average* of the sample, denoted by

$$\bar{x} = \left(\sum_{i=1}^{n} x_i \right) \Big/ n \tag{6.35}$$

EXAMPLE 6.14

A die is tossed five times with results $2, 5, 1, 5, 3$. The sample average in this case is
$$\bar{x} = (2 + 5 + 1 + 5 + 3)/5 = 3.2.$$

EXAMPLE 6.15

Lipson and Sheth (1973)[1] give the following data for wear (in units of 0.001 in.) on friction linings for wheels.

8.05	5.00
9.10	8.53
6.20	6.80
7.50	9.60
10.30	11.46

Then

$$\bar{x} = \sum_{i=1}^{n} \frac{x_i}{10} = 8.25$$

and $\bar{x}$ is an estimate of $\mu = E[X]$, the expected amount of wear per lining.

[1]C. Lipson and N. Sheth (1973), *Statistical Design and Analysis of Engineering Experiments*, McGraw–Hill, New York. Reprinted with permission of McGraw–Hill.

TABLE 6.1
Data for a Discrete Random Variable

VALUE	FREQUENCY
x_1	n_1
x_2	n_2
x_3	n_3
$\vdots$	$\vdots$
x_m	n_m
$n = n_1 + n_2 + \cdots + n_m$	

It is important to realize that μ is a fixed constant, dependent only on the density of the random variable X, while the number $\bar{x}$ used to estimate μ depends on the particular values that occur in the sample. A different set of n observations would result in a different sample $x_1, x_2, \ldots, x_n$ and, hence, would lead to a different value of $\bar{x}$ and a different estimate of μ. The estimate, in fact, is itself the value of a random variable, the averaging random variable.

When n is large, various methods exist for simplifying the calculation of the sum in (6.35).

(i) Data for Discrete Random Variables Suppose the sample contains only a few *different* numerical values, each repeated a number of times. This often might occur if the underlying random variable was discrete. In this case the sample data could be described by giving only those few values together with the number of times each is repeated, as in Table 6.1.

In this case (6.35) becomes

$$\bar{x} = \left(\sum_{i=1}^{m} n_i x_i \right) / n \qquad (6.36)$$

where $n = \Sigma n_i$.

EXAMPLE 6.16

Duncan (1965)[1] gives the following data for the number of defects found in 28 pieces of wool cloth.

No. of Defects	Frequency
0	8
1	12
2	6
3	1
4	1

[1]A. J. Duncan (1965), *Quality Control and Industrial Statistics* (3rd ed.), Richard D. Irwin, Inc., Homewood, Illinois. Reprinted with permission of Richard D. Irwin Inc.

TABLE 6.2
Grouping of Continuous Data

INTERVAL	INTERVAL MIDPOINT	FREQUENCY
$a_0 - a_1$	c_1	n_1
$a_1 - a_2$	c_2	n_2
$\vdots$	$\vdots$	$\vdots$
$a_{m-1} - a_m$	c_m	n_m

$$n = n_1 + n_2 + \cdots + n_m$$

Thus,

$$\bar{x} = \frac{8 \times 0 + 12 \times 1 + 6 \times 2 + 1 \times 3 + 1 \times 4}{28}$$

$$= 1.107$$

and 1.107 is a reasonable estimate of the value of $\mu = E[X]$, the expected number of defects per piece.

   (ii) Data for Continuous Random Variables   Suppose that the large sample contains many different numerical values. This might occur if the underlying random variable was continuous. In this case the calculation (6.35) can be simplified by *grouping* the data into intervals, see Table 6.2. If the intervals are relatively small then little error would be introduced if each value x_i was replaced by the *midpoint* c_j of the interval containing it. (Some reasonable convention should be used to classify a sample value in the unlikely case that it falls exactly on the boundary of an interval.) If this is done, then (6.35) becomes

$$\bar{x} = \left(\sum_{i=1}^{m} n_i c_i \right) / n \tag{6.37}$$

EXAMPLE 6.17

Broadbent (1949)[1] gives Table 6.3 for times between flights of a particular aphid. (See also Exercise 3.4.)
   Then

$$\bar{x} = \sum_{i=1}^{6} n_i c_i / n = \frac{55 \times 5 + 32 \times 15 + \cdots + 2 \times 115}{107}$$

$$= 15.05$$

So from this data an estimate of $E[X]$, the mean time between flights of this aphid, is 15.05.

---

[1] L. Broadbent (1949), "Factors Affecting the Activity of Alatae of the Aphids Myzus Persical (Sulzer) and Brevicoryne Bressical (L)," *Annals of Applied Biology*, **36**, pp. 40–62. Reprinted with permission of the *Annals of Applied Biology*.

TABLE 6.3
Times between Flights of an Aphid

INTERVAL	MIDPOINT	FREQUENCY
0–10	5	55
10–20	15	32
20–30	25	11
30–60	45	6
60–100	80	1
100–130	115	2

(b) **Estimating Variances from Data**   Again, in analogy with its definition (6.9) it is reasonable to estimate the variance $\sigma^2 = \text{Var}[X]$ by the corresponding sample average of the squared deviations of the observations from the sample average,

$$\sum_{i=1}^{n} (x_i - \bar{x})^2/n \tag{6.38}$$

For technical reasons that we will not discuss here, it turns out that the estimate has more desirable properties if we use the denominator $n - 1$ instead of n. Thus, our estimate of σ^2 is the *sample variance*

$$s^2 = \sum_{i=1}^{n} (x_i - \bar{x})^2/(n - 1) \tag{6.39}$$

(Clearly for large n there is little difference between (6.38) and (6.39).)

EXAMPLE 6.18

Using the data of Example 6.15 we have

$$s^2 = \left[(8.05 - 8.25)^2 + \cdots + (11.46 - 8.25)^2\right]/9$$

$$= [0.04 + \cdots + 10.3041]/9 = 3.844$$

If the term $(x_i - \bar{x})^2$ in (6.39) is expanded and some elementary algebraic manipulation is performed, an equivalent formula that is somewhat more convenient for hand calculation results.

$$s^2 = \frac{n \sum_{i=1}^{n} x_i^2 - \left(\sum_{i=1}^{n} x_i\right)^2}{n(n - 1)} \tag{6.40}$$

In the case of data with repeated values, as in part (a)(i), the sums in (6.39) and (6.40) can be simplified by inserting the counting factor multiples n_i, as in (6.36). Also, when using grouped data, as in part (a)(ii), the x_i's can be replaced by interval midpoints c_j, as in (6.37).

EXAMPLE 6.19

Using the data in Example 6.16, the two sums needed in (6.40) can be obtained.

$$\sum x_i = \sum_{i=1}^{5} n_i x_i = 8 \times 0 + 12 \times 1 + 6 \times 2 + 1 \times 3 + 1 \times 4 = 31$$

$$\sum x_i^2 = \sum_{i=1}^{5} n_i x_i^2 = 8 \times 0^2 + 12 \times 1^2 + 6 \times 2^2 + 1 \times 3^2 + 1 \times 4^2 = 61$$

Thus, (6.40) gives

$$s^2 = \frac{28 \times 61 - (31)^2}{28 \times 27} = .988$$

Clearly this is simpler than the calculation using (6.39) directly:

$$s^2 = \left[8 \times (0 - 1.107)^2 + 12 \times (1 - 1.107)^2 + \cdots + 1 \times (4 - 1.107)^2\right]/27$$

EXAMPLE 6.20

Using the Broadbent data of Example 6.17 together with (6.40) one obtains

$$\sum x_i = \sum n_i c_i = 55 \times 5 + \cdots + 2 \times 115 = 1610$$

$$\sum x_i^2 = \sum n_i c_i^2 = 55 \times 5^2 + \cdots + 2 \times 115^2 = 60450$$

$$s^2 = \frac{107 \times 60450 - (1610)^2}{107 \times 106} = 341.7$$

(c) Estimating a Correlation Coefficient To estimate the correlation coefficient ρ defined in Section 6.10 when a set of n *pairs* of data observations $(x_1, y_1), \ldots, (x_n, y_n)$ is given, we use the same method used in parts (a) and (b). Expectations are estimated as before by averages of observed data. Thus, ρ is estimated by the *sample correlation coefficient* r:

$$r = \frac{\left[\Sigma(x_i - \bar{x})(y_i - \bar{y})\right]/(n-1)}{s_x s_y} \tag{6.41}$$

Here s_x and s_y are the square roots of the variance estimates of the x-data and y-data separately, as discussed in part (b). By algebraic manipulation (6.41) can be put into a form somewhat more convenient for hand calculation.

$$r = \frac{n\Sigma x_i y_i - \left(\Sigma x_i\right)\left(\Sigma y_i\right)}{\left\{\left[n\Sigma x_i^2 - \left(\Sigma x_i\right)^2\right]\left[n\Sigma y_i^2 - \left(\Sigma y_i\right)^2\right]\right\}^{1/2}} \tag{6.42}$$

EXAMPLE 6.21

The data in Table 6.4 are final grade averages for students in their last year of high school and first year of college, as given by Rubin and Stroud (1977).[1] We

---

[1] D. B. Rubin and T. W. F. Stroud, (1977), "Comparing High Schools with Respect to Student Performance," *Journal of Educational Statistics*, **2**, pp. 121–138. Reprinted with permission of the American Educational Research Association.

TABLE 6.4
Final Year High School Grades versus First Year College Grades

LAST YEAR H.S. (x_i)	FIRST YEAR COL. (y_i)	LAST YEAR H.S. (x_i)	FIRST YEAR COL. (y_i)
62.1	49.1	71.5	70.1
66.4	54.3	72.7	58.3
66.3	61.7	75.5	67.7
67.0	65.1	76.7	57.8
67.7	55.4	76.7	69.5
68.6	51.0	77.0	64.8
69.0	57.7	77.5	69.5
69.5	51.8	79.5	53.3
69.7	65.7	81.0	64.1
71.0	67.7	84.0	76.1

wish to estimate the correlation coefficient ρ. (One might expect a high correlation, near to 1, for these grades.)

Then

$$20 \sum_{i=1}^{20} x_i y_i = 20(89,647.13) = 1,792,942.60$$

$$\left(\sum_{i=1}^{20} x_i \right) \left(\sum_{i=1}^{20} y_i \right) = 1,783,776.58$$

$$\left(\sum_{i=1}^{20} x_i \right)^2 = 2,100,760.36; \qquad \left(\sum_{i=1}^{20} y_i \right)^2 = 1,514,622.49$$

$$20 \sum_{i=1}^{20} x_i^2 = 2,113,314.40; \qquad 20 \sum_{i=1}^{20} y_i^2 = 1,536,483$$

Thus, we estimate σ by $r = .55$, which is unexpectedly far from 1.

Exercises 6

1. A factory produces 25% defectives on the average. Choose four items at random and let X = the number of defectives among them. Then X is a discrete random variable having range $0, 1, 2, 3, 4$. Compute the mean $\mu = E[X]$ and variance σ^2 of X directly from their definitions by completing the following table.

x	$p(x)$	$xp(x)$	$(x-\mu)^2 p(x)$
0			
1			
2			
3			
4			
Totals	1	$\mu =$	$\sigma^2 =$

2. Let X be a continuous random variable having a uniform density function over the interval $-1 \le x \le 1$

$$f(x) = \begin{cases} \frac{1}{2}, & \text{for } -1 \le x \le 1 \\ 0, & \text{otherwise} \end{cases}$$

Find μ and σ

3. Define the random variable Y by

$$Y = X^2$$

where X has the distribution in Exercise 2.

(a) Compute $E[Y]$ using (6.3).

(b) Compute $F_Y(y) = \Pr[Y \le y] = \Pr[X^2 \le y] = \Pr[-\sqrt{y} \le X \le \sqrt{y}]$ for $y \ge 0$.

(c) Compute $f_Y(y) = F_Y'(y)$.

(d) Compute $E[Y]$ using the definition (6.1) and compare with (a).

4. Compute $E[X]$ and $\text{Var}[X]$ for the triangular density function given in Exercise 3.15.

5. Compute $E[X]$ and $\text{Var}[X]$ for the Weibull density function given in Example 3.17.

6. If X has a uniform density function over the interval $-1 \le x \le 1$ (see Exercise 2) use the Chebyshev inequality to estimate $\Pr[|X| > 1]$. Compare it with the true value of this probability.

7. Let X and Z be random variables that differ only in origin and scale,

$$Z = aX + b$$

for some constants a and b, $a > 0$. Show that for the corresponding standardized variables $Z^* = X^*$. (See Section 6.7.)

8. Show that if X is a random variable with range $a \le x \le b$

(a) $a \le E[X] \le b$.

(b) $\text{Var}[X] \le (b-a)^2/4$.

9. Expand in terms of μ'_3, μ'_2, μ,

$$E\left[(X-\mu)^3\right]$$

10. If X is a normally distributed random variable find

$$\Pr[|X-\mu| > 3\sigma]$$

by using the normal tables. Estimate the same probability by using Chebyshev's inequality.

11. Prove Chebyshev's theorem (formula (6.11)) if the random variable X is discrete.

12. Suppose that X is a random variable with mean 0 and variance σ^2. Suppose further that $Y = aX$, $a > 0$. Show that

$$\Pr[|Y| \geq t] \leq \frac{a^2\sigma^2}{t^2}$$

13. Determine the moment-generating function for the random variable in Exercise 2.

14. Determine $E[X]$, $E[Y]$, $E[Z]$, $\mathrm{Var}[X]$, $\mathrm{Var}[Y]$, and $\mathrm{Var}[Z]$ for the random variables given in Exercise 4.10. (See Section 4.9.)

15. Determine $E[X]$, $E[Y]$, $\mathrm{Var}[X]$, $\mathrm{Var}[Y]$, and $\mathrm{Cov}[X,Y]$ for the random variables given in Exercise 4.4. ($\mathrm{Cov}[X,Y]$ means the *covariance* of X and Y.)

16. Compute $E[X]$, $E[Y]$, $\mathrm{Var}[X]$, and $\mathrm{Var}[Y]$ for the random variables given in Exercise 4.8.

17. Compute $E[X]$, $E[Y]$, $\mathrm{Var}[X]$, $\mathrm{Var}[Y]$, and $\mathrm{Cov}[X,Y]$ for random variables with the density function given in Exercise 4.13(c).

18. Show that if X and Y are jointly normally distributed random variables then they are independent if and only if $\rho = 0$.

19. Show that if X and Y are independent than $\rho = 0$.

20. Show that the standard normal density function has a moment-generating function

$$M(\theta) = e^{\theta^2/2}$$

(*Hint*: Complete the square in the exponent of the integrand.)

21. If X has the triangular distribution discussed in Exercise 3.15, find $M(\theta)$.

22. If X is a discrete random variable with the Poisson density function (Exercise 3.11), show that its probability-generating function is

$$\psi(z) = E[z^X] = e^{\lambda(z-1)}$$

From this determine $E[X]$ and $\mathrm{Var}[X]$.

23. Show that the sum of independent random variables, each one of which has a gamma density (Exercise 3.26) with the same value of β, also has a gamma density.

24. Determine the probability-generating function for the geometrically distributed random variable defined in Exercise 3.13. From this determine $\mu = E[X]$ and $\sigma^2 = \text{Var}[X]$.

25. Below are data on the number of times the word "an" appears in text segments written by Alexander Hamilton (see Exercise 3.6). Determine the sample mean and sample variance of this data.

Frequency	0	1	2	3	4	5	6	7 or more
Times appearing	77	89	46	21	9	4	1	0

26. McKendrick (1926)[1] provides the following data for the number of households in Leeckau, Germany, in which x cases of cancer had occurred in the period 1875 to 1898. (He was trying to determine whether cancer was contagious or not.)

Frequency	64	43	10	2	1
x	0	1	2	3	4

Compute the sample mean and sample variance of McKendrick's data.

27. Another word of Alexander Hamilton studied by Mosteller and Wallace was "his." The following table is their count of the frequency of his use of this word. We have adjusted their data slightly.

Frequency	0	1	2	3	4	5	6	7
Times appearing	192	18	17	7	3	2	4	3

Compute the sample mean and variance of this data. (See Exercises 3.6 and 6.25.)

28. In a study of mine accidents, Maguire, Pearson, and Wynn (1952)[2] present the following data for the time (in days) between consecutive major mine accidents. The data were collected for accidents from December 6, 1875, to May 29, 1951.

[1] A. G. McKendrick, (1926), "The Applications of Mathematics to Medical Problems," *Edinburgh Mathematical Society Proceedings*, **44**, pp. 98–130. Reprinted with permission of Edinburgh Mathematical Society and Kraus Reprint.

[2] B. A. Maguire, E. S. Pearson, and A. W. A. Wynn (1952), "The Time Intervals Between Industrial Accidents," *Biometrika*, **39**, 168–180. Reprinted with permission of the *Biometrika* Trustees.

Days between Major Mine Accidents

378	96	59	108	54	275	498	228	217	19	156
36	124	61	188	217	78	49	271	120	329	47
15	50	1	233	113	17	131	208	275	330	128
31	120	13	28	32	1205	182	517	20	312	1630
215	203	180	22	23	644	255	1613	66	171	29
11	176	345	61	151	467	195	54	291	145	217
137	55	20	78	361	871	224	326	4	75	7
4	96	81	99	312	48	566	1312	369	364	18
15	59	286	326	354	123	390	348	338	37	1357
72	315	114	275	58	457	72	745	336	19	

Compute the sample mean and sample variance for these data.

29. Fry (1928)[1] gives the following data due to Weldon (see Example 2.16). Compute the sample mean and sample variance of the data.

Number of 5's or 6's	Frequency	Number of 5's or 6's	Frequency
0	185	7	1331
1	1149	8	403
2	3265	9	105
3	5475	10	14
4	6614	11	4
5	5194	12	0
6	5067		

30. The U.S. Department of Commerce/Bureau of Industrial Economics (1981)[2] gives the following data for profit after taxes as a percentage of sales in the aerospace industry.

Year	% Profit	Year	% Profit
1968	3.2	1974	2.9
1969	3.0	1975	3.0
1970	2.0	1976	3.4
1971	1.8	1977	4.2
1972	2.4	1978	4.4
1973	2.9	1979	5.1

First compute the sample mean and sample variance of the profit data. Then plot the data with the year as the abscissa and the percentage profit as the ordinate. Based on that graph would you think that this "experiment" was conducted each year "under identical conditions"?

31. Daganzo (1975)[3] studied a model of traffic flow on a two-lane highway. He presents the following data (q = the number of vehicles in a 5-minute period

---

[1] T. C. Fry, (1928), *Probability and Its Engineering Uses*, Van Nostrand, New York. Reprinted with permission of Brooks/Cole.
[2] U.S. Department of Commerce/Bureau of Industrial Economics (1981), *U.S. Industrial Outlook for 200 Industries with Projections for 1985*. U.S. Department of Commerce, Washington, D.C.
[3] C. F. Daganzo (1975), *Two Lane Traffic: A Stochastic Model*, Ph.D. dissertation, Department of Civil Engineering, University of Michigan, Ann Arbor. Reprinted with permission of the author.

in one lane and $q_0 =$ the number of vehicles in the same period in the other, opposing lane).

q	q_0	q	q_0
45	53	36	42
58	51	30	47
51	40	52	53
23	45	36	52
31	57	24	51
54	43	36	44

What is the estimated correlation coefficient for traffic in these two lanes?

32. The following counts have been found from a random number generator on a microcomputer.

0	1	2	3	4	5	6	7	8	9
14	20	26	14	20	18	21	15	30	18

Compute the mean and variance for these counts and compare them with the results you would expect if this generator is truly generating random digits that have a uniform density function on the integers $0, 1, \ldots 9$.

33. For passengers checking in for international flights that depart between 6:00 P.M. and midnight, the following data are provided by the Air Travel Association.[1]

Minutes before flight time	120–110	110–100	100–90	90–80	80–70	70–60
Number checking in	3	4	6	9	11	14

	60–50	50–40	40–30	30–20	20–10	10–0
	15	15	15	7	1	0

Compute the sample mean time of arrival at the check-in counter for these passengers, and the sample variance.

[1]Ralph M. Parsons Company (1975), *The Apron and Terminal Building Planning Manual*, Report No. FAA-RD-75-191, prepared for the U.S. Department of Transportation, Federal Aviation Administration, Systems Research and Development Service, Washington, D.C.

7 · Special Distributions

7.1 Introduction

In many theoretical and applied problems, and under many seemingly diverse conditions, several density functions appear often enough that they are worth exploring in depth. In Chapters 3 to 6 and in the exercises to those chapters we mentioned some of these functions. In this chapter we discuss them in more detail.

The functions that we shall explore in this chapter are those that appear most frequently in later chapters. We do not pretend that this chapter exhausts the list of useful distributions. Even among the ones that we do discuss, for a particular application one of the distributions may be of paramount interest and the others of subsidiary interest. For example, in the important field of statistical hypothesis testing the normal density function is of paramount interest. In the field of reliability theory the negative exponential is of great importance. In the field of demand analysis the Poisson density function is pre-eminent. The student should not assume that any one of the distributions in this chapter is a model of all reality.

In this chapter, in addition to using the results of Chapters 3 and 4 to illustrate the ideas of Chapter 6, we shall also give some threads that tie together many of these distributions. Again, the student is warned that the threads we choose by no means exhaust the possible interrelations of these distributions. These threads merely show that some interconnections do exist.

7.2 The Bernoulli Density Function

The *Bernoulli* density function is the density function of a discrete random variable X having 0 and 1 as its only possible values. It is given by

$$p_0 = p(0) = \Pr[X = 0] = q$$
$$p_1 = p(1) = \Pr[X = 1] = p \qquad\qquad (7.1)$$
$$p_i = p(i) = \Pr[X = i] = 0, \qquad \text{for } i \neq 0, 1$$

where $p + q = 1$.

If we consider the experiment of tossing a coin, and denote "tails" by 0 and "heads" by 1, that is, X (tails) $= 0$ and X (heads) $= 1$, then X will have such a Bernoulli density function. If the coin is "honest" then $p = q = \frac{1}{2}$.

Immediately we have (Figure 7.1), using (3.6),

$$F(x) = \begin{cases} 0, & \text{for } x < 0 \\ q, & \text{for } 0 \leq x < 1 \\ 1, & \text{for } x \geq 1 \end{cases}$$

153

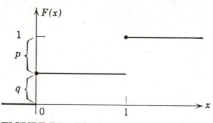

FIGURE 7.1   The Bernoulli distribution function.

For this distribution

$$\mu = E[X] = 0 \cdot q + 1 \cdot p = p \tag{7.2}$$

$$\mu_2' = E[X^2] = 0^2 \cdot q + 1^2 \cdot p = p$$

$$\sigma^2 = \mu_2' - \mu^2 = p - p^2 = p(1-p) = pq \tag{7.3}$$

The probability-generating function is

$$\psi(z) = \psi_X(z) = E[z^X] = z^0 q + z^1 p = q + zp \tag{7.4}$$

For an application of this density function see Example 3.5.

7.3   The Binomial Density Function

Consider an experiment that has two possible outcomes, "success" and "failure" (or "heads" and "tails," "hit" and "miss," "defective" and "good," et cetera). Suppose that these two outcomes have probabilities p and q, respectively, $p + q = 1$. Let us then consider the compound experiment that consists of a sequence of independent repetitions of this experiment. Equivalently, consider an independent sequence of events each one of which has the same probability p of occurring. This sequence is termed a *sequence of Bernoulli trials*.

Given a sequence consisting of a fixed number, for example, n, of Bernoulli trials, we might naturally be interested in the random variable $X =$ the number of times the given event occurs in the sequence. Thus, if the trials involve the choosing of a part from a manufacturing process and the testing of it, X might represent the number of defective parts out of n parts tested. If we consider n repeated coin tosses, X might represent the number of heads.

Clearly X is a discrete random variable with range $0, 1, 2, \ldots, n$.

Consider the sequence $X_1, X_2, \ldots, X_n$ of random variables, where X_i is 1 or 0 depending on whether the given event occurs on the ith trial or not (on whether the ith trial is a "success" or not). Then each X_i has the Bernoulli density function discussed in Section 7.2 and clearly

$$X = X_1 + X_2 + \cdots + X_n \tag{7.5}$$

since the sum on the right counts a 1 for every occurrence of the given event ("success") and, hence, simply counts the total number of occurrences. Since the events involved are independent, formulas (6.16), (6.23), and (6.31) can be applied to find the mean, the variance, and the probability-generating function of X from the mean, the variance, and the probability-generating function of the X_i given by (7.2), (7.3), and (7.4):

$$\mu = E[X] = E[X_1] + E[X_2] + \cdots + E[X_n]$$
$$= p + p + \cdots + p$$
$$= np \tag{7.6}$$

$$\sigma^2 = \text{Var}(X) = \text{Var}(X_1) + \text{Var}(X_2) + \cdots + \text{Var}(X_n)$$
$$= pq + pq + \cdots + pq$$
$$= npq \tag{7.7}$$

$$\psi(z) = E[z^X] = E[z^{X_1 + X_2 + \cdots + X_n}]$$
$$= E[z^{X_1} \cdot z^{X_2} \cdots z^{X_n}]$$
$$= E[z^{X_1}] \cdot E[z^{X_2}] \ldots E[z^{X_n}]$$
$$= \psi_{X_1}(z)\psi_{X_2}(z)\ldots\psi_{X_n}(z)$$
$$= (q + zp)(q + zp)\ldots(q + zp)$$
$$= (q + zp)^n \tag{7.8}$$

If $p_0, p_1, \ldots, p_n$ represent the density function of X, $p_i = \Pr[X = i]$, then the probability-generating function of X is

$$\psi(z) = E[z^X] = \sum_{i=0}^{n} z^i p_i \tag{7.9}$$

However, $\psi(z)$ is also given by (7.8), which can be expanded in powers of z by using the binomial theorem (Appendix A.6):

$$\psi(z) = (q + zp)^n = \sum_{i=0}^{n} \binom{n}{i} q^{n-i}(zp)^i$$

$$= \sum_{i=0}^{n} \binom{n}{i} p^i q^{n-i} z^i \tag{7.10}$$

Since (7.9) and (7.10) represent the same function of z, the coefficients of the various powers of z can be compared. Thus, we observe that the density function of X is given by

$$p_i = \binom{n}{i} p^i q^{n-i}, \qquad i = 0, 1, 2, \ldots, n \tag{7.11}$$

($p_i = 0$ otherwise).

That is, formula (7.11) gives the probability of i "successes" in n independent trials (Bernoulli trials) of an experiment that has probability p of "success" on each trial.

The fact that $\sum_{i=0}^{n} p_i = 1$ is clear, since

$$\sum_{i=0}^{n} p_i = \sum_{i=0}^{n} \binom{n}{i} p^i q^{n-i} = (q+p)^n = 1^n = 1$$

by using the binomial theorem.

Because of the presence of the binomial coefficients $\binom{n}{i}$ and also because of the extensive use of the binomial theorem in discussions, probabilities p_i given by (7.11), are said to form a *binomial density* function, and any random variable X having this density function is said to be a *binomial* random variable.

Graphs of the binomial density function for various values of n and p are sketched in Figure 7.2. For an application of this distribution see Exercises 3.27, 3.28, and 3.29. For a different derivation consult Example 2.12.

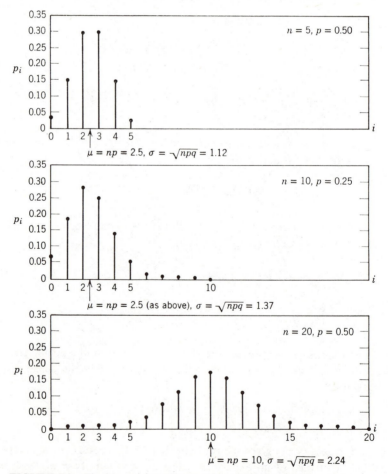

FIGURE 7.2   The binomial density function for several values of p and n.

7.4   The Geometric Density Function

Suppose that we are given a sequence of Bernoulli trials, that is, a sequence of independent events each of which has the same probability p of occurring. The binomial distribution enters when one counts the number of events that occur in a *fixed* number n of trials. Suppose instead that we count the number of trials (or events observed) until the first "failure" occurs. Let us denote this number by Y. Thus, Y is a random variable with range $1, 2, 3, \ldots$. To find its density function we note that the event $[Y = i]$ occurs if, and only if, one has a sequence of $i - 1$ "successes" followed by a "failure." Thus, we have

$$p_i = \Pr[Y = i] = p^{i-1}q \tag{7.12}$$

$i = 1, 2, 3, \ldots$, where $q = 1 - p$ is assumed to be positive. Clearly

$$\sum_{i=1}^{\infty} p_i = q + pq + p^2 q + \cdots = \frac{q}{1 - p} = \frac{q}{q} = 1$$

by using the formula for the sum of a geometric series (Appendix A.3).

Any random variable Y having range $1, 2, 3, \ldots$ and density function given by a formula of the form of (7.12) is said to have a *geometric* distribution, and (7.12) is termed a *geometric density function*.

Graphs of p_i for various values of p are sketched in Figure 7.3.

The probability-generating function $\psi(z)$ of such a random variable Y is easily determined using the formula for the sum of a geometric series.

$$\psi(z) = E[z^Y] = \sum_{i=1}^{\infty} z^i p_i$$

$$= \sum_{i=1}^{\infty} z^i p^{i-1} q$$

$$= zq(1 + zp + z^2 p^2 + \cdots)$$

$$= \frac{zq}{1 - zp} \tag{7.13}$$

The mean μ and the variance σ^2 can now be found by using formula (6.34).

$$\psi'(z) = \frac{q}{(1 - zp)^2}$$

$$\psi''(z) = \frac{2pq}{(1 - zp)^3}$$

$$\mu = \psi'(1) = \frac{q}{(1 - p)^2} = \frac{q}{q^2} = \frac{1}{q} \tag{7.14}$$

$$\mu_2' = \psi''(1) + \psi'(1) = \frac{2p + q}{q^2}$$

$$\sigma^2 = \mu_2' - \mu^2 = \frac{p}{q^2} \tag{7.15}$$

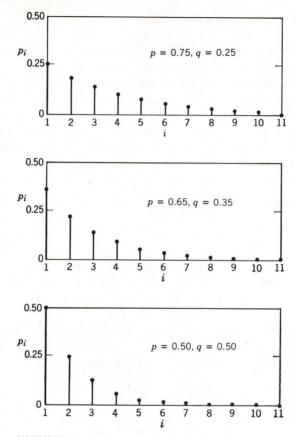

FIGURE 7.3 The geometric density function for several values of p.

A random variable X is said to have a *modified geometric* density function if it has range $0, 1, 2, \ldots$ and density function

$$\Pr[X = i] = (1 - p) p^i, \qquad i = 0, 1, 2, \ldots \tag{7.16}$$

where p is some constant with $0 \le p < 1$. If we set

$$Y = X + 1$$

we notice that Y has range $1, 2, 3, \ldots$ and density function

$$\Pr[Y = i] = \Pr[X = i - 1] = (1 - p) p^{i-1}$$

Thus, by comparison with (7.12), we observe that $Y = X + 1$ has a geometric distribution with $p = p$ and $q = 1 - p$. Since $X = Y - 1$, we have

$$E[X] = E[Y - 1] = E[Y] - 1 = \frac{p}{1 - p} \tag{7.17}$$

$$\text{Var}[X] = \text{Var}[Y] = \frac{p}{q^2} = \frac{p}{(1 - p)^2} \tag{7.18}$$

using (6.5) and (6.12).

Random variables with modified geometric distribution will be important in our later work. For applications of this density function see Exercises 3.10 and 3.13.

To discuss another important property, let us go back to the situation at the beginning of this section where Y represents the number of trials until the first "failure" in a sequence of Bernoulli trials. Now suppose that we have observed a fixed number n of these trials and found them all to be "successes." We are concerned with the number Z of *additional* trials that must be observed until the first "failure." Thus, $Z = Y - n$. We wish to study *conditional* probabilities involving Z, *given that* $Y > n$ (the first n trials observed were successes so that the number of trials until the first "failure" must be greater than n). Let us compute the following probabilities

$$q_i = \Pr[Z = i \mid Y > n] = \Pr[Y - n = i \mid Y > n]$$

$$= \Pr[Y = n + i \mid Y > n]$$

$$= \frac{\Pr[Y = n + i \text{ and } Y > n]}{\Pr[Y > n]}$$

by using the definition of conditional probability. For $i = 1, 2, 3, \ldots,$ $Y = n + i$ implies that $Y > n$. Thus, the event $[Y = n + i \text{ and } Y > n]$ is the same as the event $[Y = n + i]$. It follows that

$$q_i = \frac{\Pr[Y = n + i]}{\Pr[Y > n]}$$

$$= \frac{p_{n+i}}{\sum\limits_{j=n+1}^{\infty} p_j}$$

$$= \frac{p^{n+i-1} q}{\sum\limits_{j=n+1}^{\infty} p^{j-1} q}$$

$$= \frac{p^{n+i-1} q}{p^n q / (1 - p)}$$

$$= p^{i-1} q = p_i \tag{7.19}$$

Thus, we observe that, conditional on $Y > n$, the number of trials remaining until the first failure, $Z = Y - n$, has the *same* density function as Y had originally. This is termed the *forgetfulness property* of the geometric density function.

If a sequence of "successes" is observed in a sequence of Bernoulli trials, one need not "remember" how long the sequence is to determine probabilities for the number of additional trials needed until the first "failure."

It is important to note that the converse is also true here. If Y is any discrete random variable with range $1, 2, 3 \ldots$ and having this forgetfulness property, then Y must have a geometric probability density function. To see this we observe that the forgetfulness property implies that $p_i = \Pr[Y = i]$ satisfies

$$p_i = \Pr[Y = i] = \Pr[Y - n = i \mid Y > n] \tag{7.20}$$

for each integer n and i in the range $1, 2, 3 \ldots$. By the same reasoning that we used in deriving (7.19) we notice that the right side of (7.20) is given by

$$\frac{p_{i+n}}{\sum\limits_{j=n+1}^{\infty} p_j}$$

Thus,

$$p_i = \frac{p_{i+n}}{\sum\limits_{j=n+1}^{\infty} p_j}$$

Now set $n = 1$ and rearrange.

$$\frac{p_{i+1}}{p_i} = \sum\limits_{j=2}^{\infty} p_j = 1 - p_1 = p$$

where the constant $1 - p_1$ is denoted by p. Thus, $p_{i+1} = p_i p$. Now p_i can be expressed as $p_{i-1} p$, et cetera, and eventually we obtain

$$p_i = p_1 p^{i-1} = (1 - p) p^{i-1} = p^{i-1} q, \qquad \text{for } i = 1, 2, 3, \ldots$$

where $q = 1 - p$. Thus, we observe that p_i satisfies formula (7.12) and that Y has a geometric density function.

7.5 The Poisson Density Function

If X is a discrete random variable with range $0, 1, 2, \ldots$ and

$$p_i = \Pr[X = i] = \frac{\lambda^i e^{-\lambda}}{i!}, \qquad i = 0, 1, 2, \ldots$$
$$p_i = 0, \qquad\qquad\qquad \text{otherwise} \tag{7.21}$$

then the set $\{p_i\}$ is said to be a *Poisson* density function. Here λ represents some given positive constant (parameter). A graph of the Poisson density function for several values of λ is given in Figure 7.4. Applications were given in Exercises 3.6 and 3.11.

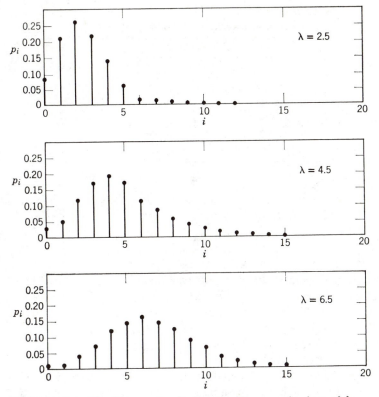

FIGURE 7.4 The Poisson density function for several values of λ.

This function occurs in many guises. First, notice that the binomial density function, $\binom{n}{i}p^i q^{n-i}$, can be written as

$$\frac{n(n-1)(n-2)\ldots(n-i+1)}{i!} \cdot p^i q^{n-i}$$

If we now multiply and divide by n^i to get

$$\frac{1\left(1-\frac{1}{n}\right)\left(1-\frac{2}{n}\right)\cdots\left(1-\frac{i-1}{n}\right)}{i!}(np)^i\left(1-\frac{np}{n}\right)^{n-i}$$

and let $n \to \infty$ in such a way that $np = \lambda$ remains constant, each term in the product $1(1-1/n)\ldots[1-(i-1)/n]$ approaches 1, while $(np)^i$ is λ^i. Also

$$\left(1-\frac{np}{n}\right)^{n-i} = \left(1-\frac{\lambda}{n}\right)^n\left(1-\frac{\lambda}{n}\right)^{-i} \to e^{-\lambda}(1) = e^{-\lambda}$$

Hence, in the limit as $n \to \infty$ with $np = \lambda$ (and as $p = \lambda/n \to 0$)

$$\binom{n}{i}p^i q^{n-i} \to \frac{\lambda^i e^{-\lambda}}{i!} \tag{7.22}$$

TABLE 7.1
The Poisson as an Approximation to the Binomial p_i

	BINOMIAL		POISSON
i	$n = 5, \ p = 0.20$	$n = 20, \ p = 0.05$	$\lambda = 1$
0	.328	.359	.368
1	.410	.377	.368
2	.205	.189	.184
3	.051	.060	.061
$\vdots$	$\vdots$	$\vdots$	$\vdots$

Thus the Poisson density function can be used as a convenient approximation to the binomial density function in the case of large n and small p.

$$\binom{n}{i} p^i q^{n-1} \approx \frac{e^{-\lambda} \lambda^i}{i!}, \qquad \text{where } \lambda = np$$

In Table 7.1 the density functions for two binomial random variables, one with $n = 5$, $p = 0.20$ and the other with $n = 20$, $p = 0.05$, are compared with one another and with that of a Poisson random variable with $\lambda = 1$. Since $np = 1$ in each of these cases, the same Poisson should provide an approximation to each of these binomials. We observe that the approximation is better in the case of the larger n and the smaller p.

In Exercise 3.4 we can show that, under reasonable assumptions, if X represents the *number* of flights during a period of time of length t, instead of the *time between* flights, then

$$\Pr[X = i] = \frac{(\lambda t)^i e^{-\lambda t}}{i!}, \qquad i = 0, 1, 2, \ldots \tag{7.23}$$

with $\lambda = 1/15$. Following Example 3.8 we can also show that the number of mine accidents up to t has (7.23) for its density function with $\lambda = 1/240$. Thus, X has a Poisson distribution with the constant λ being replaced by a new constant λt that is proportional to the time of observation.

In certain waiting-line type problems that we discuss later we find that the number of parts produced, the number of customers served, or the number of messages passed through a communication center in some fixed interval of time will have a density function that has approximately the Poisson form.

We must now show that the $\{ p_i \}$ given by the Poisson density function is, in fact, a valid density function. That is (see Section 3.6),

$$0 \leq p_i \leq 1$$

$$\sum_{i=0}^{\infty} p_i = 1$$

We easily determine that $p_i \geq 0$, since each term in the formula for p_i is positive.

It suffices, then, to show that

$$\sum_{i=0}^{\infty} p_i = \sum_{i=0}^{\infty} \lambda^i \frac{e^{-\lambda}}{i!} = 1$$

But this is true since the sum can be written as

$$\sum_{i=0}^{\infty} p_i = e^{-\lambda} \sum_{i=0}^{\infty} \frac{\lambda^i}{i!}$$

$$= e^{-\lambda} \left[1 + \frac{\lambda}{1} + \frac{\lambda^2}{2!} + \frac{\lambda^3}{3!} + \cdots \right]$$

The sum in the brackets is the infinite series expansion of e^{λ} (Appendix A.5). Hence

$$\sum_{i=0}^{\infty} p_i = e^{-\lambda} \sum_{i=0}^{\infty} \frac{\lambda^i}{i!} = e^{-\lambda} e^{\lambda} = 1$$

The probability-generating function can also be readily found:

$$\psi(z) = E[z^X] = \sum_{i=0}^{\infty} z^i p_i$$

$$= \sum_{i=0}^{\infty} z^i \frac{\lambda^i e^{-\lambda}}{i!} = e^{-\lambda} \sum_{i=0}^{\infty} \frac{(z\lambda)^i}{i!}$$

$$= e^{-\lambda} \cdot e^{z\lambda}$$

again using Appendix A.5. Thus,

$$\psi(z) = e^{-\lambda(1-z)} \tag{7.24}$$

Again (6.34) can be used to compute μ, μ_2', and σ^2:

$$\psi' = (z) = \lambda e^{-\lambda(1-z)}, \qquad \psi''(z) = \lambda^2 e^{-\lambda(1-z)}$$

$$\mu = \psi'(1) = \lambda \tag{7.25}$$

$$\mu_2' = \psi''(1) + \psi'(1) = \lambda^2 + \lambda$$

$$\sigma^2 = \mu_2' - \mu^2 = (\lambda^2 + \lambda) - \lambda^2 = \lambda \tag{7.26}$$

Thus, a random variable having a Poisson density function has equal mean and variance.

7.6 The Uniform Density Function

If X is a discrete random variable with a *finite* range $x_1, x_2, \ldots, x_N$, one of the simplest densities to consider is one for which each point in the range has equal probability. See Example 2.3. In this case, the density function is given by

$$p(x_i) = \frac{1}{N}, \qquad \text{for } x_i \text{ in the range}$$
$$p(x) = 0, \qquad \text{otherwise} \tag{7.27}$$

where N represents the number of points in the range. A random variable that has this distribution is said to have a *uniform* density. These distributions occur frequently in applications. For example, if, in a sequence of Bernoulli trials, the second "success" is known to occur on the $N + 1$st trial, a simple computation shows that the number of the trial on which the first "success" occurs will have this uniform density.

If we attempt to extend this concept to the case of a discrete random variable that has an infinite range $x_1, x_2, x_3, \ldots$ we soon arrive at an impasse. Since $\Sigma p(x_i) = 1$, it is impossible that all the terms $p(x_i)$ are the same. An infinite series of equal *positive* numbers would diverge (and hence could not equal 1), while if each $p(x_i) = 0$ the sum would always be 0, never 1. Thus, a discrete random variable with infinite range cannot have a uniform density function.

Paradoxically, the definition *does* extend to the case of a continuous random variable X having as a range some finite interval $a \le x \le b$. Such a random variable is said to have a *uniform* density if its density function is constant within the range:

$$f(x) = \begin{cases} c, & a \le x \le b \\ 0, & \text{otherwise} \end{cases} \tag{7.28}$$

Since $\int_{-\infty}^{\infty} f(x)\, dx = \int_a^b f(x)\, dx = \int_a^b c\, dx = 1$, it follows that $c = 1/(b - a)$ and that

$$f(x) = \begin{cases} \dfrac{1}{b - a}, & a \le x \le b \\ 0, & \text{otherwise} \end{cases} \tag{7.29}$$

In this case, the distribution function $F(x) = \int_{-\infty}^x f(t)\, dt$ will be given by

$$F(x) = \begin{cases} 0, & x < a \\ \dfrac{x - a}{b - a}, & a \le x \le b \\ 1, & x > b \end{cases} \tag{7.30}$$

The functions $f(x)$ and $F(x)$ are sketched in Figure 7.5. An application is given in Exercise 5.5.

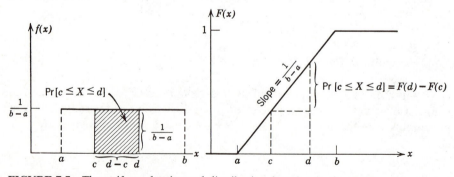

FIGURE 7.5 The uniform density and distribution function in the continuous case.

The term "at random" is frequently used to refer to such a random variable. If we say that a value is chosen *at random* in the interval $a \leq x \leq b$, it means that a value is observed of a random variable that has a uniform density over the interval $a \leq x \leq b$.

The mean, the variance, and the moment-generating functions of a random variable that has a uniform density function are easily computed:

$$\mu = E[X] = \int_a^b \frac{x}{b-a} \, dx = \frac{a+b}{2} \tag{7.31}$$

$$\sigma^2 = E[X^2] - \mu^2 = \int_a^b \frac{x^2}{b-a} \, dx - \left(\frac{a+b}{2}\right)^2 = \frac{(b-a)^2}{12} \tag{7.32}$$

$$M(\theta) = E[e^{X\theta}] = \int_a^b \frac{e^{x\theta}}{b-a} \, dx = \frac{e^{b\theta} - e^{a\theta}}{\theta(b-a)} \tag{7.33}$$

EXAMPLE 7.1

In a particular computer storage system, information is retrieved by entering the system at a random point and systematically checking through all the items in the system at a uniform speed. If it takes b seconds to run through the entire system, the time taken to reach any given item of information is a random variable that has a uniform distribution over the interval from time 0 to time b. If it takes a seconds to start the search, the time required will have a uniform distribution from time a to time $a + b$.

7.7 The Negative Exponential Density Function

A continuous random variable having range $0 \leq x < \infty$ is said to have a *negative exponential distribution* if it has a density function of the form

$$f(x) = \begin{cases} \lambda e^{-\lambda x}, & 0 \leq x < \infty \\ 0, & x < 0 \end{cases} \tag{7.34}$$

where λ is some positive constant. The corresponding distribution function is

$$F(x) = \int_{-\infty}^x f(t) \, dt = \begin{cases} 1 - e^{-\lambda x}, & 0 \leq x < \infty \\ 0, & x < 0 \end{cases} \tag{7.35}$$

Applications of this density function were given in Example 3.8 and Exercise 3.4.

The graph of $f(x)$ is sketched for several different values of λ in Figure 7.6.

The mean, the variance, and the moment-generating function of a random variable X that has this distribution are easily calculated:

$$\mu = E[X] = \int_0^\infty x\lambda e^{-\lambda x} \, dx = 1/\lambda \tag{7.36}$$

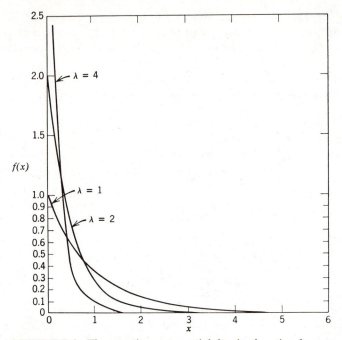

FIGURE 7.6   The negative exponential density function for several values of λ.

using integration by parts.

$$E[X^2] = \int_0^\infty x^2 \lambda e^{-\lambda x} \, dx = 2/\lambda^2$$

$$\text{Var}[X] = \frac{2}{\lambda^2} - \left(\frac{1}{\lambda}\right)^2 = \frac{1}{\lambda^2} \tag{7.37}$$

$$M(\theta) = E[e^{X\theta}] = \int_0^\infty e^{x\theta} \lambda e^{-\lambda x} \, dx = \left(1 - \frac{\theta}{\lambda}\right)^{-1} \tag{7.38}$$

provided that $\theta < \lambda$.

We notice that

$$M(\theta) = \frac{\lambda}{\lambda - \theta} = \frac{1}{1 - (\theta/\lambda)} = 1 + \frac{\theta}{\lambda} + \frac{\theta^2}{\lambda^2} + \cdots + \frac{\theta^n}{\lambda^n} + \cdots \tag{7.39}$$

provided that $|\theta/\lambda| < 1$ or $-\lambda < \theta < \lambda$. Formula (6.32) for moment-generating functions can now be applied and we observe that the nth moment of X, $E[X^n]$, is the coefficient of $\theta^n/n!$ in the series expansion of $M(\theta)$. If we write the nth term of (7.39) as

$$\frac{n!}{\lambda^n} \cdot \frac{\theta^n}{n!}$$

we find that

$$E[X^n] = \int_0^\infty x^n \lambda e^{-\lambda x} \, dx = \frac{n!}{\lambda^n} \tag{7.40}$$

an integration formula related to that in Appendix A.10.

The negative exponential density occurs frequently in applications. It frequently enters as the density of random variables that represent the lifetimes of certain types of equipment or as the time interval between events that occur in random sequence (orders to a wholesaler, telephone calls at an exchange, demands for computer service, successive arrivals of cosmic rays, or clicks on a Geiger counter, et cetera). The primary reason for the great importance of this distribution is that a random variable with this density has a certain "forgetfulness" property analogous to that of the geometric density considered in Section 7.4.

Suppose that X has this negative exponential density and suppose we know that X exceeds some given value s, $X > s$. For example, if we interpret X as the lifetime of a certain piece of equipment, suppose we have observed that this equipment has already been operating for s hours. We may then be interested in the distribution of the *remaining* lifetime, $X' = X - s$. Suppose that we denote by $G(x)$ the *conditional* distribution function of X', given that $X > s$. Thus, for $x \geq 0$, we have

$$G(x) = \Pr\left[X' \leq x | X > s\right] = \Pr\left[X - s \leq x | X > s\right]$$

$$= \Pr\left[X \leq x + s | X > s\right] = \frac{\Pr\left[X \leq x + s \text{ and } X > s\right]}{\Pr\left[X > s\right]}$$

$$= \frac{\Pr\left[s < X \leq x + s\right]}{\Pr\left[X > s\right]} \tag{7.41}$$

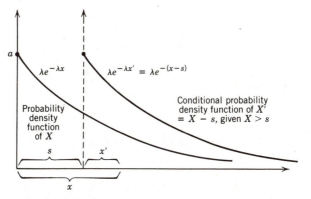

FIGURE 7.7 The comparison between densities of a negative exponentially distributed lifetime X and the remaining lifetime $X' = X - s$, given that the lifetime exceeds s.

by using the definition of conditional probability (2.4). Thus (see Figure 7.7),

$$G(x) = \frac{\int_s^{x+s} f(t)\, dt}{\int_s^\infty f(t)\, dt} = \frac{\int_s^{x+s} \lambda e^{-\lambda t}\, dt}{\int_s^\infty \lambda e^{-\lambda t}\, dt} = 1 - e^{-\lambda x} \qquad (7.42)$$

If we compare (7.42) with (7.35) we see that $G(x)$ is identical with the distribution function $F(x)$ of the original random variable X without any conditional term. In words, this says that *the distribution of the remaining life of the equipment does not depend on how long the equipment has been operating.* The equipment "forgets" how long it has been running and its eventual breakdown is the result of some suddenly appearing failure, not of gradual deterioration.

The negative exponential density is the only continuous nonnegative density that has this property. To show this, assume that X is a continuous nonnegative random variable for which the conditional distribution function of $X' = X - s$, given that $X > s$, equals the initial distribution function, $F(x)$.

Thus, by (7.41)

$$G(x) = \frac{\int_s^{s+x} f(t)\, dt}{\int_s^\infty f(t)\, dt} = F(x)$$

or

$$\frac{F(x+s) - F(s)}{1 - F(s)} = F(x) \qquad (7.43)$$

Since $F(0) = 0$ (X being nonnegative), we can arrange the terms of (7.43) as follows, dividing each side by x:

$$\frac{F(x+s) - F(s)}{x} = [1 - F(s)]\left[\frac{F(x) - F(0)}{x}\right]$$

Now let $x \to 0+$ and use the definition of a derivative:

$$F'(s) = [1 - F(s)] F'(0)$$

Let $U(s) = 1 - F(s)$ and $F'(0) = \lambda = $ constant. Thus,

$$U'(s) = -F'(s) = -U(s)\lambda$$

This simple differential equation can be solved to give

$$U(s) = Ce^{-\lambda s}$$

Since $U(0) = 1 - F(0) = 1$, it follows that $C = 1$. Thus,

$$U(s) = e^{-\lambda s} = 1 - F(s)$$

$$F(s) = 1 - e^{-\lambda s}$$

for $s > 0$. Thus, we observe that X must have the distribution function of a negative exponentially distributed random variable.

The preceding discussion can be summed up in the following theorem.

Theorem

Let X be a continuous random variable with range $0 \leq x < \infty$. For each fixed $s > 0$, assume that, conditional on X being greater than s, the density function of $X - s$ is the same as the unconditional density function of X. Then X has a negative exponential density function. Conversely, any random variable having a negative exponential density function has this property.

7.8 The Gamma Density Function

A useful generalization of the negative exponential density function is the gamma density. A continuous random variable X with range $0 < x < \infty$ is said to have a *gamma density* if its density function has the form

$$f(x) = Cx^{b-1}e^{-\lambda x}, \qquad 0 < x < \infty \tag{7.44}$$

where λ and b are positive constants and $C = \lambda^b / \Gamma(b)$. (See Appendix A.10 for the definition of $\Gamma(b)$ and the identities needed to derive this value for C.)

Notice that the negative exponential density is a special case of the gamma density with $b = 1$.

Another important special case is $\lambda = \frac{1}{2}$ and $b = n/2$, where n is a positive integer. A random variable that has this density is said to have a *chi-squared distribution with n degrees of freedom*. These distributions have important applications in statistics. Note also that the chi-squared with one degree of freedom appeared in Example 5.3 as the distribution of the square of a standard normal random variable.

Some gamma density functions for various values of λ and b are sketched in Figure 7.8. The moment-generating function of a random variable having a

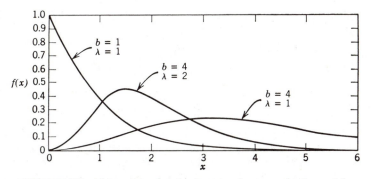

FIGURE 7.8 The gamma density function for several values of λ and b.

gamma density is easily computed:

$$M(\theta) = E[e^{X\theta}] = C \int_0^\infty e^{x\theta} x^{b-1} e^{-\lambda x} \, dx \qquad (7.45)$$

This integral will converge provided that $\lambda - \theta > 0$ and $\theta < \lambda$. If we substitute

$$x = \frac{\lambda}{\lambda - \theta} \cdot y, \qquad dx = \frac{\lambda}{\lambda - \theta} \, dy$$

we obtain

$$M(\theta) = \left(\frac{\lambda}{\lambda - \theta} \right)^b \qquad (7.46)$$

since $C \int_0^\infty y^{b-1} e^{-\lambda x} \, dy = 1$ by Appendix A.10. Thus,

$$M(\theta) = \left(1 - \frac{\theta}{\lambda} \right)^{-b} = 1 + \frac{b}{\lambda} \cdot \theta + \frac{b(b+1)}{\lambda^2} \cdot \frac{\theta^2}{2} + \cdots \qquad (7.47)$$

using the binomial theorem (Appendix A.6). From (6.32) we observe that

$$\mu = E[X] = \frac{b}{\lambda} \qquad \text{and} \qquad \mu_2' = E[X^2] = \frac{b(b+1)}{\lambda^2}$$

X has mean and variance

$$\mu = \frac{b}{\lambda}, \qquad \sigma^2 = \mu_2' - \mu^2 = \frac{b}{\lambda^2} \qquad (7.48)$$

An example of the application of the gamma density function is given in Example 5.5.

By comparing formulas (7.46) and (7.38) we observe that the moment-generating function of a gamma distributed random variable is the bth power of the moment-generating function of a negative exponentially distributed random variable. We recall that, when independent random variables are added, their moment-generating functions multiply (formula (6.31)). Thus, if b is an integer, we have the following important theorem.

Theorem

If $X_1, X_2, \ldots, X_b$ are independent random variables, each having a negative exponential density with mean $1/\lambda$, then $X = X_1 + \cdots + X_b$ will have a gamma density of the form (7.44).

EXAMPLE 7.2

The lifetime of a certain electronic tube is known to be a negative exponentially distributed random variable with a mean of 1000 hours. Three of these tubes are available (the original and two spares) and when one burns out it is replaced by a spare. Determine the distribution of the time X until the last spare burns out.

By (7.36), $\mu = 1/\lambda = 1000$ and $\lambda = 1/1000$. Thus, the lifetimes X_1, X_2, and X_3 of the three tubes will be independent random variables, since they do not interact with each other, and each will have density function $(1/1000)e^{-x/1000}$, $x > 0$. The time X until the last burn-out is given by $X = X_1 + X_2 + X_3$. By the property above, X must have a gamma density with $\lambda = 1/1000$, $b = 3$.

7.9   The Normal Density Function

The normal density function was introduced in Example 3.18 and an application was given in Exercise 3.21. A continuous random variable is said to have a normal distribution if its range is all real numbers and its density function has the form

$$f(x) = \frac{1}{\sqrt{2\pi}\,\sigma} e^{-(x-\mu)^2/2\sigma^2}, \qquad -\infty < x < \infty \qquad (7.49)$$

Here μ and σ are arbitrary parameters, $\sigma > 0$, that we shall show represent the mean and standard deviation of the random variable. A sketch of the graph of $f(x)$ is given in Figure 3.19. In the case $\mu = 0$, $\sigma = 1$, its distribution function is tabulated in Table B.1, Appendix B, and the use of this table to compute general normal probabilities was illustrated in Example 3.18 and Exercises 3.18 to 3.21.

To show that $f(x)$ given by (7.49) is a true probability density function we need the following identity (see Appendix A.8).

$$\int_{-\infty}^{\infty} e^{-z^2/2}\,dz = \sqrt{2\pi} \qquad (7.50)$$

The moment-generating function of a random variable X that has this distribution is also easily obtained: ·

$$M(\theta) = E[e^{X\theta}] = \frac{1}{\sqrt{2\pi}\,\sigma} \int_{-\infty}^{\infty} e^{x\theta} \cdot e^{-(x-\mu)^2/2\sigma^2}\,dx$$

We again make the substitution $z = (x - \mu)/\sigma$, $dz = dx/\sigma$, $x = \sigma z + \mu$ to obtain

$$M(\theta) = e^{\mu\theta + \sigma^2\theta^2/2} \qquad (7.51)$$

(See Exercise 6.20.)

The mean and variance can be computed by expanding (7.51) in powers of θ:

$$M(\theta) = e^{\mu\theta} \cdot e^{\sigma^2\theta^2/2}$$

$$= \left(1 + \mu\theta + \frac{\mu^2\theta^2}{2} + \cdots\right)\left(1 + \frac{\sigma^2\theta^2}{2} + \cdots\right)$$

$$= 1 + \mu\theta + (\mu^2 + \sigma^2)\frac{\theta^2}{2} + \cdots$$

Since the coefficients of the powers of θ give the moments, using (6.32) we obtain

$$E[X] = \mu$$

$$E[X^2] = \mu^2 + \sigma^2$$

$$\mathrm{Var}[X] = E[X^2] - (E[X])^2 = \mu^2 + \sigma^2 - \mu^2 = \sigma^2$$

and we notice that the parameters μ and σ^2 do indeed represent the mean and variance here.

The normal distribution is the most important probability distribution for use in statistics. In this book, however, it is used for the most part only as a convenient approximation to certain other distributions. Thus, we do not give an extended discussion here.

7.10 Central Limit Theorems

The laws of large numbers (Section 6.11), while extremely important for tying together the relative frequency approach to probability with the axiomatic definitions, still leave something to be desired. Those results tell us nothing about the limiting distribution of the averaging random variable S_n/n. To obtain such a result one turns to some of the most important types of theorems in probability theory, called Central Limit Theorems. There are many versions of these theorems of increasing generality; we will describe one that is adequate for most applications.

To set up our problem we let $X_1, X_2, \ldots, X_n, \ldots$ be a sequence of mutually independent, identically distributed random variables. Furthermore, we assume that both

$$E[X_i] = \mu \quad \text{and} \quad \text{Var}[X_i] = \sigma^2$$

exist and that each X_i has a moment-generating function that exists for all θ in some interval about 0. Since the X_i are identically distributed, none of these quantities depend on i.

Next we consider the sum

$$S_n = X_1 + X_2 + \cdots + X_n \tag{7.52}$$

The Central Limit Theorem states that for large n the distribution of this sum is approximately *normal*, regardless of the form of the distribution of the individual X_i's. The importance of such a statement is clear, since such sums of random variables appear repeatedly in probability and statistics.

To make this statement precise, we note that by (6.16) and (6.23) S_n will have mean $E[S_n] = n\mu$ and variance $\text{Var}[S_n] = n\sigma^2$. Introduce the *standardized* variable

$$Z_n = \frac{S_n - n\mu}{\sqrt{n}\,\sigma} = \frac{(X_1 - \mu) + (X_2 - \mu) + \cdots + (X_n - \mu)}{\sqrt{n}\,\sigma} \tag{7.53}$$

that will have mean 0 and variance 1.

Theorem   (Central Limit Theorem)

If $F_n(x)$ is the distribution function of Z_n, then

$$\lim_{n \to \infty} F_n(x) = \lim_{n \to \infty} \Pr[Z_n \le x] = \frac{1}{\sqrt{2\pi}} \int_{-\infty}^{x} e^{-u^2/2}\, du$$

(On the right we have the standard normal distribution function.)

PROOF We will give only the main ideas of the proof.

Let $M(\theta)$ be the moment-generating function of the standardized random variable $(X_i - \mu)/\sigma$. Thus, by (6.32),

$$M(\theta) = E\left[e^{\theta(X_i-\mu)/\sigma}\right]$$

$$= 1 + \mu_1\theta + \mu_2\frac{\theta^2}{2!} + \mu_3\frac{\theta^3}{3!} + \cdots$$

$$= 1 + \frac{\theta^2}{2} + \mu_3\frac{\theta^3}{3!} + \cdots = 1 + \frac{\theta^2}{2} + \theta^3 R(\theta)$$

$$= 1 + \frac{\theta^2}{2}\left[1 + 2\theta R(\theta)\right] \tag{7.54}$$

since $\mu_1 = 0$ and $\mu_2 = 1$ for a standardized random variable. Here $\theta^3 R(\theta)$ represents the sum of all the terms with powers θ^3 or higher, with θ^3 factored out. Under our assumptions $R(\theta)$ is continuous and bounded near $\theta = 0$.

Now let $M_n(\theta)$ be the moment-generating function of Z_n

$$M_n(\theta) = E\left[e^{\theta Z_n}\right] = E\left[e^{\theta[(X_1-\mu)+(X_2-\mu)+\cdots+(X_n-\mu)]/(\sqrt{n}\,\sigma)}\right]$$

$$= E\left[e^{\theta(X_1-\mu)/(\sqrt{n}\,\sigma)}\right] \cdot E\left[e^{\theta(X_2-\mu)/(\sqrt{n}\,\sigma)}\right] \cdot \ \cdots \ \cdot E\left[e^{\theta(X_n-\mu)/(\sqrt{n}\,\sigma)}\right]$$

using the independence of the X_i's. Comparing with (7.54) we have

$$M_n(\theta) = M(\theta/\sqrt{n})M(\theta/\sqrt{n})\ldots M(\theta/\sqrt{n})$$

$$= \left[M(\theta/\sqrt{n})\right]^n$$

$$= \left[1 + \frac{\theta^2}{2n}\left(1 + \frac{2\theta}{\sqrt{n}}R(\theta/\sqrt{n})\right)\right]^n \tag{7.55}$$

For each fixed θ the term $(2\theta/\sqrt{n})R(\theta/\sqrt{n})$ is small relative to 1 and approaches 0 as $n \to \infty$, since $2\theta R(\theta/\sqrt{n})$ is bounded and, in fact, approaches $2\theta R(0)$ while $\sqrt{n} \to \infty$. Let us temporarily assume that it is small enough to be neglected and write

$$M_n(\theta) = \left[1 + \frac{\theta^2}{2n}\right]^n \tag{7.56}$$

Since, from elementary calculus, $\lim_{n \to \infty}(1 + x/n)^n = e^x$, (7.56) implies that

$$\lim_{n \to \infty} M_n(\theta) = e^{\theta^2/2}$$

The right side is the moment-generating function of the standard normal. (See Exercise 6.20 and formula (7.51).) The theorem then follows from part (b) of the theorem of Section 6.12.

Rigorous justification of the assumption that $(2\theta/\sqrt{n})R(\theta/\sqrt{n})$ is small enough to be neglected would require a closer study of the bounds that can be placed on this term. While the argument is not difficult, it would take us a little too far afield.

It is important to realize that in the Central Limit Theorem as we have given it there are few assumptions made about the X_i. They must satisfy some weak condition, such as the existence of a moment-generating function, but otherwise the exact distribution form is not relevant. In this way we have shown that for every reasonable collection of independent, identically distributed random variables with a finite variance, if $S_n = X_1 + X_2 + \cdots + X_n$, then the standardized version of this sum

$$Z_n = \frac{S_n - n\mu}{\sqrt{n}\,\sigma}$$

is approximately ($n \to \infty$) normally distributed with $\mu = 0$ and $\sigma = 1$.

EXAMPLE 7.3

Suppose that $\{X_1, X_2, \ldots\}$ is a sequence of independent random variables, each having a *uniform* density over the interval $-\sqrt{3} \le x \le \sqrt{3}$ (see Section 7.6). Let $S_n = X_1 + \cdots + X_n$. Since each X_i has $\mu = 0$ and $\sigma = 1$ (show this), the standardized sum $Z_n = S_n/\sqrt{n}$ in this case.

The density function of each Z_n can be calculated explicitly (see Exercise 17) and the result is sketched in Figure 7.9. While the original uniform density is far from normal, the approach to the normal as n increases is clear.

Two other important and closely related theorems are the following.

Theorem   (Normal Approximation to the Binomial)

If X has the binomial distribution of Section 7.3 then X has an approximately normal distribution with $\mu = np$ and $\sigma = \sqrt{npq}$. More precisely, the distribution

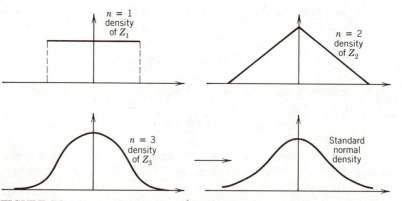

FIGURE 7.9 Approach of sums of uniform random variables to normal. $S_n = X_1 + \cdots + X_n$, X_i uniform for $-\sqrt{3} \le x \le \sqrt{3}$, $Z_n = S_n/\sqrt{n}$.

function of $Z = (X - np)/\sqrt{npq}$ approaches the standard normal distribution function as $n \to \infty$.

Theorem (Normal Approximation to the Poisson)

If X has the Poisson distribution discussed in Section 7.5 then it has approximately a normal distribution with $\mu = \lambda$ and $\sigma = \sqrt{\lambda}$. More precisely, the distribution function of $X = (X - \lambda)/\sqrt{\lambda}$ approaches the standard normal distribution function as $\lambda \to \infty$.

TABLE 7.2
The Normal Approximation to the Binomial

x	BINOMIAL $\Pr[X \le x]$	NORMAL $\Pr[X \le x]$
	$n = 10,\ p = 0.5$	
0	.0010	.0008
2	.0547	.0287
4	.3770	.2643
6	.8282	.7357
8	.9894	.9713
10	1.0000	.9992
	$n = 100,\ p = 0.5$	
40	.0284	.0228
45	.1841	.1587
50	.5398	.5000
55	.8644	.8413
60	.9824	.9772
	$n = 10,\ p = 0.05$	
0	.5987	.2327
1	.9138	.7673
2	.9884	.9854
3	.9909	.9999
4	.9919	.9999 +
	$n = 100,\ p = 0.05$	
0	.0059	.0110
2	.1183	.0838
4	.4360	.3228
6	.7660	.6772
8	.9369	.9162
10	.9885	.9890
12	.9985	.9993

TABLE 7.3
The Normal as an Approximation to the Poisson

x	POISSON $\Pr[X \le x]$	NORMAL $\Pr[X \le x]$
	$\lambda = 1$	
0	.3679	.1587
1	.7358	.5000
2	.9197	.8413
3	.9810	.9772
4	.9963	.9987
5	.9994	.9999
	$\lambda = 4$	
0	.0183	.0228
2	.2381	.1587
4	.6289	.5000
6	.8894	.8413
8	.9787	.9772
10	.9972	.9987
	$\lambda = 9$	
0	.0001	.0013
3	.0212	.0228
6	.2067	.1587
9	.5874	.5000
12	.8758	.8413
15	.9780	.9772
18	.9976	.9987

To prove the first theorem we note that since X represents the number of successes in n independent trials, one can write (see (7.5))

$$X = X_1 + X_2 + \cdots + X_n$$

where $X_i = 1$ if the ith trial is a success and $X_i = 0$ if the ith trial is a failure. Thus, the hypotheses of the Central Limit Theorem are satisfied with X replacing S_n.

The second theorem can be proved using moment-generating function methods. One simply computes the moment-generating function of Z directly, shows that it approaches the standard normal moment-generating function as $\lambda \to \infty$, and again uses the theorem of Section 6.12.

The approximations in these two theorems are illustrated by Tables 7.2 and 7.3.

We observe from Table 7.2 that the normal distribution function is a good approximation to the binomial distribution function for values of p near to $\frac{1}{2}$ and n large. Even for p small the approximation is fair for $n = 100$. For small values of n (for instance, 10) the approximation is reasonably poor and for small values of p it is very poor. We conclude that if p is near $\frac{1}{2}$ and n is large the normal serves as a good approximation to the binomial. If p is very different from $\frac{1}{2}$ the normal is a poorer approximation even for moderately large n values. However, we have observed in Section 7.5 that for p small the Poisson distribution is a good approximation to the binomial. Thus, we conclude that the binomial can be approximated by either the normal (if p is not small) or the Poisson (if p is small). In either case the approximation improves as n increases.

We observe here that the approximation is better for larger λ.

7.11   Estimating Parameters

Now we have the tools we need to assign numerical values to the unknown parameters of distributions. The idea is quite simple. We compute the mean, variance, or correlation coefficient of some distribution that we think is representative of our data (this assumed distribution is called a *model*). In Chapter 8 we will discuss how to determine if the model is a good one. For now we will just assume it is. Then we compute the *sample* mean, *sample* variance, *sample* correlation coefficient, and, if necessary, other moments of our data as done in Section 6.14. Then we argue that if the model is to represent our data it ought to have moments that have approximately the same numerical values as the data. More is required, of course, but we will defer further discussion until Chapter 8.

EXAMPLE 7.4

Suppose in Example 6.16 we assumed that $X =$ the number of defects per piece of cloth was a Poisson-distributed random variable. Then one has (see (7.25)) that $E[X] = \lambda$. Therefore, based on the data we would *estimate* that λ is 1.107 since

$\bar{x} = 1.107$. Since the Poisson density function has only one parameter, λ, this estimate is all we need to determine which member of the Poisson family has the same expected value as our sample mean and therefore is a good candidate to represent the data.

EXAMPLE 7.5

In Example 6.15 one might argue that the wear on wheel linings, X, is a uniformly distributed random variable over a range $a \le x \le b$. Then $E[X] = (a + b)/2$ (see (7.31)). The sample mean for that data was 8.25. However, $E[X]$ has two parameters, a and b, so this sample mean alone cannot estimate both. More information is needed. This is provided by $s^2 = 3.84$ (Example 6.18) and $\text{Var}[X] = (b - a)^2/12$ (see formula (7.32)). Thus, we have

$$\bar{x} = 8.25 \approx (a + b)/2$$

$$s^2 = 3.84 \approx (b - a)^2/12$$

(" $\approx$ " means "is an estimate of"). We can solve these equations simultaneously to get

$$a \approx 4.9, \qquad b \approx 11.6$$

EXAMPLE 7.6

If we assume that the record of grades (see Example 6.21) is a good representation of a bivariate normal density function, then we have the data we need to estimate all of the parameters (there are five) of that normal density (see equation (4.23)). Thus, if X is the high school grade of a senior and Y is the grade of the same student at the end of the first college year, then we assume that the pair (X, Y) is distributed with a bivariate normal density, and for that density

$$\mu_X = E[X] \approx \bar{x} = \frac{1449.40}{20} = 72.5$$

$$\mu_Y = E[Y] \approx \bar{y} = \frac{1230.70}{20} = 61.5$$

$$\sigma_X^2 = E\left[(X - E[X])^2\right] \approx s_x^2 = 33.0$$

$$\sigma_Y^2 = E\left[(Y - E[Y])^2\right] \approx s_y^2 = 57.5$$

$$\rho \approx r = .55$$

Exercises 7

1. If X and Y are two independent binomially distributed random variables with the same parameter p, show that

$$U = X + Y$$

is also binomially distributed. [*Hint*: Use formulas (7.8) and (6.31) and the theorem in Section 6.12.]

2. If X and Y are two independent Poisson-distributed random variables with the same λ, show that $U = X + Y$ is a Poisson-distributed random variable with parameter 2λ. [*Hint*: Use formulas (7.24) and (6.31) and the theorem in Section 6.12.]

3. If X and Y are two independent gamma-distributed random variables with the same parameter λ, show that $U = X + Y$ is a gamma-distributed random variable with the same λ. [*Hint*: Use formulas (7.46) and (6.31) and the theorem in Section 6.12.]

4. If X and Y are two independent normally distributed random variables, show that $U = X + Y$ is also normally distributed with $\mu_U = \mu_X + \mu_Y$ and $\sigma_U^2 = \sigma_X^2 + \sigma_Y^2$. [*Hint*: Use formulas (7.51) and (6.31) and the theorem in Section 6.12.]

5. Let $X_1, X_2, \ldots, X_n$ be mutually independent, negative exponentially distributed random variables with the same λ. Show that

$$U = X_1 + X_2 + \cdots + X_n$$

is a gamma-distributed random variable with parameters λ and $b = n$. [*Hint*: Use formulas (7.38), (7.46), and (6.31) and the theorem in Section 6.12.]

6. Let $X_1, X_2, \ldots, X_n$ be mutually independent standard normally distributed random variables. Let $U = X_1 + X_2 + \cdots + X_n$ and show that U is a normally distributed random variable with $\mu_U = 0$ and $\sigma_U^2 = n$. [*Hint*: Use formulas (7.51) and (6.31) and the theorem in Section 6.12, or Exercise 7.4.]

7. For a standard normally distributed random variable show that

$$E[X^n] = 0, \qquad \text{for } n \text{ an odd integer}$$

and

$$E[X^n] = 1 \cdot 3 \cdot 5 \cdots (n-1), \qquad \text{for } n \text{ an even integer}$$

[*Hint*: Use formula (7.51).]

8. Show that if X and Y are independent gamma-distributed random variables with parameters λ_X, b_X and λ_Y, b_Y, respectively, then $U = X + Y$ is not a gamma-distributed random variable unless $\lambda_X = \lambda_Y$.

9. Show that if X and Y are independent Poisson random variables with parameters λ_X and λ_Y, then $U = X + Y$ is not a Poisson-distributed random variable unless $\lambda_X = \lambda_Y$.

10. If X and Y are two independent, identically geometrically distributed random variables and if $U = X + Y$, determine $E[U]$ and $\text{Var}[U]$ using the moment-generating function $M_U(\theta)$.

11. If $X_1, X_2, \ldots, X_n$ are mutually independent, uniformly distributed random variables with parameters $a = 0$, $b = 1$, and $U = X_1 + X_2 + \cdots + X_n$, de-

termine $E[U]$ and Var$[U]$ using the moment-generating function $M_U(\theta)$. [*Hint*: Use formula (7.33).]

12. If X is a random variable with moment-generating function $M(\theta)$, show that $-X$ has moment-generating function $M(-\theta)$.

13. If X and Y are two independent normally distributed random variables and $U = X - Y$, show that U is a normally distributed random variable with $\sigma_U^2 = \sigma_X^2 + \sigma_Y^2$. [*Hint*: Use (7.51) and Exercise 7.12 above.]

14. Prove the Central Limit Theorem directly for $X_1, X_2, \ldots$ mutually independent, identically distributed Poisson random variables.

15. The following problem arises in the study of mechanical reliability theory. X is the strength of a part, Y is the stress induced by a load, and $U = X - Y$ is a dummy variable with the property that if $U < 0$ the part fails. Determine $M_U(\theta)$ if X and Y are independent gamma-distributed random variables with different parameters. From this determine $E[U]$ and Var$[U]$.

16. Let $X_1, X_2, \ldots, X_n$ be independent standard normally distributed random variables. Show that

$$U = X_1^2 + X_2^2 + \cdots + X_n^2$$

is a gamma-distributed random variable with parameters $\lambda = \frac{1}{2}$ and $b = n/2$. (*Hint*: See Example 5.3.)

17. In Example 7.3 determine directly the density functions of S_2 and S_3 and of Z_2 and Z_3. Verify Figure 7.9 by sketching the density functions of Z_2 and Z_3. [*Hint*: Use (5.8) and (5.2).]

18. Assume that X is a binomially distributed random variable with $n = 100$.

    (a) If $p = 0.02$, *estimate* Pr$[X \le 3]$. Compare this probability to the exact value obtained by using the binomial density.

    (b) If $p = 0.25$, *estimate* Pr$[X \le 30]$.

19. Using the Hall experiment (see Examples 4.16 and 2.17) estimate the probability that of the next 50 telephone calls 5 or less will be for ambulance service.

20. Using the same experimental results as in Exercise 19 estimate the probability that of the next 1000 calls 70 or less will be for ambulance service.

21. Suppose one throws 12 honest dice. Let X be the total number of spots showing. Estimate Pr$[25 \le X \le 40]$.

8 · Fitting Distributions to Data

8.1 Introduction

In Chapter 6 we concentrated on moment properties of distribution functions and density functions. In the first parts of Chapter 7 we concentrated on special properties of the most commonly occurring functions. The question remains as to how one knows what distribution to use in practice. Usually one has a collection of data that by itself is not in a form directly useful for probabilistic study. The question then arises: What is the most appropriate distribution to assume lies behind this data? In this chapter we will discuss some of the ways of answering this question. We will also discuss the concomitant question of how do we know whether the assumed distribution we end up with really is well represented by our data?

8.2 Estimating a Density Function

Let us start with a method that attempts to fit a discrete density function to data. The idea is very simple. Suppose we have collected n pieces of data (numbers) representing values that a discrete random variable has taken in the set $\{0, 1, 2, \dots\}$. We start by *assuming* a family of density functions that might be represented by the data. For example, let us suppose that we have theoretical reasons for assuming that the data come from a Poisson-distributed random variable. The Poisson family depends on one unknown parameter, denoted by λ in Section 7.5. Our first task is to use our data to estimate this parameter.

To accomplish this estimation we can use several methods. The one discussed in Section 7.10 was called *estimating by moments*. Recall that in Section 6.15(a) we defined the sample average $\bar{x}$ and argued that this was a "good" estimate of the mean $\mu = E[X]$. For the Poisson, the parameter $\lambda = \mu$, as in formula (7.25). Thus, the unknown parameter λ can be estimated by the number $\bar{x}$ calculated from the data. This specifies one density function out of the Poisson family that would at least agree with the data in having the same mean as the data average.

Next we wish to see whether the data really do represent this particular density function in more ways than simply having the same mean. To do this we form what is called an *empirical density function*, that is, we simply count the number of pieces of data that take the value 0. This number divided by n is a relative frequency estimate of $\Pr[X = 0]$ (see Section 2.10). (It turns out later that it is easier to work with the actual number of occurrences of 0 rather than with the

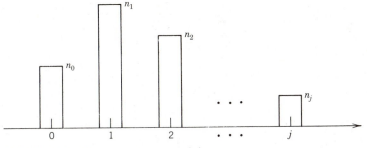

FIGURE 8.1   Histogram for a set of data.

relative frequency.) Proceeding in this same manner we can count the number of occurrences of the values $1, 2, \ldots$ in the data.

Of course n is finite and the Poisson density function takes values on all nonnegative integers so at some point we run out of data to match against the Poisson. There is a rough rule of thumb called "the rule of five" that says whenever the expected number of pieces of data at some integer, say, j, is less than five lump the data for that integer and all data counts for integers greater than j into one when making a comparison between the data and the theoretical distribution. That one count then gives a relative frequency estimate of $\Pr[X \geq j]$.

Now look at what you have. There are, say, n_0 occurrences of 0, n_1 of 1, n_2 of 2, et cetera, and the final n_j representing the counts for all integers greater than or equal to j. This display, graphed in Figure 8.1, is called a *histogram*.

8.3 Testing the Fit

Since we know $\Pr[X = k]$ for all $k = 0, 1, 2, \ldots$ from our uniquely determined density function (remember we have estimated λ) then we can match the number of actual occurrences of k in the data with what the Poisson density function says the expected number of such occurrences should be. That is, we can compare n_k with

$$n\left(\frac{\lambda^k e^{-\lambda}}{k!}\right) = m_k$$

for each $k = 0, 1, 2, \ldots, j - 1$. Remembering how we constructed the jth count we must compare n_j with

$$n\sum_{i=j}^{\infty} \frac{\lambda^i e^{-\lambda}}{i!} = m_j$$

Now we can compare n_k, $k = 0, 1, 2, \ldots, j$, with the corresponding m_k using the χ^2 test (cf. Section 4.11) to see if our Poisson density function is well represented by our data.

If $n_0, n_1, \ldots, n_j$ represent the observed counts of data in the cells and if $m_0, m_1, \ldots, m_j$ represent the expected counts using the hypothesized distribution with the estimated parameter λ, then

$$\chi^2 = \sum_{i=0}^{j} \frac{(n_i - m_i)^2}{m_i}$$

has a χ^2 distribution. Clearly χ^2 will be small if the n_i's are close to the m_i's and large if they differ substantially.

When using the χ^2 test for testing how representative data are of an assumed density function we have to slightly alter our procedures of Section 4.11. In particular the number of degrees of freedom must be modified.

A general theorem (proof omitted) states that if we have n pieces of data put into k cells and if we must estimate m parameters from our data to determine the appropriate density, then the proper number of degrees of freedom for our χ^2 testing procedures is $k - m - 1$. For the Poisson example discussed here this would be $(j + 1) - 1 - 1 = j - 1$, since we have $j + 1$ cells $\{0, 1, 2, \ldots, j\}$ and one parameter, λ, to be estimated.

If χ^2 is not too large we proceed as though our data are a good representation of the Poisson. If χ^2 *is* too large we drop our assumption that our data are a good representation of the Poisson. In the latter case we must start all over and pick a new density function to test our data against. We might even have to go to a whole new family and possibly have to construct our own density function that is unlike any of the well-known functions.

EXAMPLE 8.1

In many statistical studies of probabilistic problems one simulates the problem on a digital computer. A basic step in such a simulation is to generate on the computer a uniformly distributed random number. There are many preprogrammed methods for the approximate generation of these uniform random numbers. In Table 8.1 are 100 digits generated by a random number generator. We would like to know if they are consistent with the assumption that they come from a uniform distribution over the 10 digits $\{0, 1, 2, \ldots, 9\}$.

Of course if these numbers were truly uniformly distributed we would expect each digit to occur one-tenth of the time (the quantities in parentheses in Table

TABLE 8.1
100 Random Digits

92072	17410	58833	75974	58833
64340	74717	58524	88977	30757
73794	19516	03133	19051	48922
83001	88318	28725	68568	00812

TABLE 8.2
100 Random Digits Arranged in a Histogram

	COUNT FROM TABLE 8.1		COUNT FROM TABLE 8.1
0	10 (10)	5	10 (10)
1	11 (10)	6	4 (10)
2	7 (10)	7	14 (10)
3	12 (10)	8	16 (10)
4	7 (10)	9	6 (10)

8.2). Differences between what a uniformly distributed random variable "should" give and what the data give occur for many reasons, but the key question is: Are these differences so great that it is not tenable to assume that the data are really observations from a uniformly distributed random variable? The χ^2 test helps answer this question.

$$\chi^2 = \frac{(10-10)^2}{10} + \frac{(11-10)^2}{10} + \cdots + \frac{(6-10)^2}{10} = 12.7$$

We have $k = 10$ cells with $m = 0$ parameters to be estimated. Thus, we have $10 - 0 - 1 = 9$ degrees of freedom. From Table B.3, Appendix B, using 9 degrees of freedom we find the 95% and 99% points are given as 16.92 and 21.67, respectively. Since both of these values exceed the value of χ^2, we have no strong evidence that the data are not representative of a uniformly distributed random variable, that is, the difference between the n_i and m_i values is not great enough to force χ^2 to exceed the 95% or 99% points in the distribution.

This result is a bit surprising. The larger numbers 6–9 seem to behave differently from the smaller ones. Yet this χ^2 says that they really are not, that the deviation is not significant, and that these differences do not exceed what might be expected from random chance. Do our eyes deceive us? Often the answer is yes. Our intuition is often at variance with probabilistic behavior. One reason is that most of us have not developed keen intuition about probabilistic behavior after a life of exposure mainly to deterministic behaviors. But there is more to it than that. The χ^2 test that we used here is not a very powerful statistical test. It has trouble distinguishing "good" from "bad" data when the degrees of freedom are small. It has a tendency to imply that data are "good" when in fact they are not. For that reason one should apply more statistical analysis to data then we have done here. Other methods to enhance such studies are given in statistics courses.

EXAMPLE 8.2

In the early days of the development of probability and statistics researchers often found that the normal density fit their data. In fact there was one argument that said that the normal density was a "universal law." It came as a surprise

TABLE 8.3
Number of Prussian Soldiers Dying from Horse Kicks per Army Corps-Year

NUMBER OF DEATHS	FREQUENCY OBSERVED	POISSON EXPECTED FREQUENCY ($\lambda = 0.61$)[a]
0	109	108.7
1	65	66.3
2	22	20.2
3	3	4.1
4	1	0.6
5	0	0.1
6	0	0
	$\overline{200}$	$\overline{200}$

[a] Note that λ is estimated by $\bar{x} = 0.61$.

when in 1898 Bortkiewicz gave some naturally occurring data that were not well fit by the normal density but instead were well fit by the Poisson. Quite a controversy arose around some of the research of Bortkiewicz.

Bortkiewicz studied the records of 10 Prussian army corps over a period of 20 years. For each corps in each year he determined the number of soldiers killed by the kick of an army horse. Thus, he had 200 observations (army corps-years). The count in each observation was the number of "successes" (deaths) in a large number of "trials" (soldiers per corp) in which the probability of success on each trail was small. (Each given soldier had a very small chance of dying.) Thus, it is reasonable to assume that we have a binomial distribution with large n and small p (see Section 2.9). By the results of Section 7.5, it seems reasonable to assume that a Poisson distribution may fit the data. Bortkiewicz's data is shown in Table 8.3 (from Fry (1928)[1]).

We can then use our χ^2 test. Because of the small number of cases in the cells 3, 4, 5, and 6 we have lumped these together into one cell that will give 200 $\Pr[X \geq 3] = 4.8$. Then on these 4 cells $\{0, 1, 2, 3,$ or more$\}$ our computed χ^2 is about 0.054. The 95% and 99% values from Table B.3 in Appendix B are 5.99 and 9.21, respectively, for $4 - 1 - 1 = 2$ degrees of freedom. Thus, there is no good reason to assume that the number of Prussian soldiers dying from horse kicks is not a Poisson-distributed random variable. In fact the fit is excellent; $\chi^2 = 0.054$ is *much* smaller than 5.99 and 9.21. This was Bortkiewicz's conclusion.

So now we have a probability model for the random variable $X =$ the number of soldiers dying from the kick of a horse in a Prussian army corps during one year. We have

$$p_i = \Pr[X = i] = e^{-0.61} \frac{(0.61)^i}{i!}, \qquad i = 0, 1, 2, \ldots$$

[1] T. C. Fry (1928), *Probability and Its Engineering Uses*, Van Nostrand, New York. Reprinted with permission of Brooks/Cole.

8.4 Estimating a Distribution Function

There are some disadvantages to estimating distributions from their density functions. The histogram for a density function is often lumpy and even if the χ^2 method tells us that the data are a reasonable representation of the chosen density function our eyes are disbelieving. Because operations such as summing or integrating tend to give smoother functions, it often is better to fit distribution functions rather than density functions to data.

Let us reconsider our previous problem of fitting a Poisson distribution to some data. Recall that we compared n_k to

$$m_k = n\left(\lambda^k e^{-\lambda}/k!\right)$$

except for the end point n_j. Instead of doing that let us compute

$$n_0, n_0 + n_1, n_0 + n_1 + n_2, \ldots, \sum_{i=0}^{k} n_i, \ldots$$

The terms in this sequence, when divided by n, are denoted by $\hat{F}(0), \hat{F}(1), \hat{F}(2), \ldots, \hat{F}(k), \ldots$ and are estimates of the population distribution function

$$\hat{F}(k) = \left(\sum_{i=0}^{k} n_i\right)/n \approx F(k) = \Pr[X \le k]$$

In this way we will use our data to estimate the distribution function, not the density function.

8.5 Testing the Fit to a Distribution Function

If we do estimate the distribution function rather than the density function, there are better tools than χ^2 to test how well the data represent the distribution. The best known and most frequently used of these tools is called the Kolmogorov–Smirnov test. To use this test one computes

$$D_n = \max_{-\infty < x < \infty} \left(|\hat{F}(x) - F(x)|\right)$$

that is, one finds the maximum distance between the data $\hat{F}(x)$ and the assumed distribution function $F(x)$. If this maximum distance is not too large it is argued that the data are a good representation of the assumed distribution. If this maximum is too large it is argued that the distance of the data from the assumed curve is too great for the data to be a good representation of the assumed distribution. Therefore, the assumed distribution is dropped from further consideration and a new distribution is assumed and the whole process is repeated.

Whether D_n is "too large" or not is partly a matter of judgment. As in the case of χ^2 one says they want to be 95% "sure" of making the "right" assumption, or 99%. Then by using Table B.2 in Appendix B, with n degrees of freedom, one can find the value of D_n' such that if that tabular value is exceeded by the D_n

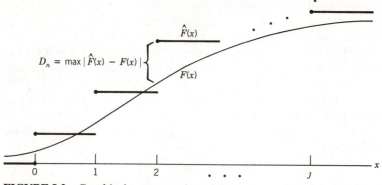

FIGURE 8.2   Graphical representation of Kolmogorov–Smirnov statistic $D_n = \max |\hat{F}(x) - F(x)|$.

computed above the data are too far from the assumed distribution to use the assumed distribution. If the D_n' of the table exceeds the D_n computed above then one is reasonably sure that the data are a good representation of the assumed distribution function.

One nice advantage of this Kolmogorov–Smirnov test is that we can use it for both discrete and continuous random variables. This can also be done with the χ^2 test used on histograms, but for continuous valued random variables the Kolmogorov–Smirnov test is easier to use. For example, one need not be concerned with the "rule of five" or how to form a histogram for this test. (See Figure 8.2.)

EXAMPLE 8.3

Let us reconsider the Bortkiewicz data on horse kicks. We repeat part of Table 8.3 in Table 8.4.

TABLE 8.4
Number of Prussian Soldiers Dying from Horse Kicks per Army Corps-Year

NUMBER	OBSERVED	$\sum_{i=0}^{k} n_i/n$	$F(k)$ POISSON DISTRIBUTION FUNCTION	DEVIATION
0	109	109/200	108.7/200	0.0015
1	65	174/200	175.0/200	0.005
2	22	196/200	195.2/200	0.004
3	3	199/200	199.3/200	0.0015
4	1	200/200	199.9/200	0.0005
5	0	200/200	200.0/200	0
6	0	200/200	200.0/200	0

The maximum deviation here is 0.005 so our computed $D_{200} = 0.005$. From Table B.2 in Appendix B, $D'_{200} = 1.36/\sqrt{200} = 0.096$ for 95% surety and $D'_{200} = 1.63/\sqrt{200} = 0.12$ for 99% surety. From this result one again concludes that there is no good reason to argue that the data are not a good representation of the Poisson distribution function.

EXAMPLE 8.4

Recall that Hall studied types of telephone calls. He concluded (Example 4.16) that the type of the nth call was independent of the type of the $(n + 1)$st call. His next task was to estimate a probability distribution for the times between consecutive calls. To do this he used 403 pieces of data, which we have summarized in Table 8.5.

There has been an enormous amount of research on telephone traffic. Work at telephone laboratories and many computer centers has indicated that the time between telephone calls probably is an exponentially distributed random variable. Hall's first attempt was then to see if his data for the time between type-0 calls were exponentially distributed with $\lambda = 0.01$.

In the tables below we compare the empirical distribution function $\hat{F}(x)$ with the exponential distribution (i.e., $F(x) = 1 - e^{-\lambda x}$) for $\lambda = 0.01$ and $x = 25, 50, \ldots, 500$.

Thus, $D_{403} = 0.08$. Then, since we have 403 degrees of freedom (the Kolmogorov–Smirnov test uses all the data), we have from Table B.2 in Appendix B that D_{403} should exceed $1.36/\sqrt{403} = 0.068$ for 95% surety or $1.63/\sqrt{403} = 0.081$ for 99% surety of rejection of the hypothesis of fit. Thus, we can see that these data do not exceed the 99% point but they exceed the 95% point. So the

TABLE 8.5[a]
Theoretical ($F(x)$) and Estimated ($\hat{F}(x)$) Distribution Function for Times between Calls for Ambulance Service

x	$\hat{F}(x)$	$F(x)$	x	$\hat{F}(x)$	$F(x)$
25	.27	.22	275	.90	.93
50	.47	.39	300	.90	.95
75	.58	.53	325	.90	.96
100	.66	.63	350	.92	.97
125	.69	.71	375	.92	.98
150	.84	.78	400	.92	.98
175	.84	.83	425	.94	.99
200	.87	.86	450	.96	.99
225	.88	.89	475	.98	.99
250	.90	.91	500	1.00	.99

[a] W. K. Hall, (1969), *A Queueing Theoretic Approach to the Allocation and Distribution of Ambulances in an Urban Area*, dissertation presented to the School of Business Adminstration, University of Michigan, Ann Arbor, Michigan. Reprinted with permission of author.

TABLE 8.6

Observed Distribution Function for Days between Major Mine Accidents ($\hat{F}(x)$) and the Negative Exponential Distribution Function ($F(x)$) with $\lambda = 1/240$

DAYS BETWEEN ACCIDENTS	$\hat{F}(x)$	$F(x)$	DAYS BETWEEN ACCIDENTS	$\hat{F}(x)$	$F(x)$
10	.037	.041	250	.661	.647
20	.128	.081	300	.725	.713
30	.165	.118	350	.826	.763
40	.201	.155	400	.881	.811
50	.239	.188	450	.881	.847
100	.404	.341	500	.908	.875
150	.514	.465	1000	.954	.984
200	.578	.565	1500	.981	1.000
			2000	1.000	1.000

data here are a bit ambivalent. We would be willing to bet at 20-to-1 odds that the distribution is *not* exponential, but not at 100-to-one odds. We could proceed with the assumption of exponentially distributed times between calls of type-0. We could search the data to see whether we had made a mistake somewhere. Hall assumed that the data were not a good representation of an exponentially distributed random variable. He tried another model and got better results.

EXAMPLE 8.5

In Exercise 6.28 we gave some data on the number of days between consecutive major mine accidents. If those data are fitted to an exponential distribution the appropriate value of λ is $1/\bar{x} = 1/240$ (see Example 3.8). To test whether these times are negative exponentially distributed we have computed Table 8.6. For this data $D'_{109} = 1.36/\sqrt{109} = 0.13$. So D_n should exceed 0.13 for 95% surety of rejection of the hypothesis of fit. It is clear from Table 8.6 that no difference between $\hat{F}(x)$ and $F(x)$ is nearly this large. So we accept that the data are representative of a negative exponentially distributed random variable with $\lambda = 1/240$ days, that is, the distribution function given in Example 3.8 is a "good" model of the time between major mine accidents for the reported data.

Exercises 8

1. Based on the data in Exercise 6.25 was Hamilton's use of the word "an" a Poisson-distributed random variable?

2. Recall the Westgren experiments on the Smoluchowski theory of Brownian motion (Exercises 2.15 and 4.17). The following table[1] compares observed

[1] From S. Chandrasekhar (1943), "Stochastic Problems in Physics and Astronomy," *Reviews of Modern Physics*, **15**, pp. 2–89. Reprinted with permission of the author and the American Physics Society.

counts with predicted counts using a Poisson density function with $\lambda = 1.428$.

Number of particles	0	1	2	3	4	5	6	7
Theory	380	542	384	184	66	19	5	2
Observed	379	563	355	174	69	26	5	2

Is there a good agreement between the observations and Smoluchowski's theory? Based on the observed data, what value of λ would you use for a Poisson comparison?

3. Return to Exercise 6.32. Using the data given there test whether that random number generator is giving numbers that one expects from a uniformly distributed random variable taking values $0, 1, \ldots, 9$.

4. In his traffic study Daganzo (see Exercise 6.32) built a theoretical model for the sizes of groups of automobiles ("platoons") proceeding on a road. From this he predicted the probability of finding a group that contains n cars, and then he observed the highway and counted group lengths over a long period of time. Table 8.7 is abbreviated from his results. (The theoretical results are obtained using a one-parameter model, with the parameter estimated from the data.) Do you think his data fit his theory well enough to accept the theory? (Be careful of the "rule of five.")

5. Using the data of Example 6.21 test whether either the students' last-year high school grades or first-year college grades are representative of data one expects to get from a normal distribution, using the Kolmogorov–Smirnov test of fit.

6. One of the first and still widely made assumptions in data analysis is that "errors" in the data (i.e., deviation from some "true" value) are normally distributed with a mean of zero. The following data are from hydrologic studies reported in Solomon (1976). They show the differences between

TABLE 8.7
Platoon Sizes in One Lane of a Two-Lane Road

PLATOON LENGTH (n)	THEORY	OBSERVATIONS	PLATOON LENGTH (n)	THEORY	OBSERVATIONS
1	60.8	66	6	5.3	4
2	34.1	19	7	3.7	9
3	20.4	26	8	2.6	4
4	12.5	11	9	1.9	4
5	8.0	6	10	1.4	4
			> 10	5.3	3

SOURCE: C. F. Daganzo (1975), *Two Lane Traffic: A Stochastic Model*, Ph.D. dissertation, Department of Civil Engineering, University of Michigan, Ann Arbor, Michigan. Reprinted with permission of the author.

TABLE 8.8
Differences between Model Predictions
and Data for Rainfall Runoff

− 2.40	− 0.49	+ 0.14	+ 0.81
− 2.05	− 0.46	+ 0.19	+ 0.95
− 1.86	− 0.43	+ 0.19	+ 1.11
− 1.71	− 0.40	+ 0.20	+ 1.12
− 1.36	− 0.33	+ 0.25	+ 1.15
− 1.05	− 0.06	+ 0.28	+ 1.18
− 0.97	− 0.02	+ 0.31	+ 1.33
− 0.86	− 0.02	+ 0.33	+ 1.55
− 0.86	0.00	+ 0.42	+ 1.62
− 0.70	+ 0.02	+ 0.65	+ 2.40
− 0.69	+ 0.10	+ 0.76	
− 0.60	+ 0.12	+ 0.79	

SOURCE: S. I. Soloman (1976), "Parameter Re-
gionalization and Network Design," in *Statistical
Approaches to Water Resources*, vol. 1 (H. W. Shen,
Ed.), H. W. Shen, Fort Collins, Colorado. Reprinted
with permission of the publisher.

rainfall runoff records and the rainfall runoff predicted by a mathematical model for 46 cases. We have arranged the data in ascending order.

The average for the data is nearly 0 and the standard deviation is nearly 1.00, using the methods of Section 6.15. Would you assume that these data are a good representation of a standard normally distributed random variable?

7. Are the data of Example 6.17 representative of what one would expect from data of an exponential density function? That is, is the resting time of that aphid an exponentially distributed random variable?

8. Reconsider the problem of Exercise 6.29 as follows. If the dice were honest and the experiments (tossing the dice) were conducted "under identical conditions" then Weldon's data should be a good representation of a binomial density function with $p = \frac{1}{3}$, $q = \frac{2}{3}$, and $n = 12$. Using a χ^2 test see whether the Weldon data are consistent with the assumptions of the binomial density function above. Can you give any reasons for such a bad fit? Now estimate p as 0.3377 and use $n = 12$ and test again. Is this a good fit to the data? See Example 2.16 to see where $p = 0.3377$ comes from.

9. In a study of hospital demand processes, Swartzman (1969) studied the times of arrivals of physical therapy patients. These patients had fixed times at which they were scheduled to come to the hospital so that the necessary equipment for their treatment would be available. However, some arrived before their scheduled time (early) and some arrived after their scheduled time (late). Table 8.9 is drawn from his data. Plot the estimated probability

TABLE 8.9
Numbers of Patients Arriving Early or Late for Physical Therapy at a Hospital

MINUTES EARLY FOR TREATMENT	NUMBER OF PATIENTS COMING THAT TIME OR EARLIER	MINUTES EARLY FOR TREATMENT	NUMBER OF PATIENTS COMING THAT TIME OR EARLIER
− 26	0	2	130
− 24	2	4	139
− 22	5	6	147
− 20	12	8	160
− 18	21	10	169
− 16	31	12	176
− 14	39	14	179
− 12	46	16	181
− 10	57	18	184
− 8	70	20	188
− 6	86	22	188
− 4	100	24	191
− 2	119	26	192

SOURCE: G. L. Swartzman (1969), *The Statistical Analysis of the Arrival Process of Three Major Types of Hospital Patients*, Ph.D. dissertation, Department of Industrial and Operations Engineering, University of Michigan, Ann Arbor, Michigan. Reprinted with permission of the author.

that the patient is x or fewer minutes away from schedule (negative x implies that the patient was early and positive x implies that the patient was late). From your graph make a guess as to what type of distribution will fit this data. Fit your guess and see whether you were shrewd enough. If so, congratulations; if not, try another guess. This is a nice problem because it illustrates how one goes from raw data to a probability model of that data.

10. Cretin (1974)[1] studied the response times of emergency medical treatment for patients suffering a myocardial infarction. One part of her data gave the times required to treat patients at the scene and conduct them by ambulance to a hospital. The resulting estimates, $\hat{F}(t)$, of the distribution function are given here.

Time (min)	$\hat{F}(t)$	Time (min)	$\hat{F}(t)$
0–3	0.004	18	0.589
6	0.016	21	0.777
9	0.045	24	0.879
12	0.175	27	0.932
15	0.357	30	1.00

[1]S. Cretin (1974), *A Model of the Risk of Death from Myocardial Infarction*, adapted from a thesis presented to the Alfred P. Sloan School of Management, Massachusetts Institute of Technology, Cambridge, Massachusetts, and from Technical Report 09-74 prepared for the National Science Foundation (RANN) under Grant GI 38004.

Is it reasonable to assume that these data are representative of an exponential distribution? (Assume sample size 1000.)

11. Here are some data collected in a bank. The data give the number of minutes that a teller requires to complete transactions with a particular type of customer. (Assume sample size 990.)

Time	Percentage taking that time or less	Time	Percentage taking that time or less	Time	Percentage taking that time or less
0.2	0.000	1.2	0.584	2.2	0.882
0.4	0.017	1.4	0.689	2.4	0.908
0.6	0.103	1.6	0.759	2.6	0.930
0.8	0.293	1.8	0.805	2.8	0.942
1.0	0.479	2.0	0.840	3.0	0.952

Is it reasonable to argue that these data are representative of what one expects if teller service times are exponentially distributed random variables? Gamma-distributed random variables? (*Note*: An initial histogram of the data is revealing.)

12. These data are part of a larger study on the times needed to perform certain duties in a hospital (e.g., take an X ray of a patient).

Time	No. of observations	Time	No. of observations
0–40	2	180–200	10
40–80	6	200–240	8
80–120	17	240–280	7
120–160	23	280–320	2
160–180	18	320 +	5

Do these data exhibit the behavior one would expect of a normally distributed random variable? Even if the data do give results expected of a normally distributed random variable, why is this not an acceptable assumption for these data?

9 · Random Processes: Introduction and Examples

9.1   Introduction

In the previous chapters we have developed much of the elementary theory of probability and exposed some of its uses in modeling phenomena as diverse as flight patterns of aphids and calls for ambulance assistance. One might call Chapters 1 to 8 a study of the statics of probability theory. In the remainder of this book we want to look at some aspects of what can be called the dynamics of probability theory.

For example, Section 4.9 was concerned with many random variables, but such collections had no particular order to them. Thus, X was some random variable, Y was some other random variable, and so on. Now we want to order these random variables so that we can speak of "the first random variable" or "the nth random variable." We also want to speak of a random variable coming "before" another random variable. We want to put the random variables into some order relation.

In doing this ordering, of course, one begins to think of some random behavior unfolding over "time" and, hence, one begins to think of phenomena occurring randomly over time. Thus, one begins to imagine "processes" not just "events." Indeed, as we shall see, these conceptualizations are extremely useful and lead to a large number of applications in nearly every field of science. One word of caution is necessary, however. One must not let our fixation on words such as "time" be the basis of our order. There are important applications to studies where, for example, "time" has no part at all but rather one is concerned with collections of random variables that are put into an order by assigning an "ordering variable" that describes "space" rather than "time." Such an example occurs in some military applications of probability where "distance" serves to order our random variables when one speaks of the target getting "closer."

Our approach is this: we start by defining a random process and then we ask what must one know to say that one has completely specified the probability behavior of this process. It turns out that one must know a very large number of probabilities to do this. It also turns out that there are several extremely important special cases of random process whose behavior allows us to completely specify probability behaviors in terms of a much smaller set of probabilities. One such special case is called a Markov process. Hence, in the remainder of this chapter we will introduce the concept of a Markov process and then we will spend several sections giving examples of such processes. In ensuing chapters we will explore properties of Markov processes in more detail and provide more examples. Finally, we will take one important class of problems, called queueing problems, and see how one can apply the theory of Markov processes to these.

9.2   Definitions

In Section 3.2 we defined a random variable as a function from the sample space to the real numbers. Now we define a *random process* as an indexed family of random variables on a probability space. Thus, "let $\{ X_t : t \in T \}$ be a random process" means that for every t in some set T, X_t is a random variable. In many applications the index "t" is interpreted as time and X_t is interpreted as "the state of the system at time t." A random process may also be called a *stochastic* process.

There are several ways that one can think intuitively about a random process. Remember from Section 3.2 that a random variable is a function whose domain is a sample space S. Thus, a random process is a function $X_t(s)$ of two variables, $s \in S$ and $t \in T$. There is nothing "random" about this function. Now if in this function (i.e., random process) we fix the sample point s, then X_t is some function of t, a real function for our purposes. The graph of this function with s fixed is called a *realization* of the random process or a *sample path*. (What one witnesses when observing a random process is part of a realization.) There is nothing "random" about this function either. Thus, one can think of a collection of functions of t (one function for each s). The random process is this collection. The "randomness" comes from the behavior due to varying the s-values.

On the other hand, if one fixes t then the two-variable function $X_t(s)$ depends only on s. It is a random variable precisely like those studied in Chapter 3. Thus, one can also think of a random process as a collection of random variables, indexed by t. There are as many random variables X_t in the collection as there are elements in T (perhaps an uncountable number). Of course if one fixes both s and t in the two-variable function $X_t(s)$, then for that s, t, $X_t(s)$ is simply some real number.

Therefore, it makes sense to ask questions such as: What is the probability that X_t at some fixed t takes the value x? For that fixed t one can answer the question precisely as in Chapter 3. One can also ask what is the probability that at the fixed values $t_1, t_2, \ldots, t_n$ the X's at these fixed t_j take the values $x_1, x_2, \ldots, x_n$? For those fixed t_j one can answer such questions precisely as in Chapter 4. Why then did we not end this book with Chapter 8? The answer is twofold. First, T may not be finite and very little that we have developed so far carries over in any obvious way to nonfinite collections of random variables, so it is reasonable to extend our discussion to these cases. Second, if we rely on the results of Chapter 4 there is an enormous amount of information that we need to know about large (possibly infinite) collections of random variables. For example, we need to know the probability distribution for each random variable X_t alone. But that is not enough if the random variables are dependent. We also need to know the joint probabilities of all pairs of random variables, if they are not pairwise independent. But that's not enough either. We will see in Section 9.3 that if we want to know the probabilistic properties of a random process we need to know the joint distributions of all possible collections of finitely many random variables. This, of course, is hopeless in any real-life application unless the random process has some simplifying structure. Thus, a second reason for our study of random processes is to study some properties of some useful and important random processes with special structures.

EXAMPLE 9.1

Let X_t be the balance in your checking account at time t, where $t = 0$ denotes the time the account was opened; then $\{X_t : t \geq 0\}$ is a random process.

EXAMPLE 9.2

Let Z_t denote the demand for electricity at time t, where $t = 0$ is midnight, July 4, 2000; then $\{Z_t : -\infty < t < \infty\}$ is a random process.

EXAMPLE 9.3

Consider a probability system with sample space $S = \{s : 0 \leq s \leq 1\}$ and $Y_i(s) = s^i$ for $s \in S$; then $\{Y_1, Y_2, \ldots\}$ is a random process.

We assume now and throughout the remainder of the text that a suitable probability system hovers in the background of any random process and is available should we need it.

9.3 A Description of a Random Process

We call T the *parameter set* of the random process. In Example 9.1, $T = \{t : 0 \leq t < \infty\}$; in 9.2, $T = \{t : -\infty < t < \infty\}$; in 9.3, $T = \{1, 2, 3, \ldots\}$. In a *continuous parameter* random process T is an interval of real numbers as in Examples 9.1 and 9.2. In a *discrete parameter* random process T is a countable set as in Example 9.3. For most of the stochastic processes in this book T is either $\{t : 0 \leq t < \infty\}$, $\{0, 1, 2, \ldots\}$, or $\{1, 2, 3, \ldots\}$.

A *state space*, E, of a random process is a set containing all possible values that the random variables can take. A state space for the random process in Example 9.1 is $E = \{\ldots, -0.02, -0.01, 0, 0.01, 0.02, \ldots, 1.00, \ldots\}$; in 9.2 E can be $\{z : 0 \leq z < \infty\}$; and in 9.3 $E = \{y : 0 \leq y \leq 1\}$. We could have used $\{x : -\infty < x < \infty\}$ as a state space for all three of Examples 9.1 to 9.3, however, the state space we selected is a more natural choice. The state space can be classified as *discrete* or *continuous* in the same way as the parameter set. Hence Example 9.1 contained the description of a discrete state, continuous parameter random process.

Thus, we can consider four types of random processes. There are those with discrete state spaces and discrete parameter spaces. There are also those with any of the other three combinations of discrete or continuous state spaces and discrete or continuous parameter spaces.

We will spend considerable time discussing certain processes with discrete state spaces and discrete parameters spaces as well as those with discrete state spaces and continuous parameter spaces. We will encounter one process with a discrete parameter space and a continuous state space (the waiting time process in Chapter 13). Special tools are required to handle the continuous state space–continuous parameter space and we will not discuss those here.

9.4   Characterizing a Random Process

To completely describe or characterize $\{X_1, \ldots, X_{10}\}$ we need the joint distribution function $F(x_1, x_2, \ldots, x_{10}) = \Pr[X_1 \leq x_1, X_2 \leq x_2, \ldots, X_{10} \leq x_{10}]$. However, if $\{X_1, \ldots, X_{10}\}$ possesses some special probabilistic structure we can specify less and still completely describe $\{X_1, \ldots, X_{10}\}$. For example, if $\{X_1, \ldots, X_{10}\}$ are independent and identically distributed, then $F(x) = \Pr[X_1 \leq x]$ completely describes $\{X_1, \ldots, X_{10}\}$ since

$$F(x_1, \ldots, x_{10}) = \Pr[X_1 \leq x_1, \ldots, X_{10} \leq x_{10}]$$
$$= \Pr[X_1 \leq x_1] \cdot \ \cdots \ \cdot \Pr[X_{10} \leq x_{10}]$$
$$= F(x_1) \cdot \ \cdots \ \cdot F(x_{10})$$

However, to characterize completely the probabilistic behavior of a general random processes $\{X_t : t \in T\}$ we need the joint distribution of $X_{t_1}, \ldots, X_{t_n}$ for every finite collection of $t_1, \ldots, t_n$ from T, i.e., we need to know

$$\Pr\left[X_{t_1} \leq x_1, \ldots, X_{t_n} \leq x_n\right]$$

for every $t_1, t_2, \ldots, t_n \in T$, $n = 1, 2, \ldots$, and for every $x_j \in E$. For example, a discrete parameter random process with $T = \{1, 2, 3, \ldots\}$ could be completely described by specifying

$$F(x_1) = \Pr[X_1 \leq x_1]$$

$$F(x_1, x_2) = \Pr[X_1 \leq x_1, X_2 \leq x_2]$$

$$\vdots$$

$$F(x_1, \ldots, x_n) = \Pr[X_1 \leq x_1, \ldots, X_n \leq x_n]$$

$$\vdots$$

which amounts to specifying an infinite number of joint distribution functions. In most of our work and in most applications the random process will have a particular structure that allows us to completely describe it with much less difficulty.

In Section 4.10 the concept of a random sequence, $\{X_1, X_2, X_3, \dots\}$, was discussed. Clearly such a sequence forms a discrete parameter random process with parameter set $T = \{1, 2, 3, \dots\}$. In Section 4.10 the important special case of *independent, identically distributed* random variables was introduced. In this case $\{X_1, X_2, \dots\}$ is completely described by $F(x) = \Pr[X_1 \le x]$, since for every $n \ge 1$

$$\Pr[X_1 \le x_1, \dots, X_n \le x_n] = \Pr[X_1 \le x_1] \cdot \dots \cdot \Pr[X_N \le x_n]$$
$$= F(x_1) \cdot \dots \cdot F(x_n)$$

A sequence of this kind can be considered as resulting from independent repetitions of the same random experiment.

We have discussed the example of telephone calls arriving for ambulance assistance several times. If $X_n = 1$ when the nth call is for an ambulance, $X_n = 0$ otherwise. In Example 4.16 we indicated that such calls had the property that X_n and X_{n+1} were independent. We did not test for independence between, say, X_n, X_{n+1}, X_{n+2}, but intuition would now lead us to believe that the general independence assumption

$$\Pr[X_1 \le j_1, X_2 \le j_2, \dots X_n \le j_n] = \Pr[X_1 \le j_1] \Pr[X_2 \le j_2] \cdot \dots \cdot \Pr[X_n \le j_n]$$

for any value of n and for each $j = 0$ or 1, might be a reasonable model for the sequence of types of telephone calls. Notice also that this independence assumption occurs rather naturally in coin-tossing experiments.

9.5   The Markov Property

An arbitrary stochastic process is usually too general to be useful in modeling a real-life system. We need to restrict ourselves to a class of stochastic processes that has enough structure with which to work yet is still flexible enough to describe a large variety of real-life systems. If we require the random variables $X_1, X_2, \dots$ to be independent and identically distributed, we have plenty of structure. Unfortunately many applications cannot, in good conscience, be modeled as independent and identically distributed random variables. For example, if X_n denotes the closing price of a share of stock on day n, X_{n+1} and X_n are obviously dependent. Similarly, in the Westgren example (Example 4.17) the number of particles in view at the nth viewing and at the $(n + 1)$st viewing are dependent. Hence, one cannot satisfactorily model this phenomenon as a sequence of independent and identically distributed random variables.

In 1907 the Russian mathematician A. A. Markov introduced a property defining a large class of random processes—a class that is flexible enough to describe many real-life systems yet possesses enough structure to be tractable. The property characterizing this class of random processes is now called the *Markov property.* The remainder of this book is devoted to the analysis of random processes satisfying the Markov property, hence, it is essential that the reader understand this property. We emphasize the Markov property for discrete state random processes, since discrete state Markov processes form a rich enough class for our purposes.

To begin, let $\{X_t : t \in T\}$ be a discrete state random process and let T be a subset of the real numbers. Thus, T might be $\{t : 0 \le t < \infty\}$, $\{1, 2, \ldots\}$, or $\{t : 28 \le t \le 54\}$. As previously mentioned, to characterize $\{X_t : t \in T\}$ we need the joint distribution function

$$\Pr\left[X_{t_1} \le x_1, \ldots, X_{t_n} \le x_n\right]$$

for every finite set $t_1, \ldots, t_n$ from T. Since $\{X_t : t \in T\}$ is a discrete state random process, we need only know the joint density function

$$\Pr\left[X_{t_1} = k_1, \ldots, X_{t_n} = k_n\right]$$

(see Section 4.9). Assume that $t_1 \le t_2 \le \cdots \le t_n$. Using equation (4.39)

$$\Pr\left[X_{t_1} = k_1, \ldots, X_{t_n} = k_n\right] = \Pr\left[X_{t_1} = k_1\right] \cdot \Pr\left[X_{t_2} = k_2 | X_{t_1} = k_1\right]$$
$$\cdot \Pr\left[X_{t_3} = k_3 | X_{t_2} = k_2, X_{t_1} = k_1\right] \cdot$$
$$\cdots \cdot \Pr\left[X_{t_n} = k_n | X_{t_{n-1}} = k_{n-1}, \ldots, X_{t_1} = k_1\right]$$

$$(9.1)$$

Note that in each of the conditional probabilities on the right-hand side of (9.1) the event is conditioned on all X_{t_i} preceding a given X_t.

DEFINITION 9.1   Let $\{X_t : t \in T\}$ be a discrete state random process and let T be a subset of the real line. $\{X_t : t \in T\}$ is said to be a *Markov process* if it possesses the following *Markov property*: For every $k_1, \ldots, k_{n+1}$ in the state space E, $t_1 \le t_2 \le \cdots \le t_{n+1}$ in T and $n = 0, 1, 2, \ldots,$

$$\Pr\left[X_{t_{n+1}} = k_{n+1} | X_{t_n} = k_n, \ldots, X_{t_1} = k_1\right] = \Pr\left[X_{t_{n+1}} = k_{n+1} | X_{t_n} = k_n\right] \quad (9.2)$$

In

$$\Pr\left[X_{t_{n+1}} = k_{n+1} | X_{t_n} = k_n, \ldots, X_{t_1} = k_1\right] \quad\quad\quad (9.3)$$

we will call the latest time that we know the exact state of the system "the present." In (9.3) we know the state of the process at times $t_1, \ldots, t_n$. Assuming $t_1 \le \cdots \le t_n$, we could call time t_n the *present*. Hence, $t_1, \ldots, t_{n-1}$ is "the past" and t_{n+1} is "the future." The Markov property states that the future is conditionally independent of the past given the present. Thus, if we know the present state, the additional knowledge of the past does not help us at all in making statements about the future in a Markov process.

A *discrete state* Markov process (satisfying Definition 9.1) is termed a *Markov chain.*

Even though we will concentrate on discrete state processes (Markov chains), the definition of the Markov property can be extended to random processes without discrete state space.

DEFINITION 9.2   Let $\{X_t : t \in T\}$ be a random process and let T be a subset of the real line. $\{X_t : t \in T\}$ is a *Markov process* if it possesses the *Markov property* that, for every $x_1, \ldots, x_{n+1}$ in the state space, $t_1 \leq t_2 \leq \cdots \leq t_{n+1}$ in T and $n = 0, 1, 2, \ldots$,

$$\Pr\left[X_{t_{n+1}} \leq x_{n+1} | X_{t_n} = x_n, \ldots, X_{t_1} = x_1\right] = \Pr\left[X_{t_{n+1}} \leq x_{n+1} | X_{t_n} = x_n\right]$$

Proposition 9.1   If $X_1, X_2, \ldots$ is a discrete state, discrete parameter random process and if for every n

$$\Pr[X_{n+1} = j | X_n = i, X_{n-1} = i_{n-1}, \ldots, X_1 = i_1] = \Pr[X_{n+1} = j | X_n = i] \quad (9.4)$$

then $\{X_1, X_2, \ldots\}$ possesses the Markov property.

EXAMPLE 9.4

Let $X_1, X_2, \ldots$ be independent Bernoulli random variables (see Section 7.2) with

$$\Pr[X_i = 1] = p_i > 0 \quad \text{and} \quad \Pr[X_i = 0] = 1 - p_i = q_i > 0$$

Define $Y_0 = 0$ and

$$Y_n = X_1 + X_2 + \cdots + X_n, \quad n = 1, 2, \ldots$$

Hence, Y_n counts the number of ones in $X_1, \ldots, X_n$. $\{Y_0, Y_1, \ldots\}$ possesses the Markov property since

$$\Pr[Y_{n+1} = k_{n+1} | Y_n = k_n, \ldots, Y_0 = 0]$$
$$= \Pr[Y_{n+1} = k_{n+1} | Y_n = k_n]$$
$$= \Pr[X_{n+1} = k_{n+1} - k_n] = \begin{cases} p_{n+1}, & \text{if } k_{n+1} - k_n = 1 \\ q_{n+1}, & \text{if } k_{n+1} - k_n = 0 \\ 0, & \text{otherwise} \end{cases}$$

EXAMPLE 9.5

We can extend the telephone example from Section 9.4, where we discussed the independence and identical distributivity of the types of telephone calls. Now let

$$Y_n = X_1 + X_2 + \cdots + X_n, \quad n = 0, 1, 2, \ldots$$

We know from Example 2.17 that

$$\Pr[X_n = 1] = .081 \quad \text{and} \quad \Pr[X_n = 0] = .919$$

Hall spent considerable time showing that, at least for his data, these probabilities do not depend on n. Thus, we have a special case of Example 9.4 and $\{Y_n : n =$

$0, 1, 2, \ldots \}$ is a Markov process. But what does this process tell us? Notice that if call j is for an ambulance $X_j = 1$, otherwise $X_j = 0$. Thus, Y_n in the total *number of telephone calls asking for ambulance assistance* up to the nth call (whatever its type) to arrive. This count of number of ambulance calls to arrive is important since it helps to determine how many ambulances must be made available to satisfy the demand for such service. But there is more here. Notice (see Section 7.3) that for this problem our Markov process Y_n is nothing more than sums of sequences of Bernoulli trials, so that we have immediately from (7.11)

$$\Pr[Y_n = j] = \binom{n}{j} p^j q^{n-j}, \qquad j = 0, 1, 2, \ldots n$$

where $p = .081$ and $q = .909$. This tells us the probability of j calls for ambulance service in the next n calls to come.

EXAMPLE 9.6

Let $X_1, X_2, \ldots$ be independent Bernoulli variables as defined in Example 9.4. Define

$$Z_0 = 1$$

$$Z_1 = 1$$

$$Z_{n+1} = Z_n + Z_{n-1} + X_n, \qquad n = 1, 2, 3, \ldots$$

Note that Z_{n+1} depends only on $X_1, X_2, \ldots, X_n$ and, thus, X_n is independent of Z_k with $k \le n$. However, $\{ Z_0, Z_1, Z_2, \ldots \}$ fails to have the Markov property. To see this, notice that

$$\Pr[Z_4 = 7 \,|\, Z_3 = 4, Z_2 = 3] = \Pr[X_3 = 0 \,|\, Z_3 = 4, Z_2 = 3] = \Pr[X_3 = 0] = q_3$$
$$(9.5)$$

since X_3 is independent of Z_2 and Z_3. But by the same argument

$$\Pr[Z_4 = 7 \,|\, Z_3 = 4, Z_2 = 2] = \Pr[X_3 = 1 \,|\, Z_3 = 4, Z_2 = 2] = \Pr[X_3 = 1] = p_3$$
$$(9.6)$$

Since in general $q_3 \ne p_3$, these two probabilities are not equal. However, if $\{ Z_0, Z_1, \ldots \}$ possessed the Markov property both (9.5) and (9.6) would have to be equal to

$$\Pr[Z_4 = 7 \,|\, Z_3 = 4]$$

Clearly this is not so.

9.6   Transition Probabilities for a Discrete Parameter Markov Chain

Let us now assume that not only is the state space discrete (a set of integers) but also that the parameter set is discrete, $T = \{0, 1, 2, 3, \ldots\}$ or $T = \{1, 2, 3, \ldots\}$.

Thus, we are dealing with a Markov chain of the form $\{X_0, X_1, \ldots\}$ or $\{X_1, X_2, \ldots\}$.

Probabilities of the form

$$\Pr[X_{n+1} = j \mid X_n = i]$$

are termed *transition probabilities*. If $\Pr[X_{n+1} = j \mid X_n = i]$ is the same for every n we say that the chain has *stationary transition probabilities* or has the *homogeneity* property. If $\{X_0, X_1, \ldots\}$ has stationary transition probabilities we call

$$p_{ij} = \Pr[X_{n+1} = j \mid X_n = i]$$

a *one-step transition probability*. Clearly it is a conditional probability.

Since, given that $X_n = i$, X_{n+1} must have *some* value, it follows that

$$\sum_j p_{ij} = 1, \qquad \text{for each } i \text{ in the state space}$$

The sum is extended over all states in the state space of the process. The reader should verify that the process $\{Y_0, Y_1, \ldots\}$ in Example 9.4 does *not* have stationary transition probabilities unless $X_1, X_2, \ldots$ are identically distributed, i.e., when $p_n = p$ and $q_n = q$ as in Example 9.5.

Frequently it is convenient to express the set of all one-step transition probabilities in matrix form, e.g.,

$$\mathbf{P} = (p_{ij}) = \begin{bmatrix} p_{00} & p_{01} & \cdots \\ p_{10} & p_{11} & \cdots \\ \vdots & \vdots & \end{bmatrix}$$

EXAMPLE 9.7

To fix some of these ideas and to anticipate our later theory let us look at a machinery breakdown example. (See Exercise 3.25.) DiMarco (1972) collected 10 years worth of data on three turbo-boiler assemblies in a power generating plant. There were 1061 breakdowns in this period. Let us number the systems 1, 2, and 3. Then let $X_n = i$ if the ith failure was due to turbo-boiler system i breaking down. Then, from Exercise 3.25, we have Table 9.1 below for the pair of random variables X_n and X_{n+1}. That is, the data of Exercise 3.25 give us the number of

TABLE 9.1
Repair Record for Three
Turbo-Boiler Systems

j / i	1	2	3
1	412	54	165
2	54	19	25
3	165	24	143

times a breakdown in system i was followed by a breakdown in system j, for i, $j = 1, 2, 3$.

In Exercise 4.23 we asked whether X_n and X_{n+1} were independent random variables. The answer is no, since our computed $\chi^2 = 46.1$ and the tabulated values in Appendix B are 9.49 or 13.28 for 95% or 99% surety with $(r - 1)(c - 1)$ = 4 degrees of freedom. (See Section 4.11.) On the basis of that analysis we are now going to *assume* that the sequence of types of breakdowns, $\{ X_n : n = 0, 1, 2, \ldots \}$, possesses the Markov property. Do be careful here. Our data analysis tells us that X_n and X_{n+1} are dependent random variables, but that analysis does not specifically tell us that X_{n+1} and X_{n-1} are independent given X_n. So our data analysis has not *proven* that we are justified in saying that $\{ X_n : n = 0, 1, 2, \ldots \}$ is a Markov process. A more detailed analysis of the raw data does provide evidence that this may be a good guess of what is going on and is likely to give us a first approximation in any case. Often a good first approximation is good enough.

Now using the DiMarco data and our assumption that this breakdown process is a Markov process we can *estimate* a matrix of transition probabilities $p_{ij} = \Pr[X_{n+1} = j | X_n = i]$. We have used the estimation procedures of Section 2.9 to help us here. The estimates are given in Table 9.2 below.

Thus, we estimate that if turbo-boiler system 1 has just broken down, then turbo-boiler system 1 will also be the next system to break down with probability .65; with probability .09 the next system to break down will be system 2; and with probability .26 the next system to break down will be system 3.

Now we can do a small computation using equation (2.8).

$$\Pr[X_{n+2} = j | X_{n+1} = k, X_n = i, X_{n-1}, \ldots, X_0] \cdot \Pr[X_{n+1} = k | X_n = i, X_{n-1}, \ldots, X_0]$$
$$= \Pr[X_{n+2} = j, X_{n+1} = k | X_n = i, X_{n-1}, \ldots, X_0]$$

That is, if someone tells us the type of breakdown for every breakdown all the way back to the initial breakdown (10 years ago for this problem), then we can determine the probability that the next breakdown is due to system k and the one after that is due to system j. But notice that by the Markov property *assumption*

$$\Pr[X_{n+2} = j | X_{n+1} = k, X_n = i, X_{n-1}, \ldots, X_0] = \Pr[X_{n+2} = j | X_{n+1} = k]$$
$$\Pr[X_{n+1} = k | X_n = i, X_{n-1}, \ldots, X_0] = \Pr[X_{n+1} = k | X_n = i]$$

TABLE 9.2
Estimated One-Step Transition Probabilities for the Turbo-Boiler Breakdown Problem

i \ j	1	2	3
1	.65	.09	.26
2	.55	.19	.26
3	.50	.07	.43

and if we further assume that this process has stationary transition probabilities (which DiMarco spent many pages trying to justify) then neither of these probabilities depend on n. Another way of stating this is to say that Table 9.2 gives us each of these probabilities for *every* $n = 0, 1, 2, \ldots$. We do not need a different table for each n.

But there is much more that we can gather from this data. Using the theorem of total probability (see equation (2.10)) we see that

$$\Pr[X_{n+2} = j | X_n = i] = \sum_{k=1}^{3} \Pr[X_{n+2} = j, X_{n+1} = k | X_n = i]$$

$$= \sum_{k=1}^{3} \Pr[X_{n+2} = j | X_{n+1} = k] \cdot \Pr[X_{n+1} = k | X_n = i]$$

$$= \sum_{k=1}^{3} p_{kj} p_{ik} = \sum_{k=1}^{3} p_{ik} p_{kj} \qquad (9.7)$$

and the terms in the sum do not depend on n. Now comes an important observation. If we call $\mathbf{P}^{(2)}$ the matrix whose elements are $p_{ij}^{(2)} = \Pr[X_{n+2} = j | X_n = i]$ for $i, j = 1, 2, 3$, then equation (9.7) is nothing more than the formula used to find the i, j th element of a matrix that is the product of two other matrices. (Recall that if $\mathbf{A}$ and $\mathbf{B}$ are two matrices and if $\mathbf{C} = \mathbf{AB}$, then $c_{ij} = \sum_k a_{ik} b_{kj}$.) Thus, we have the important result that

$$\mathbf{P}^{(2)} = \mathbf{P} \cdot \mathbf{P} = \mathbf{P}^2$$

The matrix whose elements are $\Pr[X_{n+2} = j | X_n = i]$ is simply the square of the matrix represented by Table 9.2. The elements $p_{ij}^{(2)}$ of $\mathbf{P}^{(2)}$ are termed *two-step* transition probabilities. For this problem Table 9.3 shows the following matrix for $\mathbf{P}^{(2)}$.

Thus, if we know that turbo-boiler system 3 just broke down then we would estimate that with probability .579 turbo-boiler system 1 would be the system to break down two breakdowns from now, no matter which turbo-boiler system breaks down next.

It is easy, now, to see that if $\mathbf{P}^{(3)}$ is the matrix with elements $p_{ij}^{(3)} = \Pr[X_{n+3} = j | X_n = i]$ (*three-step* transition probabilities), then $\mathbf{P}^{(3)} = \mathbf{P}^3$ and, in general, $p_{ij}^{(m)} = \Pr[X_{n+m} = j | X_n = i]$ (*m-step* transition probability) is the i, j th

TABLE 9.3
The Probabilities $\Pr[X_{n+2} = j | X_n = i]$
for the Turbo-Boiler Systems

i \ j	1	2	3
1	.602	.094	.304
2	.592	.104	.304
3	.579	.088	.333

element of the matrix $\mathbf{P}^m$ for any $m = 1, 2, \ldots$. Thus, we have computed $\mathbf{P}^5$ as

$$
\mathbf{P}^5 = \begin{bmatrix} .594 & .093 & .313 \\ .594 & .093 & .313 \\ .594 & .093 & .313 \end{bmatrix}
$$

to three decimal places. How does one interpret this matrix? If we know that turbo-boiler system 2 just went down then we would estimate that with probability .594 turbo-boiler system 1 would go down five breakdowns from now; with probability .093 turbo-boiler system 2 would go down five breakdowns from now; and with probability .313 turbo-boiler system 3 would go down five breakdowns from now.

Now there is another important observation we can make. Since $\mathbf{P}^5$ has all rows equal, correct to three decimal places,

$$
\mathbf{P}^n = \mathbf{P}^5, \qquad \text{for } n \ge 5
$$

That is, once the matrix $\mathbf{P}^m$ has all *rows* equal (as is true for $\mathbf{P}^5$ to three decimal places) then any higher power of $\mathbf{P}$ is identical to $\mathbf{P}^m$. What does this mean? Clearly, for this example, it is saying that no matter how far ahead we must estimate (beyond five breakdowns from this one) the probability of turbo-system j breaking down, given that turbo-boiler i has just broken down, we can do no better (to within three decimal places) than to use $\mathbf{P}^5$. Of course we can do better if we want to look ahead only 1, 2, 3, or 4 breakdowns from now.

But there is still more to observe. All of the rows of $\mathbf{P}^5$ are the same (not the columns). Therefore, $\Pr[X_{n+5} = j \mid X_n = i]$ does not depend on i. For $j = 1, 2, 3$ the probabilities are the same whether $i = 1$, 2, or 3. Those probabilities depend only on j. Thus, if we must look ahead five or more breakdowns from now, knowing which turbo-boiler system just broke down does not change our estimate of which turbo-boiler system will break down later.

There is still more. If we let $\pi_j^{(n)} = \Pr[X_n = j]$, then $\pi_j^{(n)}$ tells us the probability that the nth breakdown is due to turbo-boiler system j. Then using, again, the theorem of total probability (equation (2.10)) we have

$$
\pi_j^{(n)} = \sum_{i=1}^{3} \Pr[X_n = j \mid X_0 = i] \Pr[X_0 = i] \tag{9.8}
$$

If we now let $\boldsymbol{\pi}^{(n)}$ be a row vector whose elements are $\pi_j^{(n)}$, $j = 1, 2, 3$, then one recognizes (9.8) as the formula for finding the jth term in the vector that is the product of a vector and a matrix. Thus, one can write

$$
\boldsymbol{\pi}^{(n)} = \boldsymbol{\pi}^{(0)} \mathbf{P}^n
$$

where $\boldsymbol{\pi}^{(0)}$ is a vector whose elements are $\Pr[X_0 = i]$, $i = 1, 2, 3$. There is another happy property here. Since $\mathbf{P}^n$ has essentially identical rows when $n \ge 5$, the above product gives about the same $\boldsymbol{\pi}^{(n)}$ for all $n \ge 5$. In that case $\boldsymbol{\pi}^{(n)}$ does not depend on n as long as $n \ge 5$ no matter what $\boldsymbol{\pi}^{(0)}$ is. Furthermore, no matter what $n \ge 5$ is

$$
\boldsymbol{\pi}^{(n)} = (.594, .093, .313)
$$

to three decimal places, and these numbers do not depend on n or $\boldsymbol{\pi}^{(0)}$. Of course, in this example, if $n < 5$, $\boldsymbol{\pi}^{(n)}$ may depend on both n and $\boldsymbol{\pi}^{(0)}$. How shall we interpret this result? It says that the probability is .594 that turbo-boiler system 1 is the nth one to break down (if $n \geq 5$) no matter what turbo-boiler system just went down and no matter how far in the future (beyond five) we must predict.

And there is yet more, though we cannot defend the results here. $\pi_j^{(n)}$, $n \geq 5$, gives us the long-run *proportion* of all breakdowns ever to occur that were caused by turbo-boiler system j. Thus, nearly 60% (.594) of all breakdowns will be caused by turbo-boiler system 1, nearly 10% (.093) will be caused by turbo-boiler system 2, and about 30% (.313) will be caused by turbo-boiler system 3, which conforms to the proportions in any row of Table 9.3. Correspondingly $1/\pi_j^{(n)}$, $n \geq 5$, is an estimate of the mean *number* of breakdowns between breakdowns of turbo-boiler system j. This number is called the *mean lifetime* or *mean time between failures* (MTBF) of turbo-boiler system j. For our example the mean lifetime of turbo-boiler system 1 is about $\frac{5}{3}$ or $1\frac{2}{3}$ breakdowns, for turbo-boiler system 2 this mean lifetime is about 11 breakdowns, and it is about 3 for turbo-boiler system 3.

What does one conclude from all this? An obvious conclusion (which, of course, could have been made by an initial inspection of Table 9.1) is that turbo-boiler system 1 is not a very reliable system. Perhaps it should be replaced. Alternatively, since $p_{11} = .65$, each time turbo-boiler system 1 goes down it is quite likely to be the next system to go down. Perhaps it is very difficult to repair correctly the first time and maintenance procedures might need to be changed. Turbo-boiler system 2 is the best of the three. Remember that the basic data were the accumulation of 1061 breakdowns over a 10-year period. Thus, there are an average of 100 breakdowns per year and about 60 of these are due to turbo-boiler system 1. Furthermore, of the 100 breakdowns per year, "on the average" there will be one breakdown per 3.65 days, which means that turbo-boiler system 1 will go down every 5 to 6 days on the average.

This simple example illustrates many ideas that we shall discuss in detail in subsequent chapters. We shall be concerned with the following.

1. General methods of computing probabilities like $p_{ij}^{(m)}$.
2. Conditions under which $p_{ij}^{(m)}$ approaches a limiting value p_j as $m \to \infty$.
3. Conditions under which this limiting value is the same regardless of the initial state i.
4. Conditions under which the limits p_j themselves form a density function (that is, add up to 1).
5. Ways of computing the limits p_j.

9.7 The Poisson Process

The examples considered so far in this chapter have all been discrete parameter processes. Let us now use an example to introduce one continuous parameter Markov chain. Let us reconsider Example 8.5, in which we studied a process of major mine accidents. In 8.5 we showed that the times between consecutive accidents seemed to be exponentially distributed.

Instead of looking at the time between accidents, let us now consider a new, continuous parameter random process. Call it $\mathbf{N} = \{N_t : t \geq 0\}$, where for each t, N_t counts the number of accidents up to time t. To produce a model for $\mathbf{N}$ let us introduce the following assumption.

Assumption 9.1 (for a Poisson Process)

(1) $\mathbf{N}$ is a process whose sample paths proceed by jumps, where each jump is of unit magnitude, $N_0 = 0$.

(2) If N_{t+s} is the number of jumps in $\mathbf{N}$ up to time $t + s$ and N_t is the number of jumps in $\mathbf{N}$ up to t, then for any s, $t \geq 0$, $N_{t+s} - N_t$, the number of jumps in $\mathbf{N}$ (mine accidents) between time t and time $t + s$, is *independent* of N_u for $u \leq t$.

(3) For any s, $t \geq 0$ the distribution of $N_{t+s} - N_t$ may depend on s but not on t.

Are these assumptions realistic for the mine accident data? The first assumption seems to be valid, for we have no zeros in Table 8.6, which indicates that two accidents never occurred at exactly the same time. So the number of accidents increases (when it does increase) by only one accident.

The second assumption is more difficult to defend. Physically, it seems reasonable that mine accidents are independent in the sense of assumption (2). The fact that a mine accident occurred in mine 1 probably has no bearing on the occurrence or nonoccurrence of an accident in mine 2, which could be miles away, under different management, and mining different ores. So the assumption is not too farfetched.

The third assumption may depend on the total mining activity. If there were more mines operating at a particular time (say, perhaps 1880 to 1890) then there may well be more accidents during that time than at a time when a smaller number of mines were in operation. We do not have enough data here to really know how many mines were operating at any time. We could plot the average number of mine accidents per year. If that plot was a horizontal line it could indicate that at least the average number of accidents did not change over time. That would be some supporting evidence leading to assumption (3). The fact that one exponential distribution fits the data well (see Example 8.5) is a piece of evidence that the distribution of $\mathbf{N}$ satisfies (3).

The acid test of how good a model we have for $\mathbf{N}$ is to see what assumptions (1) to (3) imply about $\mathbf{N}$ and then to use our tests of goodness of fit to determine if our data really support such a model. (Recall the Daganzo Exercise 8.4.)

Thus, let us do a bit of analysis to see what assumptions (1) to (3) imply. Here is an easy implication: Let $N_0 = 0$. What is $\Pr[N_t = 0]$? Let us look at $t + s$. If $N_{t+s} = 0$ then two events must have occurred: $N_t = 0$ and $N_{t+s} - N_t = 0$. Thus,

$$\Pr[N_{t+s} = 0] = \Pr[N_t = 0] \cdot \Pr[N_{t+s} - N_t = 0] = \Pr[N_t = 0] \cdot \Pr[N_s = 0] \quad (9.9)$$

Why? By assumption (2) $N_{t+s} - N_t$ is independent of N_u for all $u \leq t$. In particular, the events $N_t = 0$ and $N_{t+s} - N_t = 0$ are independent. But by assumption (3) $\Pr[N_{t+s} - N_t = 0]$ does not depend on t. Thus, it has the same value as $\Pr[N_s - N_0 = 0] = \Pr[N_s = 0]$.

Now let $f(t) = \Pr[N_t = 0]$. Then from (9.9) we have

$$f(t + s) = f(t)f(s)$$

The only nonzero continuous function that will satisfy this equation for all $t \geq 0$ and which has the value 1 when $t = 0$ ($\Pr[N_0 = 0] = 1$) is the function

$$f(t) = \Pr[N_t = 0] = e^{-\lambda t}$$

for some $\lambda > 0$. This is a nice confirming observation. The only way $N_t = 0$ can occur is for the time between two consecutive mine accidents to be greater than t. Thus, $N_t = 0$ implies that the time between two consecutive accidents must be exponentially distributed. But we know that those intervals are exponentially distributed (with $\lambda = 1/240$) from Example 8.5. Thus, our model for **N** is not completely wrong. Other data at least support our model.

Now let us look at the event $\{N_t = n\}$. Again we consider the interval $(0, t + s]$ split into two parts $(0, t]$ and $(t, t + s]$. If there are n accidents in $t + s$ then there must be k in $(0, t]$ for *some* $k = 0, 1, 2, \ldots, n$ and $n - k$ in $(t, t + s]$. Again using assumptions (1), (2), and (3) we obtain

$$\Pr[N_{t+s} = n] = \sum_{k=0}^{n} \Pr[N_t = k]\Pr[N_s = n - k] \qquad (9.10)$$

We will make two observations to help study this probability. Remember from Section 7.8 that if X_i for $i = 1, 2, \ldots, n$ are independent, identically distributed, exponential random variables and if $Z = \sum_{i=1}^{n} X_i$, then the density function of Z is given by

$$\lambda(\lambda z)^{n-1} e^{-\lambda z}/(n-1)!, \qquad n = 1, 2, \ldots$$

which is a special case of the gamma density. Also note that if T_n is the time of the nth mine failure, then

$$T_n = T_1 + (T_2 - T_1) + \cdots + (T_n - T_{n-1})$$

That is, T_n, the time of the nth failure, can be written as the sum of n independent, identically distributed, exponential random variables where $(T_j - T_{j-1})$ is the time between the jth and $(j-1)$st failures, which we know to be exponentially distributed from Example 8.5. So the density function of T_n is given by

$$\lambda(\lambda t)^{n-1} e^{-\lambda t}/(n-1)!, \qquad n = 1, 2, \ldots$$

where λ is estimated to be $1/240$.

Now here is a very useful result: $\{T_{n+1} - T_n : n = 0, 1, 2, \ldots\}$, $T_0 = 0$, is a sequence of independent, identically distributed, exponential random variables. Let $\Pr[N_t = 0] = e^{-\lambda t}$. Choose an s "very small" (we will eventually let s go to zero). The function

$$1 - e^{-\lambda s} = \lambda s - (\lambda s)^2/2! + (\lambda s)^3/3! \mp \cdots \qquad (9.11)$$

(see Appendix A.5) has the nice property that

$$\frac{1 - e^{-\lambda s}}{s} = \lambda - \lambda^2 s/2! + \lambda^3 s^2/3! \mp \cdots$$

and for "small enough" s this is approximately λ. In more formal language, any function $g(s)$ is said to be of the *small order of s* if

$$\lim_{s \to 0} \frac{g(s)}{s} = 0$$

This property of a function is symbolized by $o(s)$. Notice that each term after the first term in (9.11) has this small order of s property. Thus, we can write

$$1 - e^{-\lambda s} = \lambda s + o(s)$$

So we have, for "small enough" s,

$$\Pr[N_s = 0] = e^{-\lambda s} = 1 - \lambda s + o(s)$$

Consider $\Pr[N_s = 1]$. The event $N_s = 1$ can occur if and only if $0 < T_1 \leq s < T_2$. The probability of exactly one mine accident in the interval $(0, s]$ is the probability of one accident occurring in $(0, s]$ and the one after this occurring after s. Thus,

$$\Pr[N_s = 1] = \Pr[0 < T_1 \leq s] \Pr[T_2 > s | T_1 \leq s]$$

But T_1 is exponentially distributed, $T_2 - T_1$ is exponentially distributed, and T_1 and $T_2 - T_1$ are independent. So using our arguments above we have

$$1 \geq \Pr[T_2 > s | T_1 \leq s] = \Pr[T_2 - T_1 > s - T_1 | T_1 \leq s]$$
$$\geq \Pr[T_2 - T_1 > s | T_1 \leq s]$$
$$= \Pr[T_2 - T_1 > s] = e^{-\lambda s} \to 1 \qquad \text{as } s \to 0$$

Thus,

$$\Pr[N_s = 1] = (1 - e^{-\lambda s}) \Pr[T_2 > s | T_1 \leq s] = \lambda s + o(s)$$

since the second term approaches 1 as $s \to 0$.

To find $\Pr[N_s = 2]$ we can use arguments similar to those that gave us $\Pr[N_s = 1]$ to show that

$$\Pr[N_s = 2] = o(s)$$

and for $k \geq 2$

$$\Pr[N_s = k] = o(s)$$

Thus, for "small" s we have

$$\Pr[N_s = 0] = 1 - \lambda s + o(s)$$
$$\Pr[N_s = 1] = \lambda s + o(s)$$
$$\Pr[N_s = k] = o(s), \qquad k \geq 2$$

The question then is what is $\Pr[N_t = k]$ where t need not be "small"? To answer this we will go back to equation (9.10).

Equation (9.10) was

$$\Pr[N_{t+s}=n]=\sum_{k=0}^{n}\Pr[N_t=k]\Pr[N_s=n-k]$$

Let s be "small." Then

$$\Pr[N_{t+s}=n]=\Pr[N_s=0]\Pr[N_t=n]+\Pr[N_s=1]\Pr[N_t=n-1]$$
$$+\sum_{m=2}^{n}\Pr[N_s=m]\Pr[N_t=n-m]$$
$$=(1-\lambda s+o(s))\Pr[N_t=n]$$
$$+(\lambda s+o(s))\Pr[N_t=n-1]$$
$$+\sum_{m=2}^{n}o(s)\Pr[N_t=n-m]$$
$$=(1-\lambda s)\Pr[N_t=n]+\lambda s\Pr[N_t=n-1]+o(s)$$

since each probability is ≤ 1. Write this as

$$\frac{\Pr[N_{t+s}=n]-\Pr[N_t=n]}{s}=-\lambda\Pr[N_t=n]+\lambda\Pr[N_t=n-1]+\frac{o(s)}{s}$$

Now let $s\to 0$ and remember your differential calculus to get

$$\frac{d}{dt}\Pr[N_t=n]=-\lambda\Pr[N_t=n]+\lambda\Pr[N_t=n-1],$$

a differential equation for the unknown function $\Pr[N_t=n]$.

Let $\Pr[N_t=n]=p_n(t)$ and rewrite the differential equation as

$$\frac{dp_n(t)}{dt}=-\lambda p_n(t)+\lambda p_{n-1}(t),\qquad n=1,2,\ldots$$

Then we have

$$p_0(t)=\Pr[N_t=0]=e^{-\lambda t}$$
$$p_1'(t)=-\lambda p_1(t)+\lambda p_0(t)$$
$$p_2'(t)=-\lambda p_2(t)+\lambda p_1(t)$$
$$\vdots$$

We can solve the equations iteratively, since we know $p_0(t)$, using elementary methods in the theory of differential equations. Each of these equations is an ordinary, linear, homogeneous, constant coefficient differential equation. Their solution is

$$p_n(t)=(\lambda t)^n e^{-\lambda t}/n!,\qquad n=0,1,2,\ldots \qquad (9.12)$$

where for the mine data $\lambda=1/240$.

So we argue that if the mine accident data satisfy assumptions (1) to (3) then the number of accidents up to time t is a Poisson-distributed random variable. The random process $\mathbf{N}=\{N_t:t\geq 0\}$ is called a *Poisson process*.

One can test whether this is a good model for the mine accident process. Simply choose t's (maybe 100 of them) in the interval from the start of the data to its end, count the number of accidents until that date, and divide that count by the total number of accidents in the whole set of data. These ratios give relative frequency estimates of $\Pr[N_t = n]$. Use a Kolmogorov test to see whether (9.12) fits the data well. If it does then we have a good (useful) model of mine accidents. Otherwise, mine accidents do not satisfy one of our assumptions. In the latter case we must build a more sophisticated model.

Here are two important theorems that we have exemplified. Indeed, our preceding discussion, if made somewhat more rigorous, would constitute a proof of these theorems.

Theorem

Any random process $\mathbf{X} = \{ X_t : t \geq 0 \}$ satisfying assumptions (1) to (3) is a Poisson process. Realizations of X are always step-functions with jumps of magnitude 1. For such processes the times between jumps are independent, identically distributed, exponential random variables.

Theorem

A random process $\mathbf{X}$ that proceeds in jumps of unit magnitude is a Poisson process satisfying assumptions (1) to (3) if and only if the time between jumps is a random process of independent, identically distributed, exponential random variables.

Exercises 9

1. Flip a coin 50 times and let $X_n = -1$ if a head appears and $X_n = +1$ if a tail appears. Plot the results of your flips as X_n versus n.

2. Repeat Exercise 1 twice and plot the three sets of flips on the same graph.

3. Identify the stochastic process for which you have three realizations in Exercises 1 and 2. Is it an independent sequence? A Markov sequence?

4. Define a new sequence for Exercise 1 by

$$Y_1 = X_1$$

$$Y_2 = \frac{X_1 + X_2}{2}$$

$$\vdots$$

$$Y_n = \frac{X_1 + X_2 + \cdots + X_n}{n}$$

$$\vdots$$

Plot Y_n versus n.

5. Repeat Exercise 4 for each sequence in Exercise 2.

6. Identify the $\{Y_n\}$ sequence. Is it an independent sequence? A Markov sequence?

7. Define new sequences for the results obtained in Exercises 1 and 2 by letting

$$Z_1 = X_1$$

$$Z_2 = \frac{X_1 + X_2}{2}$$

$$Z_3 = \frac{X_{n-1} + X_n}{2}$$

$$\vdots$$

$$Z_n = \frac{X_{n-1} + X_n}{2}, \qquad n = 2, 3, \dots, 50$$

Plot graphs of the $\{Z_n\}$ sequence for each set of data in Exercises 1 and 2.

8. Give reasons why $\{Z_n\}$ of Exercise 7 is not Markovian.

9. Consider an urn containing 50 red balls and 50 blue balls. One ball at a time is withdrawn without replacement. Let X_n denote the number of red balls remaining in the urn after the nth ball is drawn.

   (a) Does $\{X_1, X_2, \dots, X_{100}\}$ possess the Markov property?
   (b) Compute $\Pr[X_{10} = 47 | X_9 = 48]$.
   (c) Compute $\Pr[X_{20} = 4 | X_{19} = 48]$.
   (d) Compute $\Pr[X_{n+1} = j | X_n = i]$.
   (e) Does $\{X_1, X_2, \dots, X_{100}\}$ have stationary transition probabilities?

10. Consider Exercise 9, with the difference that each time a ball is withdrawn a new ball is inserted in the urn. The new ball is red with probability $\frac{1}{2}$. Now answer (a) through (e) for the new situation.

11. Suppose that

$$X_0 = \begin{cases} 1, & \text{with probability } p \\ -1, & \text{with probability } 1 - p \end{cases}$$

For $n = 0, 1, 2, \dots$

$$X_{2n} = \begin{cases} X_0, & \text{if } n \text{ is even} \\ -X_0, & \text{if } n \text{ is odd} \end{cases}$$

$$X_{2n+1} = 0$$

   (a) List all of the possible realizations for $\{X_0, X_1, X_2, \dots\}$.
   (b) Does $\{X_0, X_1, \dots\}$ possess the Markov property? Explain your answer.

12. The probability is small that we would receive more than 10 calls for ambulance service in the next 100 calls. Estimate this probability. (*Hint*: See Sections 7.5 and 7.10.)

13. Let Y and Z be independent random variables. Decide whether each of the following random processes possesses the Markov property.

   (a) $\{X_t : t \geq 0\}$ where $X_t = tY$.

   (b) $\{X_t : t \geq 0\}$ where $X_t = t + Z$.

   (c) $\{X_t : t \geq 0\}$ where $X_t = tY + Z$.

   (d) $\{X_0, X_1, \ldots\}$ where $X_n = nY$.

   (e) $\{X_0, X_1, \ldots\}$ where $X_n = n + Z$.

   (f) $\{X_0, X_1, \ldots\}$ where $X_n = nY + Z$.

   (g) $\{X_t : t \geq 0\}$ where $X_0 = 0$ and X_t is the greatest integer less than or equal to tY.

   (h) $\{X_1, X_2, \ldots\}$ where X_n is the greatest integer less than or equal to Y/n.

14. In the development of the Poisson process in Section 9.7 we asserted that $\Pr[N_s = 2] = o(s)$. Show that this is true. Also present some argument that leads you to conclude $\Pr[N_s \geq 2] = o(s)$.

15. In the development of the Poisson process in Section 9.7 we asserted (Formula 9.12) that

$$p_n(t) = \Pr[N_t = n] = (\lambda t)^n e^{-\lambda t}/n!, \qquad n = 0, 1, 2, \ldots$$

but we only proved the case for $n = 0$. Show that this result is obtained for any n. (*Hint*: An inductive proof is quite useful here.)

16. Using the results given for the Poisson process in Section 9.7 determine the following.

   (a) The probability that the second accident occurs between $t = 3$ and $t = 4$.

   (b) The probability that the second accident has not occurred by $t = 5$.

   (c) $\Pr[N_{t+s} - N_t = k | N_u, u \leq t]$. (You'll need assumptions (1) to (3) here explicitly.)

17. Return to the Broadbent problem (Exercise 8.7). Build a model to show that if one counts the number of resting periods of the aphid in $(0, t]$ then $\mathbf{N} = \{N_t : t \geq 0\}$ is a Poisson process. Here $N_t = $ the number of resting periods in $(0, t]$ for the aphid.

10 · Discrete Parameter Markov Chains

10.1 Introduction

In Chapter 9 we introduced the concept of a Markov chain as a discrete state random process possessing the Markov property. (See Definition 9.1 and the following discussion.) In this chapter we will restrict our discussion to Markov chains having a discrete parameter space $T = \{0, 1, 2, \ldots\}$ as well as a discrete state space E. Thus, we will deal with Markov chains of the form $\mathbf{X} = \{X_0, X_1, X_2, \ldots\}$.

At the end of Section 9.6 we posed five questions about the behavior of Markov chains. In this chapter we will explore the answers to these questions. We will also formalize and extend some of the computations discussed in Example 9.7. However, before we can answer the questions we must digress, for the answers to these types of questions depend very much on particular properties of the state space and the way in which its elements are interconnected by the random process. Thus, in this chapter we must go on a seeming sidetrack into a discussion of classes of states before we can understand some of the important questions about Markov chains.

10.2 Preliminary Properties

Since every discrete set can be put into a one-to-one correspondence with the integers, without any real loss in generality we can assume that the discrete state space $E = \{0, 1, 2, \ldots\}$, where this set may be finite or infinite.

The statement "$X_n = i$" can be read as "the process is in state i at the nth step." In Example 9.5 we reintroduced the problem of types of incoming telephone calls. In that usage "$X_n = i$" can be interpreted as "the nth telephone call to arrive is of type i," where $i = 1$ means the nth call to arrive was for an ambulance. Also in Example 9.5 we extended this model to the counting process $\mathbf{Y} = \{Y_0, Y_1, \ldots\}$. Notice that this Y process is a discrete parameter, discrete state process with the Markov property. We can say that "$Y_n = i$" is to be interpreted there to mean "the number of calls for ambulance assistance among the first n calls is i." In Example 9.7, "$X_n = i$" can be interpreted as "the ith system is the one causing the nth failure." Many other interpretations are possible for particular applications.

213

Now we know from Section 9.4 that to completely specify the probability structure of a discrete parameter random process we need to know

$$\Pr\left[X_{n_1} = j_1, X_{n_2} = j_2, \ldots, X_{n_k} = j_k \right]$$

for every $n_1, n_2, \ldots, n_k \in T$, $k = 0, 1, 2, \ldots$, and every $j_k \in E$. We also know that in general applications obtaining such knowledge is hopeless. But suppose $\mathbf{X}$ is a Markov chain. Then, for example, we can use equation (9.4) as follows.

$$\Pr[X_0 = j_0, X_1 = j_1, \ldots, X_k = j_k] =$$
$$\Pr[X_k = j_k | X_{k-1} = j_{k-1}, \ldots, X_1 = j_1, X_0 = j_0] \cdot$$
$$\Pr[X_{k-1} = j_{k-1} | X_{k-2} = j_{k-2}, \ldots, X_1 = j_1, X_0 = j_0] \cdot$$
$$\cdots \cdot \Pr[X_1 = j_1 | X_0 = j_0] \Pr[X_0 = j_0]$$

But since $\mathbf{X}$ is a Markov chain, each of these conditional probabilities has the form

$$\Pr[X_{n+1} = j_{n+1} | X_n = j_n, \ldots, X_1 = j_1, X_0 = j_0] = \Pr[X_{n+1} = j_{n+1} | X_n = j_n]$$

for any $n = 1, 2, \ldots$. X_0 presents a special problem that we will pick up later. The important point to observe, however, is that every finite-dimensional density function can be obtained if we know the transition probabilities $\Pr[X_{n+1} = j_{n+1} | X_n = j_n]$ and the initial state probability $\Pr[X_0 = j_0]$. For, by multiplication as above, we can obtain the joint density for any sequence $X_0, X_1, \ldots, X_k$ and by summation we can find the joint density for any finite collection, not just $X_0, X_1, \ldots, X_k$. (See Section 4.9, especially equations (4.36) and (4.37).) Furthermore, by summing these marginal densities (see Section 4.6) we can obtain the distribution functions for all finite collections of random variables in $\mathbf{X}$. Thus, according to our Definition 9.1, we completely know the probability structure of any Markov chain once we know the transition probabilities and the initial state probabilities. It is this property (plus, perhaps, one more below) that makes Markov chains so useful. There is some hope of determining the probabilities necessary to determine everything needed to satisfy Definition 9.1 simply by examining this much smaller set of probabilities.

But even now we have complications. The probabilities, $\Pr[X_{n+1} = j_{n+1} | X_n = j_n]$, depend on which pair of random variables (X_{n+1}, X_n) we are talking about in a (possibly) countably infinite collection of random variables $\mathbf{X}$. In some applications these conditional probabilities have a nice structure as a function of n. Perhaps they decrease monotonically with increasing n. Nonetheless, the point is that we have reduced one difficult problem to a considerably less difficult problem.

Now we make one more assumption that is quite common. *We assume* $\Pr[X_{n+1} = j_{n+1} | X_n]$ *does not depend on n.* In a considerable number of applications this is a reasonable assumption and we will make it throughout our discussion because it is useful and makes many messy calculations easy and unmessy. The assumption is called *the assumption of time homogeneity* or, if there is no confusion, simply *the homogeneity property*.

Then for Markov chains with the homogeneity property we have

$$\Pr[X_{k+1}=j|X_k=i] = \Pr[X_{n+1}=j|X_n=i]$$

for every k, $n \in T$. Let us denote this probability by p_{ij}, that is,

$$p_{ij} = \Pr[X_{n+1}=j|X_n=i]$$

for every i, $j \in E$, and has the same value for every $n \geq 0$.

To completely define the probability structure of a Markov chain with the homogeneity property we need only know the set of *one-step transition probabilities* p_{ij}. In addition, we still need to know $\Pr[X_0=i]$ for every $i \in E$.

Unless explicitly stated otherwise we will assume that when we talk about Markov chains we mean Markov chains with the homogeneity property.

In addition to the one-step transition probabilities we must also know the *initial state probabilities*

$$p_j = \Pr[X_0=j]$$

for every $j \in E$. In applications these quantities p_{ij} and p_j must be either given, determined from theoretical considerations, or estimated from data.

Now here is a nice space saver that we first saw in Example 9.7. Put p_{ij} into a matrix, with i designating the *row* of the matrix and j designating the *column* of the matrix. Denote this matrix of one-step transition probabilities by

$$\mathbf{P} = \begin{bmatrix} p_{00} & p_{01} & p_{02} & \cdots \\ p_{10} & p_{11} & p_{12} & \cdots \\ p_{20} & p_{21} & p_{22} & \cdots \\ \vdots & \vdots & & \vdots \end{bmatrix}$$

Also define a row vector

$$\mathbf{p}^{(0)} = (p_0, p_1, p_2 \cdots)$$

as the *vector of initial probabilities*. The number of entries in $\mathbf{P}$ and $\mathbf{p}^{(0)}$ can be either finite or infinite, depending on whether the state space E is finite or infinite.

From our previous discussion, we need two pieces of information to completely specify the probability structure of a discrete parameter Markov chain with the homogeneity property: the matrix of one-step transition probabilities and the vector of initial probabilities.

10.3 Examples

Some examples of Markov chains were given in Chapter 9. A few others follow.

EXAMPLE 10.1

Two players, A and B, play a game that consists of a sequence of independent trials for stakes of $1 per trial. Let the probability of A winning any given trial be

p and the probability of B winning be q. If tie games are possibly allowed then $p + q \leq 1$, and the probability of a tie is $r = 1 - p - q$. In a tie no money changes hands.

Suppose that A starts with k dollars, B starts with $N - k$ dollars, and thus a total of N dollars are available for play. We are interested in the behavior of A's fortunes.

Let X_n = the amount of money that A has after n trials ($X_0 = k$). Clearly if $X_n = x$, then X_{n+1} must equal $x + 1$, $x - 1$, or x, with probabilities p, q, and r, regardless of the values of X_i for $i < n$.

Thus $\{ X_n \}$ forms a Markov chain with one-step transition probabilities given by

$$P_{x, x+1} = \Pr[X_{n+1} = x + 1 \mid X_n = x] = p$$

$$P_{x, x-1} = \Pr[X_{n+1} = x - 1 \mid X_n = x] = q$$

$$P_{x, x} = \Pr[X_{n+1} = x \mid X_n = x] = r$$

for $0 < x < N$, $p_{00} = p_{NN} = 1$, and all other $p_{ij} = 0$. Clearly the game terminates if A's winnings ever reach 0 or N.

The initial probabilities are given by

$$p_k = \Pr[X_0 = k] = 1$$

$$p_i = \Pr[X_0 = i] = 0, \qquad \text{for } i \neq k$$

The matrix **P** of one-step transition probabilities has the following form.

	X_{n+1}:	0	1	2	3	$\ldots$	$N-1$	N
(state now) X_n								
0		1	0	0	0	$\ldots$	0	0
1		q	r	p	0	$\ldots$	0	0
2		0	q	r	p	$\ldots$	0	0
$\vdots$		$\vdots$	$\vdots$	$\vdots$	$\vdots$		$\vdots$	$\vdots$
$N-1$		0	0	0	0	$\ldots$	r	p
N		0	0	0	0	$\ldots$	0	1

(state next play) over the columns; **P** = labels the matrix.

The initial probability vector is

$$\mathbf{p}^0 = (0, 0, \ldots, 0, 1, 0, \ldots, 0)$$

where the 1 occupies the kth position (counting the initial term as the zeroth position).

EXAMPLE 10.2

In a certain production process items pass through three manufacturing stages. At the end of each stage the items are inspected and are either junked, sent through the stage again for reworking, or passed on to the next stage. Suppose that at each stage the probability is p that the item will be junked, q that it will be reworked, and r that it will be passed on, $p + q + r = 1$. The process has five states that can be indexed as follows.

> 0: item junked
> 1: item completed
> 2: item in first manufacturing stage
> 3: item in second manufacturing stage
> 4: item in third manufacturing stage

Let us assume that the process can be modeled by a Markov chain $X_0, X_1, X_2, \ldots,$ in which X_n gives the state of the item after n inspection points. Thus, $X_4 = 3$ means that the item is in the second manufacturing stage after four inspections. Clearly once an item is junked or is completed it cannot shift into another state. The transition matrix is

$$\mathbf{P} = \begin{pmatrix} 1 & 0 & 0 & 0 & 0 \\ 0 & 1 & 0 & 0 & 0 \\ p & 0 & q & r & 0 \\ p & 0 & 0 & q & r \\ p & r & 0 & 0 & q \end{pmatrix}$$

If we assume that the item begins in the first manufacturing stage the initial probability vector will be

$$\mathbf{p}^{(0)} = (0, 0, 1, 0, 0)$$

EXAMPLE 10.3

A two-state process has been used as a simple model for weather changes. If the weather is classified as State 0: *dry* or State 1: *wet*, then the transition matrix might, for example, have the form

$$\mathbf{P} = \begin{pmatrix} \frac{1}{4} & \frac{3}{4} \\ \frac{1}{2} & \frac{1}{2} \end{pmatrix}$$

The actual entries would have to be estimated by using meteorological records. Thus, if an examination of these records showed that out of 1000 days 600 were dry and that 150 of the 600 dry days were followed by other dry days, we would estimate p_{00} by $150/600 = \frac{1}{4}$.

EXAMPLE 10.4

A Markov model has been used to study the behavior of satellite vehicles. Here we suppose that at a sequence of points in time a vehicle reentering from space is given corrective signals. These signals are successful or not. The transition probability matrix can be structured as in Example 10.3, where the states are now 0 for a "Success" and 1 for a "Failure."

EXAMPLE 10.5

Suppose that in some establishment customers are served for one hour, beginning on the hour, and at most one customer can be served. Let X_n denote the number of customers in the system just before the beginning of the nth hour and let Y_n denote the number of arrivals during the nth hour. Then the number of customers in the system just before the beginning of the $(n+1)$st hour is

$$X_{n+1} = (X_n - 1)^+ + Y_n$$

where

$$(X_n - 1)^+ = \begin{cases} 0, & \text{if } X_n - 1 \leq 0 \\ X_n - 1, & \text{if } X_n - 1 > 0 \end{cases}$$

Assume that $Y_1, Y_2, \ldots$ are independent and identically distributed random variables

$$\Pr[Y_n = k] = q_k$$

Then $\{X_1, X_2, \ldots\}$ is a Markov chain with transition matrix

	0	1	2	3	$\cdots$		
0	q_0	q_1	q_2	q_3	$\cdots$		
1	q_0	q_1	q_2	q_3	$\cdots$		
2	0	q_0	q_1	q_2	q_3	$\cdots$	
3	0	0	q_0	q_1	q_2	$\cdots$	
$\vdots$	$\vdots$	$\vdots$	$\vdots$				

EXAMPLE 10.6

An interesting two-state Markov process arises in learning theory. Here we suppose that at the nth trial either a correct response or an incorrect response is given. Which response occurs depends only on the previous response. Learning theory attempts to study the pattern of (correct, incorrect) responses.

10.4   *n*-step Transition Probabilities

In most uses one is not content with knowing the one-step transition probabilities. In addition, one needs to find probabilities such as

$$\Pr[X_n = j \mid X_0 = i]$$

or

$$\Pr[X_{n+m} = j \mid X_m = i]$$

for any n, $m \in T$. By our homogeneity assumption the second probability here does not depend on m and is the same as the first. Therefore, we might just as well set $m = 0$.

In the case $n = 2$, the probability

$$p_{ij}^{(2)} = \Pr[X_2 = j \mid X_0 = i]$$

is termed a *two-step transition probability*. The question is how does one determine such a probability? Remember all we have to work with are the one-step transition probabilities and the initial state probabilities. We have seen how to compute these two-step transition probabilities in the special case of Example 9.7. In general we reason as follows:

Consider a path that the process might take to get from state i to state j in two steps (see Figure 10.1). Suppose that k is the state (possibly even i or j) that the process visits on the first step. Then, using the Markov property for this path, the probability of going from i to k to j in exactly two steps is

$$p_{ik}p_{kj} = \Pr[X_1 = k \mid X_0 = i]\Pr[X_2 = j \mid X_1 = k] = \Pr[X_2 = j;\ X_1 = k \mid X_0 = i]$$

More generally, since the chain has the homogeneity property, this is the probability of going from i to k to j in any two consecutive steps (not necessarily the first two).

Since in the first step the process must visit *some* state k, we observe that (see Figure 10.1)

$$p_{ij}^{(2)} = \sum_k p_{ik}p_{kj} \tag{10.1}$$

In general, the quantities

$$p_{ij}^{(n)} = \Pr[X_n = j \mid X_0 = i] = \Pr[X_{n+m} = j \mid X_m = i]$$

are termed *n-step transition probabilities*.

Let us denote the matrix whose entries are $p_{ij}^{(2)}$ by $\mathbf{P}^{(2)}$ and, more generally, let us denote the matrix whose entries are $p_{ij}^{(n)}$ by $\mathbf{P}^{(n)}$. By using the well-known formula for computing the elements of a product of two given matrices, we observe that (10.1) is the formula for determining the elements of the product matrix. Formula (10.1) can be written in matrix form as

$$\mathbf{P}^{(2)} = \mathbf{P} \cdot \mathbf{P} = \mathbf{P}^2 \tag{10.2}$$

where $\mathbf{P}$ is the matrix with entries p_{ij}. That is, the matrix of two-step transition probabilities is formed by squaring the matrix of one-step transition probabilities. It easily follows that, in general,

$$\mathbf{P}^{(n)} = \mathbf{P}^n \tag{10.3}$$

that is, the matrix of n-step transition probabilities can be formed by multiplying

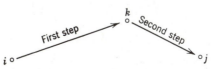

FIGURE 10.1 A possible path from state i to state j in two steps.

the matrix of one-step transition probabilities by itself n times. Thus, the problem of finding n-step transition probabilities reduces to the determination of powers of a given matrix.

From the matrix relation $\mathbf{P}^{n+m} = \mathbf{P}^n \mathbf{P}^m$, we can deduce also an important relation about the elements of $\mathbf{P}^{(n)}$:

$$\mathbf{P}^{(n+m)} = \mathbf{P}^{(n)} \mathbf{P}^{(m)} \tag{10.4}$$

or, in terms of the transition probabilities,

$$p_{ij}^{(n+m)} = \sum_k p_{ik}^{(n)} p_{kj}^{(m)} \tag{10.5}$$

If we take $p_{ik}^{(0)} = 0$ for $i \neq k$ and $p_{kk}^{(0)} = 1$, that is, if we define $\mathbf{P}^{(0)}$ to be the identity matrix $\mathbf{I}$, (10.5) holds true for all integers m, $n \geq 0$. Equation (10.4), or its equivalent form (10.5), is termed the *Chapman–Kolmogorov Equation* and is one of the basic relations of Markov process theory.

One interpretation of (10.5) is as follows. To go from state i to state j in $(n+m)$ steps, one must go from i to some intermediate state k in n steps, and then proceed to j from k in m steps. Because of the Markov property, the probability of going from i to k in n steps is simply $p_{ik}^{(n)}$. Likewise, because of the Markov property, the probability of going from k to j does not depend on which state i formed the "first" state. Hence the probability of going from k to j in m steps is $p_{kj}^{(m)}$. Furthermore, the events "go from i to k in n steps" and "go from k to j in m steps" are independent. Hence the probability of going from i to j in $(n+m)$ steps by following the path i, k, j is $p_{ik}^{(n)} p_{kj}^{(m)}$. The probability of going from i to j no matter what path is followed is then $\sum_k p_{ik}^{(n)} p_{kj}^{(m)}$ and, of course, this is $p_{ij}^{(n+m)}$.

EXAMPLE 10.3 (continued)

For Example 10.3 we have

$$\mathbf{P} = \begin{pmatrix} \dfrac{3}{4} & \dfrac{1}{4} \\[2mm] \dfrac{1}{2} & \dfrac{1}{2} \end{pmatrix}$$

Then by direct multiplication we have

$$\mathbf{P}^{(3)} = \mathbf{P}^3 = \begin{pmatrix} \dfrac{43}{64} & \dfrac{21}{64} \\[2mm] \dfrac{21}{32} & \dfrac{11}{32} \end{pmatrix}$$

$$\mathbf{P}^{(5)} = \mathbf{P}^2 \mathbf{P}^3 = \begin{pmatrix} \dfrac{683}{1024} & \dfrac{341}{1024} \\[2mm] \dfrac{341}{512} & \dfrac{171}{512} \end{pmatrix} = \begin{pmatrix} .6689 & .3330 \\ .6660 & .3339 \end{pmatrix}$$

$$\mathbf{P}^{(10)} = \mathbf{P}^5 \mathbf{P}^5 = \begin{pmatrix} .666 & .333 \\ .666 & .333 \end{pmatrix}$$

Thus, if it is dry today ($X_0 = 0$) we would say that the probability of rain tomorrow is $\frac{1}{4}$, ($\Pr[X_1 = 1 | X_0 = 0] = \frac{1}{4}$). On the other hand, if it is dry today, then the probability of rain 10 days from now is .333.

EXAMPLE 10.7

Suppose that we have a piece of equipment that can be in one of three conditions: (a) running, (b) under repair, or (c) idle—awaiting more work. We observe this equipment only when it changes states. Let $X_n = 0$ if the nth change of state puts it into a running condition, $X_n = 1$ if the nth change of state puts it into a repair condition, and $X_n = 2$ if the nth change of state puts it into an idle condition. One possible assumption is that if the equipment is down for repair, the next change of state must be to a running condition. (If no work is waiting, the repair crew is taken off the job, so that no transition from under repair to idle is possible.) If the equipment is idle, the next change of state has it in a running condition. But, if it is running, it can change to an idle condition or it can break down and be placed under repair. For simplicity, suppose that the last two possibilities are equally likely. It is clear that $\{ X_n \}$ is a Markov chain and that

$$\mathbf{P} = \begin{pmatrix} 0 & \frac{1}{2} & \frac{1}{2} \\ 1 & 0 & 0 \\ 1 & 0 & 0 \end{pmatrix}$$

From this we find that

$$\mathbf{P}^{(2)} = \begin{pmatrix} 1 & 0 & 0 \\ 0 & \frac{1}{2} & \frac{1}{2} \\ 0 & \frac{1}{2} & \frac{1}{2} \end{pmatrix}$$

$$\mathbf{P}^{(3)} = \begin{pmatrix} 0 & \frac{1}{2} & \frac{1}{2} \\ 1 & 0 & 0 \\ 1 & 0 & 0 \end{pmatrix}$$

and we observe, in general, that

$$\mathbf{P}^{(2n)} = \begin{pmatrix} 1 & 0 & 0 \\ 0 & \frac{1}{2} & \frac{1}{2} \\ 0 & \frac{1}{2} & \frac{1}{2} \end{pmatrix}, \qquad n = 1, 2, \ldots$$

$$\mathbf{P}^{(2n-1)} = \begin{pmatrix} 0 & \frac{1}{2} & \frac{1}{2} \\ 1 & 0 & 0 \\ 1 & 0 & 0 \end{pmatrix}, \qquad n = 1, 2, \ldots$$

We notice that this chain exhibits rather peculiar behavior. If we start in state 0, we return to state 0 only after an even number of steps. We will discuss such behavior in Section 10.5.

EXAMPLE 10.8

In Example 9.5 we looked at the random process X where each X_n was 1 if the nth call was for ambulance assistance and $X_n = 0$ otherwise. We had

$$\mathbf{P} = \begin{bmatrix} .919 & .081 \\ .919 & .081 \end{bmatrix}$$

One-step transition matrices with a structure in which each row is identical to every other row have an interesting property. Notice, for example,

$$p_{00}^{(2)} = .919 \times .919 + .081 \times .919$$
$$= .919 = p_{00}^{(1)}$$

In general, for such matrices

$$\mathbf{P}^n = \mathbf{P}$$

We will comment in more detail on such behavior later. It is telling us a great deal about the random process.

EXAMPLE 10.9

In Example 9.5 we continued the above example and looked at Y, where each Y_n gave the number of ambulance calls received in the first n calls. Since each random variable, Y_n, can take values from 0 to n and since n can be arbitrarily large, it is useful to take the state space E here as the infinite set $\{0, 1, 2, \dots\}$. Then

$$\mathbf{P} = \begin{bmatrix} .919 & .081 & 0 & 0 & 0 & \cdots \\ 0 & .919 & .081 & 0 & 0 & \cdots \\ 0 & 0 & .919 & .081 & 0 & \cdots \\ 0 & 0 & 0 & \ddots & \ddots & \\ \vdots & \vdots & \vdots & & & \end{bmatrix}$$

So $\mathbf{P}$ is an infinite-dimensional matrix with .919 on the major diagonal representing $\Pr[Y_n = i \mid Y_{n-1} = i]$. Physically this means that given that i calls have occurred for ambulance service in the first $n - 1$ calls, then the probability that i will occur in the first n calls is just the probability that the next call (the nth one) is not for an ambulance and this probability is .919. By a similar argument one finds that the first super-diagonal element is .081 and all other entries are 0.

   Note

(a) $p_{00}^{(2)} = (.919)^2$ and $p_{00}^{(n)} = (.919)^n$.

(b) $p_{01}^{(2)} = (.919)(.081) + (.081)(.919) = 2(.081)(.919)$.

(c) $p_{02}^{(2)} = (.081)^2$.

More generally, as is seen for $n = 2$,

$$p_{0j}^{(n)} = \frac{n!}{j!(n-j)!}(.081)^j(.919)^{n-j}, \qquad \text{for each } j$$

In Section 7.3 we saw this same binomial density function. It is easy to see then that for each fixed j, as $n \to \infty$,

$$p_{0j}^{(n)} \to 0$$

since $(.919)^{n-j} \to 0$. More generally one can show $p_{ij}^{(n)} \to 0$ in this case. That is, as $n \to \infty$, $\mathbf{P}^n$ approaches a matrix all of whose elements are 0. What this is saying is that for large n the probability of having exactly j calls for ambulance assistance approaches 0 for all fixed j. We will briefly comment on this behavior later. Certainly it is reasonable, since as $n \to \infty$ clearly the demand eventually will *exceed* any fixed value of j.

10.5 *n*th-step State Probabilities

Just as we often want to know the conditional probability

$$\Pr[X_n = j \mid X_0 = i]$$

so too we often want to know the unconditional probability

$$\Pr[X_n = j]$$

This latter probability is called an *n*th *step state probability* and is denoted by $p_j^{(n)}$.

Let us define $\mathbf{p}^{(n)} = (p_0^{(n)}, p_1^{(n)}, \dots)$ as the *vector of state probabilities after n transitions*. The element $p_k^{(1)}$ of the state vector $\mathbf{p}^{(1)}$ is the *probability of being in state k after 1 step*. It follows that

$$p_k^{(1)} = \Pr[X_1 = k] = \sum_j \Pr[X_0 = j]\Pr[X_1 = k \mid X_0 = j] = \sum_j p_j p_{jk} \quad (10.6)$$

In words, the left-hand side is the probability of being in state k after 1 transition. The right-hand side can be interpreted simply as the probability of being in some state j to start, and in one step moving from j to k. Since we can start in any state j (with probability $p_j = \Pr[X_0 = j]$), we must add all the ways that we can be in state k, regardless of the initial state, to determine $p_k^{(1)}$.

A more formal derivation of (10.6) comes by noting that

$$\Pr[X_1 = k, X_0 = j] = \Pr[X_0 = j]\Pr[X_1 = k \mid X_0 = j]$$

by the definition of conditional probabilities. The left-hand side is the joint probability of being in state k at time 1 (after 1 step) and being in state j at time 0 (to start). The marginal probability $\Pr[X_1 = k]$ can then be obtained from this joint probability as

$$\Pr[X_1 = k] = \sum_j \Pr[X_1 = k, X_0 = j] = \sum_j \Pr[X_0 = j]\Pr[X_1 = k \mid X_0 = j]$$

Formula (10.6) can be written in vector and matrix form

$$\mathbf{p}^{(1)} = \mathbf{p}^{(0)}\mathbf{P}$$

where the elements of the vector $\mathbf{p}^{(1)}$ are the $p_k^{(1)}$ in (10.6).

If we continue this process for $n = 2$, it follows that

$$p_k^{(2)} = \Pr[X_2 = k] = \sum_j p_j^{(1)} p_{jk}$$

or, in vector and matrix form, that

$$\mathbf{p}^{(2)} = \mathbf{p}^{(1)} \mathbf{P} \tag{10.7}$$

By induction it follows easily that

$$\mathbf{p}^{(n)} = \mathbf{p}^{(n-1)} \mathbf{P}$$

Thus, the components of the vector $\mathbf{p}^{(n)}$ satisfy the relation

$$p_j^{(n)} = \sum_k p_k^{(n-1)} p_{kj} \tag{10.8}$$

We also observe that, if (10.6) is introduced into (10.7) for $\mathbf{p}^{(1)}$, we then have

$$\mathbf{p}^{(2)} = \mathbf{p}^{(0)} \mathbf{P}^2$$

Here $\mathbf{P}^2 = \mathbf{P} \cdot \mathbf{P}$ is the square of the matrix $\mathbf{P}$. In general

$$\mathbf{p}^{(n)} = \mathbf{p}^{(0)} \mathbf{P}^n, \qquad n = 1, 2, 3, \ldots \tag{10.9}$$

Thus, we notice that $\mathbf{p}^{(n)}$, the vector of state probabilities after n steps, is easily found in terms of the vector of initial conditions $\mathbf{p}^{(0)}$ and the nth power of the matrix of one-step transition probabilities.

Observe the distinction between $p_{ij}^{(n)}$ and $p_j^{(n)}$. The former is a conditional probability and the latter is a marginal probability. That is, $p_{ij}^{(n)}$ is the probability of being in state j *given* that the process was in state i, n steps previously. On the other hand, $p_j^{(n)}$ is just the probability of being in state j after n steps, regardless of the initial state. Formula (10.9), which can be written in component form as

$$p_j^{(n)} = \sum_k p_k p_{kj}^{(n)}$$

can be interpreted as follows. The probability of being in state j after n steps is the weighted average of the probabilities of going to j in n steps, given that the process was in k at the start, weighted by the probability of being in state k to start.

EXAMPLE 10.3   (continued)

In Example 10.3 (continued) we saw how $\mathbf{P}^n$ behaves for several values of n. Now suppose tomorrow is day 0 and the probabilities for $n = 0$ are given by

$$\mathbf{p}^{(0)} = (3/4, 1/4)$$

Then

$$\mathbf{p}^{(1)} = (3/4, 1/4) \begin{bmatrix} 3/4 & 1/4 \\ 1/2 & 1/2 \end{bmatrix}$$

$$= (.688, .312)$$

$$\mathbf{p}^{(3)} = (3/4, 1/4) \begin{bmatrix} 43/64 & 21/64 \\ 21/32 & 11/32 \end{bmatrix}$$

$$= (.668, .332)$$

$$\mathbf{p}^{(10)} = (3/4, 1/4) \begin{bmatrix} .666 & .333 \\ .666 & .333 \end{bmatrix}$$

$$= (.666, .333)$$

(We have done some rounding here to get everything to three decimal places.) So whether it is rainy or dry tomorrow, it will be dry 3 days after tomorrow (4 days from now) with probability .668 and 10 days after tomorrow the probability of rain is .333. Now recall from our discussion of Example 10.8 that once a matrix $\mathbf{P}^n$ achieves identical rows then further powers do not change it. Notice that $\mathbf{P}^{10}$ in the above example has identical rows. Thus, for all $n \geq 10$, $\mathbf{P}^n = \mathbf{P}^{10}$. And because $\mathbf{p}^{(0)}$ doesn't depend on n we have

$$\mathbf{p}^{(n)} = (.666, .333)$$

for all $n \geq 10$.

Note also that this result will be the same for *any* initial probability vector $\mathbf{p}^{(0)} = (p_0, p_1)$, not just for $(3/4, 1/4)$. All that is needed is that $p_0 + p_1 = 1$ and $\mathbf{p}^{(n)} = \mathbf{p}^{(0)} \mathbf{P}^n = (.666, .333)$. This implies that if we try to predict the weather using this Markov model we are led to predict the probability of rain to be .333 whenever we are predicting more than 9 days into the future, regardless of which day after 9 days we are trying to predict and regardless of any assumption about the initial probabilities.

EXAMPLE 10.8   (continued)

Recall that in Example 10.8, $\mathbf{P}$ had identical rows. So $\mathbf{P}^n = \mathbf{P}$ for all $n \geq 1$. Then using the same argument as in the preceding example we can argue that

$$\mathbf{p}^{(n)} = (.919, .081)$$

for every $n \geq 1$ and for any $\mathbf{p}^{(0)}$. Thus, the probability of a call for ambulance service is .081 no matter which call in the future we want to consider (which X_n) and no matter what type of phone call we just received (X_0). But, of course, this analysis only supports our earlier analysis that $\mathbf{X}$ is a sequence of independent, identically distributed random variables. What is useful to observe, however, is that

$$\mathbf{p}^{(n)} = \mathbf{p}^{(0)} \mathbf{P}^n = \mathbf{p}^{(0)} \mathbf{P}$$

implies that $\mathbf{p}^{(n)}$ does not depend on n. Thus, if we drop the superscripts on $\mathbf{p}$ we have

$$\mathbf{p} = \mathbf{p}\mathbf{P}$$

for every n. A vector $\mathbf{p}$ satisfying this equation is called a *stationary vector*. We will have much to say about Markov chains and stationary vectors later.

EXAMPLE 10.7   (continued)

The interesting consequence of Example 10.7 is that if

$$\mathbf{p}^{(0)} = (a, b, c)$$

where $a + b + c = 1$, then direct calculation using the matrices $\mathbf{P}^{2n}$ and $\mathbf{P}^{2n-1}$ in that example gives

(i) $\mathbf{p}^{(2n)} = (a, (b + c)/2, (b + c)/2)$

(ii) $\mathbf{p}^{(2n-1)} = (b + c, a/2, a/2)$

for all n. Thus, for state 0, $\Pr[X_n = 0] = a$ if n is even and $b + c$ if n is odd. The sequence $p_0^{(1)}, p_0^{(2)}, p_0^{(3)}, \ldots$ is an alternating sequence, $b + c, a, b + c, a, \ldots$. We know from first principles that such an alternating sequence does not in general have a limit. Thus, if we were interested in

$$\lim_{n \to \infty} p_0^{(n)}$$

we would have to conclude that no such limit exists.

10.6   Classification of States

It frequently happens that when we are studying a system its response is somewhat erratic early in its life due to start-up conditions. However, after some time it settles down to a response with some stability. We say it has achieved a steady state or it has reached equilibrium. This also occurs in randomly behaving systems, in which *probabilities* may exhibit an erratic behavior to start with but eventually these probabilities reach a steady state or an equilibrium behavior. Since we mostly deal with these equilibrium systems, it is important to know how the system behaves in equilibrium. Almost always when one uses terms such as "equilibrium" what is meant is that one wishes to know how the system might operate for large values of the parameter. In most engineering studies one asks how the system operates after it has been broken in or after it has seen a long "time." In systems that behave randomly we can also ask how the probabilities of the system behave after a long time. Mathematically we model the behavior of the system as $n \to \infty$ or $t \to \infty$ to represent "a long time." In this section we want to study the probability behavior of Markov chains as $n \to \infty$.

Be careful here. In systems with random behavior, which we are going to examine, such limiting quantities are $\lim_{n \to \infty} \Pr[X_n = j]$. That is, we are interested in limiting ("long run") behaviors of *probabilities*. Physically, the system itself may behave quite erratically even if the probabilities behave quite nicely in the limit. We are discussing the behavior of the *probability* of the system in the long run not the behavior of the system in the long run. The distinction is important.

We saw in Examples 10.3, 10.7, 10.8, and 10.9 that limiting probabilities can be quite odd. In Example 10.7 such limits do not even exist. In Example 10.8 such limits exist, are probabilities, and are the same as $\mathbf{p}^{(0)}$. In Example 10.9 the limits exist but are zero for every state. In Example 10.3 the limits exist, are probabilities, and $\mathbf{p}^{(n)}$ converges to them. These different behaviors are caused by different *types* of states. So to understand how the limiting probabilities behave we must first establish a classification of the states.

To classify the states of a Markov chain, we first need to define terms such as *recurrent* (Examples 10.3 and 10.8) and *transient* (Example 10.9), as well as the terms *periodic* (Example 10.7) and *aperiodic*. We will define these terms and then give techniques that simplify the classification of states.

Roughly, state j is *recurrent* if once the Markov chain visits state j, it will visit state j again in the future with probability one. *Transient* implies that the Markov chain may not return to state j in the future.

To define these terms precisely, let the random variable T_j be the "time" (i.e., the trial number) of the first visit to state j after time zero, unless state j is never visited in which case we set $T_j = \infty$. Thus, if $T_j = 28$, then $X_1 \neq j, X_2 \neq j, \ldots, X_{27} \neq j$ but $X_{28} = j$. T_j is a discrete random variable taking values in $\{1, 2, 3, \ldots, \infty\}$. Let us compute the probability density function of T_j conditioned on X_0. Define

$$f_{ij}^{(k)} = \Pr\left[T_j = k \,|\, X_0 = i\right]$$

It is simple to compute $f_{ij}^{(1)}$, since

$$f_{ij}^{(1)} = p_{ij}. \tag{10.10}$$

To compute $f_{ij}^{(k)}$ for $k > 1$, note that if the Markov chain goes from i to j in k steps, the first step must take the chain from i to some state l where $l \neq j$ (or else we would be in j for the first time on the first step not the kth). Now after that first step to l we have $k - 1$ steps left and the chain must get to j, from l, on the last of those steps. That is, the first visit to j must occur on the $(k - 1)$st step starting now in state l. Thus, we must have

$$f_{ij}^{(k)} = \sum_{l \neq j} p_{il} f_{lj}^{(k-1)}, \qquad k = 2, 3, 4, \ldots \tag{10.11}$$

Equation (10.11) provides a recursive scheme for determining $f_{ij}^{(k)}$. From (10.10) we can determine $f_{ij}^{(1)}$. From $f_{ij}^{(1)}$ and (10.11) we can determine $f_{ij}^{(2)}$. From $f_{ij}^{(2)}$ and (10.11) we can determine $f_{ij}^{(3)}$, and so on.

There is an easier way to compute $f_{ij}^{(k)}$ if you know how to handle matrices. First, pick a state j that you are interested in studying. Now fix that j from now on. Since you have that j fixed, $f_{ij}^{(k)}$ has only two parameters to assign, i and k.

For each k define a *column vector* whose elements are $f_{ij}^{(k)}$, $i = 1, 2, \ldots$. Call this vector $\mathbf{f}^{(k)}$. Now $\mathbf{f}^{(1)}$ is the jth column of $\mathbf{P}$, that is, $\mathbf{f}^{(1)}$ is the vector of probabilities of going from any i to the fixed j in just one step. Now define a new matrix and call it $\mathbf{Q}$. $\mathbf{Q}$ is obtained from $\mathbf{P}$ by setting the jth column of $\mathbf{P}$ equal to 0 and leaving the remainder of the matrix alone. Then we have by (10.11)

$$\mathbf{f}^{(k)} = \mathbf{Q}\mathbf{f}^{(k-1)}, \qquad k = 2, 3, \ldots$$

Now

$$\mathbf{f}^{(2)} = \mathbf{Q}\mathbf{f}^{(1)}$$
$$\mathbf{f}^{(3)} = \mathbf{Q}\mathbf{f}^{(2)} = \mathbf{Q} \cdot \mathbf{Q}\mathbf{f}^{(1)}$$
$$= \mathbf{Q}^2 \mathbf{f}^{(1)}$$

and by iteration

$$\mathbf{f}^{(k)} = \mathbf{Q}^{k-1}\mathbf{f}^{(1)} \tag{10.12}$$

In this way one can obtain $f_{ij}^{(k)}$ for every i and k for each fixed j. Of course we still have to compute the powers of $\mathbf{Q}$. (A computer might be helpful.)

Why does this algorithm work? The answer lies in something we've already noticed, and the algorithm is just a formalization. $\mathbf{f}^{(1)}$ is the vector of probabilities of getting to j from each i in one step. Because we set the jth column of $\mathbf{P}$ equal to zero when we formed $\mathbf{Q}$, it is not possible to get to j from any i using the $\mathbf{Q}$ matrix. Thus $q_{il} = \Pr[X_n = l | X_{n-1} = i]$ for $l \neq j$ and $q_{ij} = 0$. $\mathbf{Q}$ is a matrix of probabilities for paths that take us from i to l avoiding j on the way. But since each step using $\mathbf{Q}$ avoids j, $\mathbf{Q}^n$ avoids j for n steps. Then

$$\mathbf{Q}^n \mathbf{f}^{(1)}$$

is the probability of avoiding j for n consecutive steps and then on the $(n + 1)$st step making a transition into j from whatever state we were in on the nth step. But since we avoided j for n steps and landed there on the $(n + 1)$st step, the $(n + 1)$st step must be the first time that we got to j. But that is precisely what $\mathbf{f}^{(n+1)}$ is.

EXAMPLE 10.3   (continued)

Let

$$\mathbf{P} = \begin{bmatrix} 1/4 & 3/4 \\ 1/2 & 1/2 \end{bmatrix}$$

and let us compute $f_{01}^{(1)}$, $f_{01}^{(2)}$, and $f_{01}^{(3)}$ using our algorithm on this example. Remember to choose j. In the above we will choose $j = 1$. Then make up the column vector that is the jth column of $\mathbf{P}$.

$$\mathbf{f}^{(1)} = \begin{bmatrix} f_{01}^{(1)} \\ f_{11}^{(1)} \end{bmatrix} = \begin{bmatrix} 3/4 \\ 1/2 \end{bmatrix}$$

Then define $\mathbf{Q}$ by setting the jth column of $\mathbf{P}$ to 0.

$$\mathbf{Q} = \begin{bmatrix} 1/4 & 0 \\ 1/2 & 0 \end{bmatrix}$$

Then

$$\mathbf{f}^{(2)} = \mathbf{Q}\mathbf{f}^{(1)} = \begin{bmatrix} 3/16 \\ 3/8 \end{bmatrix}$$

$$\mathbf{f}^{(3)} = \mathbf{Q}^2\mathbf{f}^{(1)} = \begin{bmatrix} 1/16 & 0 \\ 1/8 & 0 \end{bmatrix} [3/4 \quad 1/2] = \begin{bmatrix} 3/64 \\ 3/32 \end{bmatrix}$$

Continuing this iteration it is easy to show that

$$\mathbf{f}^{(n)} = \begin{bmatrix} (3/4)(1/4)^{n-1} \\ (3/8)(1/4)^{n-2} \end{bmatrix}$$

If the only term we want is $f_{01}^{(n)}$ all we need do is pick out the term corresponding to state 0 from $\mathbf{f}^{(n)}$, that is,

$$f_{01} = \tfrac{3}{4}(1/4)^{n-1}$$

Then we have that, starting in state 0, the probability of getting to state 1 for the first time on exactly the nth step is $(3/4)(1/4)^{n-1}$, that is,

$$\Pr[T_j = n \mid X_0 = i] = \tfrac{3}{4}(1/4)^{n-1}.$$

EXAMPLE 9.7 (continued)

In Example 9.7 we let $X_n = i$ if the nth turbo-boiler system to break down was the ith one, where $i = 1, 2, 3$. Let us now ask that if turbo-boiler system 1 just broke down, what is the probability it will next break down n system breakdowns from now? If we return to Example 9.7 we see that

$$\mathbf{P} = \begin{bmatrix} .65 & .09 & .26 \\ .55 & .19 & .26 \\ .50 & .07 & .43 \end{bmatrix}$$

So using our algorithm

$$\mathbf{f}^{(1)} = \begin{bmatrix} .65 \\ .55 \\ .50 \end{bmatrix}; \qquad \mathbf{Q} = \begin{bmatrix} 0 & .09 & .26 \\ 0 & .19 & .26 \\ 0 & .07 & .43 \end{bmatrix}$$

Then by matrix multiplication

$$\mathbf{Q}^2 = \begin{bmatrix} 0 & .035 & .135 \\ 0 & .054 & .161 \\ 0 & .043 & .203 \end{bmatrix}; \qquad \mathbf{Q}^3 = \begin{bmatrix} 0 & .016 & .067 \\ 0 & .022 & .083 \\ 0 & .023 & .099 \end{bmatrix}$$

$$\mathbf{Q}^4 = \begin{bmatrix} 0 & .008 & .033 \\ 0 & .010 & .042 \\ 0 & .011 & .048 \end{bmatrix}; \qquad \mathbf{Q}^5 = \begin{bmatrix} 0 & .004 & .016 \\ 0 & .005 & .020 \\ 0 & .006 & .024 \end{bmatrix}$$

where we have rounded to three decimal places. And

$$\mathbf{f}^{(2)} = \begin{bmatrix} .180 \\ .235 \\ .254 \end{bmatrix}; \quad \mathbf{f}^{(3)} = \begin{bmatrix} .087 \\ .110 \\ .125 \end{bmatrix}; \quad \mathbf{f}^{(4)} = \begin{bmatrix} .042 \\ .054 \\ .062 \end{bmatrix}; \quad \mathbf{f}^{(5)} = \begin{bmatrix} .010 \\ .013 \\ .015 \end{bmatrix}$$

Thus, $f_{11}^{(1)} = .650$, $f_{11}^{(2)} = .180$, $f_{11}^{(3)} = .087$, $f_{11}^{(4)} = .042$, and $f_{11}^{(5)} = .010$, to three decimal places.

DEFINITION State j is said to be *recurrent* if

$$\Pr[T_j < \infty \mid X_0 = j] = 1$$

That is, starting from state j, the probability of eventual return to state j is one. State j is said to be *transient* if

$$\Pr[T_j < \infty \mid X_0 = j] < 1$$

That is, the system may never return to state j.

$\Pr[T_j < \infty \mid X_0 = j]$ can be computed since

$$\Pr[T_j < \infty \mid X_0 = j] = f_{jj}^{(1)} + f_{jj}^{(2)} + f_{jj}^{(3)} + \cdots$$

As an example, suppose that we have a Markov chain with state space $\{0, 1\}$ and transition matrix $\mathbf{P}$ given by

$$\mathbf{P} = \begin{array}{c} \\ 0 \\ 1 \end{array} \begin{array}{cc} 0 & 1 \\ \begin{bmatrix} 1 & 0 \\ 1/2 & 1/2 \end{bmatrix} \end{array}$$

State 0 is recurrent, since

$$\begin{aligned} \Pr[T_0 < \infty \mid X_0 = 0] &= f_{00}^{(1)} + f_{00}^{(2)} + \cdots \\ &= 1 + 0 + 0 + \cdots \\ &= 1 \end{aligned}$$

However, state 1 is transient, since

$$\begin{aligned} \Pr[T_1 < \infty \mid X_0 = 1] &= f_{11}^{(1)} + f_{11}^{(2)} + \cdots \\ &= \tfrac{1}{2} + 0 + 0 + \cdots \\ &= \tfrac{1}{2} < 1 \end{aligned}$$

A recurrent state may be *null recurrent* or *positive recurrent*.

DEFINITION Let state j be recurrent. State j is *positive recurrent* if

$$E[T_j \mid X_0 = j] < \infty$$

and state j is *null recurrent* if

$$E[T_j \mid X_0 = j] = \infty$$

Thus, state j is positive recurrent if the *expected* time until the first return to state j is finite, given that we started in state j. Similarly, state j is null recurrent if the *expected* time until the first return to state j is infinite, given that the Markov chain started in j. Of course,

$$E\left[T_j \mid X_0 = j\right] = f_{jj}^{(1)} + 2f_{jj}^{(2)} + 3f_{jj}^{(3)} + \cdots$$

At first, the null recurrent case may seem paradoxical since with probability one the Markov chain returns to state j but the expected number of steps until the Markov chain returns is infinite. Roughly, null recurrence is between transience and positive recurrence. If state j is transient, the Markov chain may never return to j. If state j is positive recurrent, the Markov chain will surely return to j and the expected length of time until the return is finite.

To continue the classification of states, we must now decide whether each of the recurrent states is *periodic* or *aperiodic*. Roughly, a state is *periodic* with period δ if the chain can only return to state j on some multiple of δ steps, $\delta \geq 2$.

DEFINITION   Let state j be given. Let δ be the largest common divisor of

$$\left\{ n : p_{jj}^{(n)} > 0 \right\}$$

If $\delta > 1$, state j is called *periodic with period* δ. If $\delta = 1$, state j is *aperiodic*.

Example 10.7 is a Markov chain with all states periodic with period 2. Example 10.3 is one in which all states are aperiodic.

EXAMPLE 10.10

For each of the following transition matrices state 0 is periodic with period 2.

$$\text{(i)} \quad \begin{array}{c} \\ 0 \\ 1 \end{array} \begin{array}{cc} 0 & 1 \\ \left[\begin{array}{cc} 0 & 1 \\ 1 & 0 \end{array}\right] \end{array}$$

$$\text{(ii)} \quad \begin{array}{c} \\ 0 \\ 1 \\ 2 \end{array} \begin{array}{ccc} 0 & 1 & 2 \\ \left[\begin{array}{ccc} 0 & 1/2 & 1/2 \\ 1 & 0 & 0 \\ 1 & 0 & 0 \end{array}\right] \end{array}$$

$$\text{(iii)} \quad \begin{array}{c} \\ 0 \\ 1 \\ 2 \\ 3 \end{array} \begin{array}{cccc} 0 & 1 & 2 & 3 \\ \left[\begin{array}{cccc} 0 & 0 & 1/2 & 1/2 \\ 0 & 0 & 1/2 & 1/2 \\ 1/2 & 1/2 & 0 & 0 \\ 1/2 & 1/2 & 0 & 0 \end{array}\right] \end{array}$$

$$\text{(iv)} \quad \begin{array}{c} \\ 0 \\ 1 \\ 2 \\ 3 \end{array} \begin{array}{cccc} 0 & 1 & 2 & 3 \\ \left[\begin{array}{cccc} 0 & 1 & 0 & 0 \\ 0 & 0 & 1 & 0 \\ 0 & 0 & 0 & 1 \\ 1/2 & 0 & 1/2 & 0 \end{array}\right] \end{array}$$

In the first three

$$\left\{ n : p_{00}^{(n)} > 0 \right\} = \{2, 4, 6, 8, \dots\}$$

Clearly the largest common divisor, that is, the largest integer that divides into each number evenly, is 2. In (iv)

$$\{n: p_{00}^{(n)} > 0\} = \{4, 6, 8, \dots\}$$

Again 2 is the largest common divisor.

EXAMPLE 10.11

For each of the following transition matrices state 0 is aperiodic.

(i) $\begin{bmatrix} 0 & 1 \\ 1/2 & 1/2 \end{bmatrix}$

(ii) $\begin{bmatrix} 0 & 1/2 & 1/2 \\ 1/2 & 0 & 1/2 \\ 1 & 0 & 0 \end{bmatrix}$

(iii) $\begin{bmatrix} 0 & 0 & 1/2 & 1/2 \\ 0 & 0 & 1/2 & 1/2 \\ 1/2 & 1/2 & 0 & 0 \\ 1/2 & 1/4 & 0 & 1/4 \end{bmatrix}$

(iv) $\begin{bmatrix} 0 & 1 & 0 & 0 \\ 0 & 0 & 1 & 0 \\ 0 & 0 & 0 & 1 \\ 1/2 & 1/2 & 0 & 0 \end{bmatrix}$

In the first three

$$\{n: p_{00}^{(n)} > 0\} = \{2, 3, 4, 5, \dots\}$$

In the fourth,

$$\{n: p_{00}^{(n)} > 0\} = \{4, 7, 8, 10, \dots\}$$

The largest common divisor of both of these sets $\delta = 1$.

10.7   Classes of States

There are several results that simplify the process of classifying states, particularly for finite Markov chains. First we need to introduce some more definitions: State j is *accessible* or *can be reached* from state i if $p_{ij}^{(n)} > 0$ for some n. Hence, if j is accessible from i, which we denote by the symbol $i \to j$, then it is possible eventually to be in j after leaving i. If $i \to j$ and $j \to i$, states i and j are said to *communicate*.

Let C be a set of states. C is said to be a *closed* set if no state outside of C is accessible from a state in C. Hence, once the Markov chain enters C it will always stay in C. If, in addition, each pair of states in C communicates, then C is called a *closed communicating class*. If all states in the state space communicate, then the Markov chain is said to be *irreducible*. If $p_{jj} = 1$, state j is said to be *absorbing*, that is, once state j is reached it is never left.

The simplest way to determine whether a finite state space Markov chain is irreducible or not, to locate the closed communicating classes, and to check for periodicities is to examine the *transition diagram*. A *transition diagram* is a line diagram with a vertex corresponding to each state and an arrow between two vertices i and j if and only if $p_{ij} > 0$. In such a diagram if one can move from i to j by a path following the arrows, then $i \to j$.

EXAMPLE 10.12

The transition diagram corresponding to the transition matrix

$$\mathbf{P} = \begin{array}{c} \\ 0 \\ 1 \\ 2 \\ 3 \\ 4 \end{array} \begin{array}{ccccc} 0 & 1 & 2 & 3 & 4 \\ \left[\begin{array}{ccccc} 1/3 & 1/3 & 0 & 0 & 1/3 \\ 0 & 1 & 0 & 0 & 0 \\ 0 & 0 & 0 & 1/2 & 1/2 \\ 0 & 0 & 0 & 0 & 1 \\ 0 & 0 & 1 & 0 & 0 \end{array}\right] \end{array}$$

is shown in Figure 10.2.

Note that states 2 and 4 communicate. (There's an arrow from 2 to 4 and from 4 to 2.) States 3 and 4 also communicate. In fact, states 2, 3, and 4 form a closed communicating class. (There are no arrows leading out of this set.) The state space $E = \{0, 1, 2, 3, 4\}$ is not irreducible, since states 0 and 4 do not communicate. There are no arrows from 4 to 0. State 4 is accessible from 0, but not vice versa. One might suspect that state 3 is periodic with period 3 since the least number of steps to get back to state 3 is three. However, one can also get back in five steps by going from 3 to 4 to 2 to 4 to 2 to 3. Hence, state 3 must be aperiodic. Note that state 1 is absorbing. (There are arrows into 1 but none out of 1.)

In applying the following results be careful to make sure that the assumptions are satisfied. In particular, make sure that you do not assume that results for finite Markov chains necessarily apply to Markov chains with infinite state spaces.

Theorem 10.1

The state space E of a finite Markov chain can be partitioned uniquely into sets $T, C_1, C_2, \ldots, C_m$, where each of the C_i are closed communicating classes and T is the set of states that do not belong to any proper closed communcating class. (If the chain is irreducible, $E = C$.)

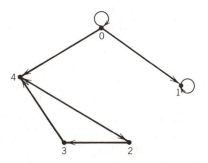

FIGURE 10.2   Transition diagram for Example 10.12.

Theorem 10.2

In a finite Markov chain, any state belonging to a closed communicating class is positive recurrent; any state not belonging to a closed communicating class is transient. Finite Markov chains do not have null recurrent states.

Theorem 10.3

Any two states that communicate have the same classification: positive recurrent, recurrent null, or transient; aperiodic or periodic with the same period.

With these three results we can classify the states of a finite Markov chain. First, we find the closed communicating classes. Any state not belonging to a closed communicating class is transient. The states belonging to closed communicating classes are positive recurrent. Then for each positive recurrent state, we need to determine whether it is aperiodic or periodic, and if periodic, we also need the period. From the last Theorem it suffices to determine periodicity for only one state in each closed communicating class.

The simplest way to locate the closed sets and to determine whether they are periodic is by examining the transition diagram.

EXAMPLE 10.12   (continued)

By inspection of the transition diagram the closed communicating classes are $\{1\}$ and $\{2,3,4\}$. Hence state 0 is transient and state 1 is positive recurrent. In addition, since $p_{00} > 0$, state 0 is aperiodic (see Exercise 10.42). States 2, 3, and 4 are positive recurrent. To complete the classification, we need to examine the periodicity of only one state in $\{2,3,4\}$. From state 4 it is possible to return to 4 in two steps or three steps. Since the greatest common diviser is one, we know that state 4 is aperiodic. Consequently, states 2 and 3 are also aperiodic.

EXAMPLE 10.3   (further continued)

Recall the weather forecasting in Example 10.3. There **P** was of the form

$$\mathbf{P} = \begin{pmatrix} a & 1-a \\ b & 1-b \end{pmatrix}$$

If $a \neq 0, 1$ and if $b \neq 0, 1$, then the Markov chain is irreducible. Each state can be reached from the other. On the other hand, if $a = 1$, state 1 cannot be reached from state 0. Hence, 0 is a closed set. Furthermore, since the closed set contains only one state, this state is an absorbing state. It has the property that once a dry day is encountered the weather remains dry forever.

EXAMPLE 10.2   (continued)

The transition matrix for the production item's progress through manufacturing stages (Example 10.2) had the form

$$\mathbf{P} = \begin{pmatrix} 1 & 0 & 0 & 0 & 0 \\ 0 & 1 & 0 & 0 & 0 \\ p & 0 & q & r & 0 \\ p & 0 & 0 & q & r \\ p & r & 0 & 0 & q \end{pmatrix}$$

This matrix can be decomposed into blocks, as indicated by the dotted lines, to take the form

$$\mathbf{P} = \begin{pmatrix} \mathbf{A}_1 & \mathbf{0} & \mathbf{0} \\ \mathbf{0} & \mathbf{A}_2 & \mathbf{0} \\ \mathbf{B}_1 & \mathbf{B}_2 & \mathbf{B}_3 \end{pmatrix}$$

where

$$\mathbf{A}_1 = (1), \qquad \mathbf{A}_2 = (1), \qquad \mathbf{B}_1 = \begin{pmatrix} p \\ p \\ p \end{pmatrix}, \qquad \mathbf{B}_2 = \begin{pmatrix} 0 \\ 0 \\ r \end{pmatrix}, \qquad \mathbf{B}_3 = \begin{pmatrix} q & r & 0 \\ 0 & q & r \\ 0 & 0 & q \end{pmatrix}$$

and each **0**-symbol means a matrix of zeros. The sets $\{0\}$ (item junked) and $\{1\}$ (item completed) are each closed sets. Since each contains only one state, 0 and 1 are absorbing states. Clearly the chain is not irreducible, and the set of states $\{2, 3, 4\}$ is not closed.

EXAMPLE 10.5   (continued)

This waiting-line example is an irreducible chain with an infinite number of states. All states communicate and so all states have the same class. They are obviously aperiodic. But whether they are all transient, null recurrent, or positive recurrent is not clear since Theorem 10.2 does not apply to infinite Markov chains.

EXAMPLE 10.9   (continued)

The ambulance calling example for the Markov chain Y is an example of a chain with an infinite number of states and in which every state is transient. We will prove the transient result later.

We haven't given sufficient tools for completely classifying infinite Markov chains, as in Example 10.9. We can usually determine whether the chain is irreducible and its periodicity if any. However, we cannot determine whether the

states are positive recurrent, null recurrent, or transient. To get an idea of the
difficulty consider the following infinite transition matrix.

$$\mathbf{P} = \begin{array}{c} \\ 0 \\ 1 \\ 2 \\ 3 \\ \vdots \end{array} \begin{array}{cccccc} 0 & 1 & 2 & 3 & \cdots \\ \left[\begin{array}{ccccc} p & 1-p & 0 & 0 & \cdots \\ p & 0 & 1-p & 0 & \cdots \\ 0 & p & 0 & 1-p & \cdots \\ 0 & 0 & p & 0 & \cdots \\ \vdots & \vdots & \vdots & & \end{array}\right] \end{array}$$

where $0 < p < 1$.

Clearly all states communicate. (Why?) Since $p_{00} > 0$, state 0 is aperiodic, and
hence all states are aperiodic. It turns out that all states are transient if $p < \frac{1}{2}$, null
recurrent if $p = \frac{1}{2}$, and positive recurrent if $p > \frac{1}{2}$. To get an intuitive feeling, note
that if $p < \frac{1}{2}$ the Markov chain tends to drift toward the right. However, if $p > \frac{1}{2}$
the Markov chain tends to return to the origin. In the next two sections two
results will be given that allow us to determine if the states of any irreducible
Markov chain are positive recurrent.

10.8   Limiting Distributions

Frequently, we wish to study a Markov chain that has been operating for some
time, as discussed in the beginning of Section 10.6. Thus, we focus attention on
the behavior of the state probabilities $p_j^{(n)}$ for large n. More precisely, let us
consider the quantities

$$v_j = \lim_{n \to \infty} p_j^{(n)} \tag{10.13}$$

The quantities v_j are sometimes referred to as *steady-state probabilities* or limiting
probabilities. In designing a physical system there are often "start-up" effects that
are different from what can be expected in the "long run." A designer might be
mainly interested in the "long-run" behavior that is described by the limiting
probabilities v_j.

From another point of view, v_j might be interpreted as the *long-run proportion*
of time the process spends in state j.

We can reasonably expect that, over a long period of time, the influence of the
initial state in which the process started may die out and, thus, that the limiting
probabilities v_j may not depend on the initial state. If this is the case, then v_j can
also be interpreted as the limit of the n-step transition probabilities $p_{ij}^{(n)}$,
$v_j = \lim_{n \to \infty} p_{ij}^{(n)}$, since $p_{ij}^{(n)}$ is simply the probability of the process being in state
j after n steps, given that it was initially in state i. Thus, if each v_j does not
depend on the initial state, the matrix $\mathbf{P}^{(n)} = (p_{ij}^{(n)})$ will converge to a matrix $\mathbf{V}$ as

$n \to \infty$, each row of which is identical to the vector $\mathbf{v}$ with components v_j,

$$\mathbf{P}^{(n)} \to \mathbf{V} = \begin{pmatrix} \mathbf{v} \\ \mathbf{v} \\ \vdots \end{pmatrix}, \qquad \text{as } n \to \infty$$

We have already seen this behavior in Examples 10.3 (continued) and 10.8. We discussed some of its implications in Example 9.7.

Clearly we must be concerned with the following questions: Do the limits defining the v_j's exist? If they exist, do they form a probability distribution, that is, do they add to 1, $\Sigma v_j = 1$? How can the v_j's be computed?

If the limits $v_j = \lim_{n \to \infty} p_j^{(n)} = \lim_{n \to \infty} p_{ij}^{(n)}$ exist and do not depend on the initial state, then by letting $n \to \infty$ in the identity (10.8)

$$p_j^{(n)} = \sum_k p_k^{(n-1)} p_{kj}$$

we obtain

$$v_j = \sum_k v_k p_{kj}, \qquad j = 0, 1, 2, \ldots \tag{10.14}$$

or, equivalently,

$$\mathbf{v} = \mathbf{vP} \tag{10.15}$$

Formula (10.14) (or formula (10.15)) gives a system of linear equations that must be satisfied by the quantities v_j.

Any vector $\mathbf{x} = (x_0, x_1, x_2, \ldots)$ with $x_j \geq 0$, $\Sigma x_j = 1$, that satisfies the system (10.14) or (10.15)

$$x_j = \sum_k x_k p_{kj} \qquad \text{or} \qquad \mathbf{x} = \mathbf{xP} \tag{10.16}$$

is termed a *stationary probability vector* of the process.

The following theorems are stated without proof.

Theorem 10.4

In any aperiodic Markov chain all the limits $v_j = \lim_{n \to \infty} p_j^{(n)}$ exist. (See the discussion at the beginning of Section 10.5 for examples.)

Theorem 10.5

In any irreducible, aperiodic Markov chain the limits $v_j = \lim_{n \to \infty} p_j^{(n)} = \lim_{n \to \infty} p_{ij}^{(n)}$ do not depend on the initial distribution.

Theorem 10.6

In any finite, irreducible, aperiodic Markov chain the limit vector $\mathbf{v} = (v_0, v_1, \ldots, v_j, \ldots)$ is the unique stationary probability vector of the process.

These theorems generally are sufficient to allow us to discuss and to determine the limiting probabilities in finite Markov chains encountered in practice.

Theorem 10.5 implies that, for a finite, aperiodic, irreducible Markov chain, the matrix $\mathbf{P}^{(n)} = (p_{ij}^{(n)})$ approaches a matrix that has every row equal, each row being identical with the stationary vector $\mathbf{v} = (v_0, v_1, \dots)$

$$\lim_{n \to \infty} \mathbf{P}^{(n)} = \begin{pmatrix} \mathbf{v} \\ \mathbf{v} \\ \vdots \end{pmatrix} = \begin{pmatrix} v_0, v_1, v_2, \dots \\ v_0, v_1, v_2, \dots \\ \vdots \quad \vdots \quad \vdots \end{pmatrix} \tag{10.17}$$

EXAMPLE 10.3   (again)

In Example 10.3 (continued) we discussed an irreducible aperiodic process with one-step transition matrix

$$\mathbf{P} = \begin{pmatrix} \frac{3}{4} & \frac{1}{4} \\ \frac{1}{2} & \frac{1}{2} \end{pmatrix}$$

By computation we found that

$$\mathbf{P}^{(10)} = \begin{pmatrix} .666 & .333 \\ .666 & .333 \end{pmatrix}$$

If we compare this with (10.17), we surmise that the limit vector is $\mathbf{v} = (\frac{2}{3}, \frac{1}{3})$. To check this, we use Theorem 10.6 and determine the stationary probability vector that satisfies (10.16): $\mathbf{x} = \mathbf{xP}$. Thus,

$$\mathbf{x} = \mathbf{x} \begin{pmatrix} \frac{3}{4} & \frac{1}{4} \\ \frac{1}{2} & \frac{1}{2} \end{pmatrix}$$

or

$$x_0 = x_0\left(\tfrac{3}{4}\right) + x_1\left(\tfrac{1}{2}\right)$$
$$x_1 = x_0\left(\tfrac{1}{4}\right) + x_1\left(\tfrac{1}{2}\right)$$

Each of these equations is equivalent to $x_0 = 2x_1$, so the system gives only one condition on the stationary vector (x_0, x_1):

$$x_1 = \tfrac{1}{2}x_0$$

Another condition comes from the fact that this is a probability vector, $\Sigma x_i = 1$. Thus,

$$x_0 + x_1 = 1$$

Thus, we find that $x_0 = \frac{2}{3}$ and $x_1 = \frac{1}{3}$. By Theorem 10.6, the limiting vector is given by

$$\mathbf{v} = \mathbf{x} = \left(\tfrac{2}{3}, \tfrac{1}{3}\right)$$

EXAMPLE 9.7   (again)

In Example 9.7 we tried to extract a large amount of information concerning system breakdown from the DiMarco data. In particular, we computed $\pi^{(n)} = \Pr[X_n = j]$ for $j = 1, 2, 3$. We also gave a heuristic argument based on numerical computation and the assertion that if $\mathbf{P}^n$ approaches a matrix with all rows equal then $\pi^{(n)}$ approaches a vector of probabilities where the rows of $\mathbf{P}^n$ were copies of $\lim_{n \to \infty} \pi^{(n)}$. If all we were interested in was $\lim_{n \to \infty} \pi^{(n)}$, this approach creates a lot of make-work. One must compute $\mathbf{P}^n$ for large values of n (we stopped at $n = 5$) and then hope that whenever we stopped we had a $\mathbf{P}^n$ with identical rows. (We quit when the rows agreed to the third decimal place, but Example 9.7 (continued) shows that some of the problems converge slowly numerically. In that example the $\mathbf{f}^{(k)}$ sums should give a vector all of whose elements are 1 and, even after five matrix multiplications, those sums differ from 1 in the second decimal.) What we have discussed in this section is a much easier way to get $\lim_{n \to \infty} \pi^{(n)}$ when it exists. Let us now use this method to see how good our numbers were in Example 9.7.

The $\mathbf{P}$ matrix in 9.7 was

$$\mathbf{P} = \begin{bmatrix} .65 & .09 & .26 \\ .55 & .19 & .26 \\ .50 & .07 & .43 \end{bmatrix}$$

Thus, this Markov chain is finite, irreducible, and aperiodic. By Theorem 10.6 above, the vector $\mathbf{v}$ exists, is a probability vector, is the unique stationary vector, and is also the limiting vector (the vector of $\lim_{n \to \infty} \pi^{(n)}$). The only problem then is to compute this vector. Since it is a stationary vector it satisfies equations such as

$$\mathbf{v} = \mathbf{v}\mathbf{P}$$

or in component form

$$v_1 = .65v_1 + .55v_2 + .50v_3$$

$$v_2 = .09v_1 + .19v_2 + .07v_3$$

$$v_3 = .26v_1 + .26v_2 + .43v_3$$

Since $\mathbf{P}$ has each row summing to 1, one can show that the above equations are not independent. One (any one) of the equations is redundant, so we can eliminate it from our consideration. Any v_1, v_2, and v_3 satisfying the remaining equations will automatically satisfy the equation thrown away. So let us compute with

$$v_1 = .65v_1 + .55v_2 + .50v_3$$

$$v_3 = .26v_1 + .36v_2 + .43v_3$$

or

$$.35v_1 = .55v_2 + .50v_3$$

$$-.26v_1 = .36v_2 - .57v_3$$

(Notice that we have three unknowns but only two equations. Thus, we cannot get a unique solution here. The best we can do is to express two of the unknowns in terms of the third and this is what we shall do.)

We get

$$v_1 = 1.8953v_3, \qquad v_2 = .297v_3$$

Now the point is this: Any vector of the form $v = (1.8953v_3, .297v_3, v_3)$ will satisfy equations (10.15). But that is not good enough for a stationary probability vector. To be the latter one also needs $v_i \geq 0$ for $i = 1, 2, 3$. Thus, we need to have $v_3 \geq 0$ to satisfy this. But, moreover, $\Sigma v_j = 1$. In this example we must then require that

$$1.8953v_3 + .2970v_3 + v_3 = 1$$

With this requirement there is only one value that v_3 can be. It must be .3132 (to four decimal places). But if $v_3 = .3132$ then $v_1 = 1.8953$ (.3132) and $v_2 = .2970$ (.3132). So we have

$$v_1 = .5936, \qquad v_2 = .0930, \qquad v_3 = .3132$$

Thus, correct to four decimal places, the unique stationary probability vector is

$$\mathbf{v} = (.5936, .0930, .3132)$$

and by Theorem 10.6 this vector is also the limiting ($\lim_{n \to \infty} p_j^{(n)}$) probability vector for this process. A comparison with $\pi^{(n)}$ for $n \geq 5$ in Example 9.7 shows good agreement in the first three decimal places. So our computations in Example 9.7, while tedious, are at least reasonably accurate. Therefore, we can argue that after only five failures this system is operating as it always will if no other changes occur (e.g., if a new turbine or boiler is not purchased to replace the unreliable system 1).

In Theorem 10.6 we indicated that, for a finite (irreducible, aperiodic) Markov chain, the transition probabilities $p_{ij}^{(n)}$ always approach limits v_j as $n \to \infty$, which form a probability vector. In the infinite state case, it might happen that $X_n \to \infty$ as $n \to \infty$, with probability 1, in which case $p_{ij}^{(n)} \to 0$ for every fixed i and j (see Example 10.9).

Theorems 10.4 and 10.5 apply equally well to the infinite state case. Thus, in an aperiodic, irreducible, infinite Markov chain we know that all the limits

$$v_j = \lim_{n \to \infty} p_{ij}^{(n)}$$

exist and do not depend on the initial state i. In the infinite state case the possibility remains, however, that these limits might be 0 and thus $\mathbf{v}$ will not be a stationary *probability* vector since $\Sigma v_j \neq 1$.

Theorem 10.7

(a) An irreducible, aperiodic Markov chain is positive recurrent if, and only if, there exists a unique stationary probability vector $\mathbf{x}$; that is, if there exists a unique solution to the equations (10.16) that satisfy $x_j \geq 0$, $\Sigma x_j = 1$. In this case the components of $\mathbf{x}$ give the limiting probabilities, $v_j = x_j$ for each j.

(b) If an irreducible, aperiodic Markov chain is not positive recurrent, then

$$v_j = \lim_{n \to \infty} p_{ij}^{(n)} = 0$$

for every j.

In an irreducible Markov chain v_j has another interpretation that is useful in some applications: v_j is the *long-run proportion of time* that the process spends in state j. Hence, if there is a cost c_j incurred each time state j is visited, then the long-run average cost per unit time is

$$\sum c_j v_j.$$

EXAMPLE 10.13

In Example 10.5 we considered a waiting-line process in which X_n represented the number of individuals waiting in line at the time when the nth individual completed his service. The transition matrix had the form

$$\mathbf{P} = \begin{pmatrix} p_{00} & p_{01} & p_{02} & \cdots \\ p_{10} & p_{11} & p_{12} & \cdots \\ 0 & p_{21} & p_{22} & \cdots \\ 0 & 0 & p_{32} & \cdots \\ \vdots & \vdots & \vdots & \end{pmatrix}$$

The zero terms below the dotted line are indicative of the fact that during a single service operation the length of the waiting line cannot decrease by more than one. Here $\{X_n\}$ is an irreducible process with possible states $0, 1, 2, \ldots$. The study of these processes is included in the extensive and useful branch of applied probability termed queueing theory (see Chapter 13).

In one important special case the terms in the matrix $\mathbf{P}$ have the form

$$\mathbf{P} = \begin{pmatrix} 1-t & (1-t)t & (1-t)t^2 & \cdots \\ 1-t & (1-t)t & (1-t)t^2 & \cdots \\ 0 & 1-t & (1-t)t & \cdots \\ 0 & 0 & 1-t & \cdots \\ \vdots & \vdots & \vdots & \end{pmatrix}$$

or $p_{ij} = (1-t)t^{j-i+1}$ for $i = 1, 2, \ldots$ and $j = i-1, i, i+1, \ldots$; $p_{0j} = (1-t)t^j$ for $j = 0, 1, 2, \ldots$; and $p_{ij} = 0$ otherwise. Here t, $0 < t < 1$, represents a parameter that is related to the relative speed of the server compared to the rate of arrivals to the system.

Direct calculation of the n-step transition probabilities is difficult. Since $\mathbf{P}$ is irreducible and aperiodic we can apply Theorem 10.7 and attempt to solve the system (10.16), which becomes

$$x_0 = (1-t)x_0 + (1-t)x_1$$
$$x_1 = (1-t)tx_0 + (1-t)tx_1 + (1-t)x_2$$
$$\vdots$$
$$x_j = (1-t)t^j x_0 + (1-t)\sum_{i=1}^{j+1} t^{j+1-i} x_i, \qquad \text{for } j = 1, 2, 3, \ldots$$

This system can be solved recursively in terms of x_0 to give

$$x_1 = \left(\frac{t}{1-t}\right)x_0$$

$$x_2 = \left(\frac{t}{1-t}\right)^2 x_0$$

$$\vdots$$

$$x_j = \left(\frac{t}{1-t}\right)^j x_0, \qquad \text{for } j = 0, 1, 2, \ldots$$

(The student should check this.) In order that $\Sigma x_j = 1$ we must have

$$x_0 \sum_{j=0}^{\infty} \left(\frac{t}{1-t}\right)^j = 1 \tag{10.18}$$

The condition that the geometric series in (10.18) converges is that $t/(1-t) < 1$, which is equivalent to $t < 1 - t$ or $t < \frac{1}{2}$. If this holds true, we have

$$x_0 \cdot \frac{1}{1 - \left(\dfrac{1}{1-t}\right)} = 1$$

$$x_0 = \frac{1-2t}{1-t}$$

$$x_j = \frac{1-2t}{1-t}\left(\frac{t}{1-t}\right)^j, \qquad \text{for } j = 0, 1, 2, \ldots \tag{10.19}$$

Thus, by our theorem, if $t < \frac{1}{2}$ the system is positive recurrent and the limiting probabilities v_j are given by (10.19). Formula (10.19) gives the probability that j individuals are waiting in line after the process has been going on for a long time.

If $t \geq \frac{1}{2}$, then $t/(1-t) \geq 1$ and the series in (10.18) *diverges*. Thus, no solution to the system (10.16) exists that forms a probability vector. It follows from our theorem that the system is not positive recurrent and that $v_j = \lim_{n \to \infty} P_{ij}^{(n)} = 0$ for all j. For $t \geq \frac{1}{2}$ the waiting line will tend to grow indefinitely, and the probability of there being any fixed number j in the line will approach zero. If $t = \frac{1}{2}$ the Markov chain is null recurrent but the limiting probabilities are still 0 for all finite j.

EXAMPLE 10.14

Return now to Example (10.9), where

$$\mathbf{P} = \begin{bmatrix} .919 & .081 & 0 & 0 & 0 & \cdots \\ 0 & .919 & .081 & 0 & 0 & \cdots \\ 0 & 0 & .919 & .081 & 0 & \cdots \\ & & & \ddots & \ddots & \end{bmatrix}$$

That **Y** process is irreducible and aperiodic. The state space is infinite. Now if we try to find the limiting probability vector we try to solve the equations

$$\mathbf{x} = \mathbf{x}\mathbf{P}$$

In component form, the first few of these are

$$x_0 = .919x_0$$
$$x_1 = .081x_0 + .919x_1$$
$$x_2 = .081x_1 + .919x_2$$
$$\vdots$$

So we have $x_0 = 0$, $x_1 = 0$, $x_2 = 0, \ldots$, and the only solution to this system is the null solution. Thus, the states are not positive recurrent. One interpretation is that in the long run ($n \to \infty$) an infinite number of telephone calls will arrive. There will also be an infinite number of calls for ambulance service. Thus, the probability that there are a *finite* number of telephone calls for ambulance service, any finite number, is zero.

To close this section, notice that the theorems that we have used all require the Markov chain to be aperiodic. The very useful Theorem 10.7, for example, requires this. To see what happens if the chain is not aperiodic recall Example 10.7. Let us unwittingly set up the equations in Theorem 10.6. This chain is finite and irreducible (but periodic). Then using equations (10.15) we obtain

$$\mathbf{v} = \mathbf{v}\mathbf{P}$$

or in component form, using **P** from Example 10.7,

$$v_1 = v_2 + v_3$$
$$v_2 = v_1/2$$
$$v_3 = v_1/2$$

So we obtain $\mathbf{v} = (v_1, v_1/2, v_1/2)$ is a stationary vector for any v_1. To be a stationary probability vector we still must require that $v_i \geq 0$ and $\Sigma v_i = 1$. Now $v_i \geq 0$ if $v_1 \geq 0$ and $\Sigma v_i = 1$ if $v_1 = \frac{1}{2}$, $v_2 = \frac{1}{4}$, and $v_3 = \frac{1}{4}$. Thus,

$$\mathbf{v} = \left(\tfrac{1}{2}, \tfrac{1}{4}, \tfrac{1}{4}\right)$$

is a stationary probability vector. But we argued in Example 10.7 (continued) that no limiting vector existed because $p_j^{(n)}$ was an alternating series in n.

This is what is going on: Both Theorems 10.6 and 10.7 say that for irreducible *aperiodic* Markov chains any stationary probability vector is also a limiting probability vector (i.e., its elements are $\lim_{n \to \infty} p_{ij}^{(n)}$). However, the above example shows that if we drop the aperiodic requirement then we can find stationary probability vectors *but* these vectors are *not* limiting probability vectors. So for periodic Markov chains (which are also irreducible) there can be stationary probability vectors even when no limiting probability vector exists. That's why the word "aperiodic" is crucial to our theorems.

We have thrown a lot together so far in this chapter. Let us summarize where we stand. First we had to define the basic building blocks of Markov chains and show that these gave us complete knowledge of the probability structure of a Markov chain. These preliminaries required all of Section 10.2. Section 10.3 was simply a collection of examples illustrating the ideas of Section 10.2. In Section 10.4 the idea of nth-step transition was introduced and several examples were carried out. Section 10.5 studied nth-step state probabilities and gave some examples illustrating why one needs to classify states in a Markov chain. Section 10.6 then developed ideas on the classification of states. We saw that there really are only two types of states in a Markov chain: states are either transient or recurrent. We gave a means of distinguishing these through the random variable T_j—the time of first visit to state j. Unfortunately, there are two types of recurrent states: positive recurrent and null recurrent. (Luckily there is only one type of transient state.) We have a condition to distinguish positive recurrence from null recurrence (look at $E[T_j|X_0=j]$). Again, unfortunately, there are two more types of recurrent states: aperiodic and periodic. We gave a criterion to distinguish these.

Thus, we have the following results.

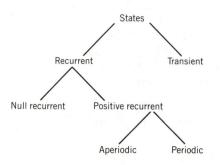

We saw in Section 10.7 that we can form groups of states (classes of states) according to whether the states communicate or not. The most important result was Theorem 10.3, which says essentially that all states in the same class have the same properties regarding recurrence or transience, positive recurrence or null recurrence, aperiodicity or periodicity. That theorem can be a great time saver. From it one studies large classes of states (those that communicate) by looking at just one state in the class. It happens in many applications that the chain is irreducible (all states are in the same class because they all communicate). From Theorem 10.3 we can then classify all the states at once by looking at the properties of just one state.

Sections 10.2 and 10.5 tell us how to compute the n-step probabilities and Sections 10.6 and 10.7 tell us what types of states we have. Sections 10.6 and 10.7 are not terribly important if all we want are n-step probabilities for finite n. However, as we argued in the beginning of Section 10.6, it is often desirable to find limiting probabilities (also called "steady-state probabilities" or "equilibrium probabilities") since for many systems the system "settles down" to its equilibrium behavior and stays there forever (unless the system experiences some major change). To determine these limiting probabilities the class structure for states in Sections 10.6 and 10.7 is crucial. Thus, in this section we gave four major theorems for the behavior of and determination of these limiting probabilities. In summary the following chart is useful.

Classification of States

Positive Recurrent	Transient, Null Recurrent
$v_j = \lim_{n \to \infty} p_{ij}^{(n)} = \lim_{n \to \infty} p_j^{(n)}$	$\lim_{n \to \infty} p_{ij}^{(n)} = \lim_{n \to \infty} p_j^{(n)} = 0$

v_j satisfies

$$v_j \geq 0, \qquad \sum v_j = 1$$

$$\mathbf{v} = \mathbf{v}\mathbf{P}$$

10.9 Passage to a Closed Set

From the previous sections we know how to handle many finite step probabilities. For each closed communicating class we know how to handle limiting probability results whether those classes are of positive recurrent or null recurrent or aperiodic or periodic states. We also know how to handle limiting probabilities for transient states alone. There is but one major problem left. Suppose i is a transient state and $j \in C_1$ and $k \in C_2$, each a closed set of states. If there is a path leading us from i to j and another leading us from i to k it is important to know the probabilities of getting to k or j for if the chain gets to k first it will never get to j since C_1 and C_2 are each closed.

Thus, to complete our discussion of limiting distributions we will look at nonirreducible but *finite* Markov chains. From Theorem 10.1 we know that the state space can be partitioned into mutually exclusive sets $C_1, \ldots, C_s$ and T, where each of the C_k are closed communicating classes and T is the set of transient states. Clearly the limiting distribution depends on the initial state. If $X_0 \in C_m$, the Markov chain is always in C_m. If $X_0 \notin C_m$ and if the process ever gets to C_m it remains there, and within C_m it acts like an irreducible process, to which the discussion of Section 10.8 applies. Thus, its behavior within each closed communicating class can be analyzed separately (and, of course, if the class consists of just one absorbing state, this analysis is trivial). If the process starts in T, then eventually it must enter one of the C_m's. (Why?) To analyze the long-run behavior of the process we must compute the probabilities f_i that a process starting in the transient state i eventually enters the closed communicating class C_m, for each such C_m:

$$f_i = \Pr[\, X_n \text{ in } C_m \text{ eventually} \mid X_0 = i \,] \tag{10.20}$$

for each transient state i.

The computation of (10.20) is easy if we return to Section 10.6, where we defined $f_{ij}^{(n)}$. Recall that $f_{ij}^{(k)}$ was the probability of going from i to j *for the first time on the kth step.* $f_{ij}^{(k)}$ satisfied the following equations.

(a) $f_{ij}^{(1)} = p_{ij}$.

(b) $f_{ij}^{(k)} = \sum_{l \neq j} p_{il} f_{lj}^{(k-1)}$, $k = 2, 3, \dots$.

Now we wanted $i \in T$ so these formulas, in this case, can use any state i as long as $i \in T$. However, if $l \notin T$ then in one step we leave the set T and therefore are in some C_n, maybe the one we want (C_m), maybe not. The only way to guarantee that we get to C_m for the first time on the kth (for $k > 1$) step is to make sure that the first step keeps us in T. Thus, we must amend (b) above as

(c) $f_{ij}^{(k)} = \sum_{l \in T} p_{il} f_{lj}^{(k-1)}$, $i \in T$, $j \in C_m$, $k = 2, 3, \dots$.

Now comes the crucial observation: to get to C_m *eventually*, we must get there for the first time on the 1st or 2nd or 3rd or … step. But the events "the chain gets to state j starting from state i on exactly the kth step" are mutually exclusive when considered as events on k. Thus, the probability of eventually getting to j from i is simply the sum of (a) and the sum of terms in (c) summed over all k. It is not hard to verify that the following result is obtained.

(d) $f_{ij} = p_{ij} + \sum_{l \in T} p_{il} f_{lj}$, $i \in T$, $j \in C_m$.

If we now sum (d) over all states in C_m and define

$$s_i = \sum_{j \in C_m} p_{ij}$$

we obtain

$$f_i = s_i + \sum_{l \in T} p_{il} f_l, \qquad i \in T \tag{10.21}$$

is the probability of eventually getting to closed set C_m, when starting in transient state i.

EXAMPLE 10.1   (continued)

The gambling example (Example 10.1) is a nonirreducible process with state space $E = \{0, 1, 2, \dots, N\}$ that contains two closed communicating classes each of which consists of only a single absorbing state, $\{0\}$ and $\{N\}$. Thus the transient states are $T = \{1, 2, \dots, N-1\}$ and, by the definition of transient states, if the system starts in one of these states it eventually must pass into state 0 (player ruined) or into state N (opponent ruined).

Let us fix our attention on the "ruin" absorbing state 0. The probabilities p_{ij} are given by $p_{i,i+1} = p$ if $i = 1, 2, \dots, N-1$, $p_{i,i-1} = q$ if $i = 1, 2, \dots, N-1$, and $p_{ij} = 0$ otherwise. The probability s_i is the probability of ruin (entering state 0) in one step from state i. Thus, $s_1 = q$ and $s_i = 0$ for $i = 2, 3, \dots, N-1$. Hence, the system (10.21) becomes

$$f_1 = q + pf_2$$
$$f_i = 0 + qf_{i-1} + pf_{i+1}, \qquad \text{for } i = 2, 3, \ldots, N-2$$
$$f_{N-1} = 0 + qf_{N-2}$$

$f_0 = 1$ and $f_N = 0$ can be interpreted to mean that ruin is certain starting from state 0 and that ruin is impossible starting from state N. These equations can be solved using standard difference equation methods.

EXAMPLE 10.2   (continued)

Return to Example 10.2. For that example we asked for the probability that an item is junked before it is completed. But this question is equivalent to the question: Find the probability that the Markov chain ever reaches state 0. The reasoning here is quite simple. If the chain ever reaches state 0 it must never before have gone to state 1. So it reached state 0 before it reached state 1.

Then using equation (10.21) we have the set

$$f_2 = p + qf_2 + rf_3$$
$$f_3 = p + qf_3 + rf_4$$
$$f_4 = p + qf_4$$

These three equations in three unknowns can be solved easily for f_2, f_3, and f_4 (e.g., $f_4 = p/(p+r)$). Hence the probability of the item being scrapped before it is completed is $p/(p+r)$ if the item is now in production stage 3.

Here is a nice trick for *finite* Markov chains. Remember (see Section 10.6) that we gave an algorithm for computing $\mathbf{f}^{(k)}$ that was the vector of probabilities for the Markov chain moving from state i to some fixed state j for the first time on the kth step. We argued in (d) above that the probability of eventually getting to j from i was simply the sum of the $f_{ij}^{(k)}$. Now we can put all of this into matrix form. Let $\mathbf{A}$ be the matrix obtained from $\mathbf{P}$ by deleting all rows and columns except those for the set T. Let $\mathbf{f}^{(1)} = (p_{ij}$ for $i \in T$, $j \in C_m$ fixed). Then it is easy to see that

$$\mathbf{f}^{(k)} = \mathbf{A}^{k-1}\mathbf{f}^{(1)}, \qquad k = 2, 3, \ldots$$
$$\mathbf{f}^{(1)} = \text{The } j\text{th column of } \mathbf{P} \text{ after throwing away all rows and columns except those in } T \text{ (i.e., the first column of } \mathbf{A})$$

Then

$$\mathbf{f} = \sum_{k=1}^{\infty} \mathbf{f}^{(k)} = \mathbf{f}^{(1)} + \sum_{k=2}^{\infty} \mathbf{A}^{(k-1)}\mathbf{f}^{(1)}$$

or

$$\mathbf{f} = (\mathbf{I} + \mathbf{A} + \mathbf{A}^2 + \cdots)\mathbf{f}^{(1)}$$

It is shown in courses on matrix analysis that the term $(\mathbf{I} + \mathbf{A} + \mathbf{A}^2 + \cdots) = (\mathbf{I} - \mathbf{A})^{-1}$ when the Markov chain is finite. (This doesn't work for infinite chains because $(\mathbf{I} - \mathbf{A})^{-1}$ may not be unique, so we need another means to find the correct matrix to use in that case.) We have the useful computing formula

$$\mathbf{f} = (\mathbf{I} - \mathbf{A})^{-1}\mathbf{f}^{(1)}$$

EXAMPLE 10.2   (again)

For Example 10.2 we can use our algorithm since that chain is finite.

$$\mathbf{A} = \begin{bmatrix} q & r & 0 \\ 0 & q & r \\ 0 & 0 & q \end{bmatrix}$$

$$\mathbf{f}^{(1)} = \begin{pmatrix} p \\ p \\ p \end{pmatrix}$$

$$(\mathbf{I} - \mathbf{A})^{-1} = \begin{bmatrix} \dfrac{1}{p+r} & \dfrac{r}{(p+r)^2} & \dfrac{r^2}{(p+r)^3} \\ 0 & \dfrac{1}{p+r} & \dfrac{r}{(p+r)^2} \\ 0 & 0 & \dfrac{1}{p+r} \end{bmatrix}$$

And thus $\mathbf{f}$ is easy to compute for all $i \in T$ and $j = \{0\}$ in this example. For example, $f_4 = p/(p+r)$ as previously.

This algorithm works if C_m is a set of states with more than one state in the set. Then if we are only interested in the probability of ever reaching the set, there is no need to distinguish states in the set. Therefore, we can replace the whole set C_m with one single new state, say, j', make j' an absorbing state, and continue with the above algorithm.

Exercises 10

1. Suppose one die is thrown many times. Let X_n = the number showing on the nth toss. Is $\mathbf{X} = \{ X_n : n = 1, 2, \ldots \}$ a Markov chain? If so give a one-step transition probability matrix for it assuming the die is honest.

2. In Exercise 1 compute the following.

   (a) $\Pr[X_3 = 2 | X_0 = 1]$.
   (b) $\Pr[X_3 = 2, \ X_2 = 3, \ X_1 = 6 | X_0 = 1]$.
   (c) $\Pr[X_{75} = 5 | X_0 = 1]$.
   (d) $\Pr[X_3 = 2 | X_0 = 1, \ X_1 = 2, \ X_2 = 6]$.
   (e) $\Pr[X_3 = 2 | X_1 = 2]$.

3. Suppose we throw a die many times and total the number of points accumulated until the nth toss. Let Y_n be this total. Is $\mathbf{Y} = \{Y_n : n = 0, 1, \ldots\}$ a Markov chain? If it is, what is its one-step transition probability matrix $\mathbf{P}$?

4. Suppose we flip a penny many times. Let $X_n = 0$ if the nth flip turns up tails and $X_n = 1$ otherwise. Now define a new random process $\mathbf{Y}$ by taking $Y_1 = (X_1, X_2)$, $Y_2 = (X_2, X_3)$, $Y_3 = (X_3, X_4)\ldots$. Give a calculation to show that $\mathbf{Y}$ is a Markov process.

5. Give an example of a Markov chain that has infinitely many states.

6. Let $\mathbf{X}$ be a Markov chain with state space $E = \{0, 1, 2,\}$, initial probability vector $\mathbf{p}^{(0)} = (\frac{1}{4}, \frac{1}{2}, \frac{1}{4})$, and one-step transition matrix

$$\mathbf{P} = \begin{bmatrix} 1/4 & 3/4 & 0 \\ 1/3 & 1/3 & 1/3 \\ 0 & 1/4 & 3/4 \end{bmatrix}$$

(a) Compute $p(0, 1, 1) = \Pr[X_0 = 0,\ X_1 = 1,\ X_2 = 1]$.

(b) Show that $\Pr[X_1 = 1$ and $X_2 = 1 | X_0 = 0] = p_{01} p_{11}$.

(c) Compute $p_{01}^{(2)}$.

7. Let $X = \{X_n : n = 1, 2, 3, \ldots\}$ be a Markov chain with state space $E = \{0, 1\}$, initial probability vector $\mathbf{p}^{(0)} = (a, b)$, and one-step transition matrix

$$\mathbf{P} = \begin{bmatrix} 2/3 & 1/3 \\ 1/2 & 1/2 \end{bmatrix}$$

(a) Compute $\Pr[X_0 = 0,\ X_1 = 1,\ X_2 = 1]$.

(b) Compute $\Pr[X_n = 0,\ X_{n+1} = 1,\ X_{n+2} = 1]$ for $n = 1, 2, 3$.

(c) Compute $\Pr[X_n = 0,\ X_{n+2} = 1]$ for $n = 1, 2, 3$.

(d) Compute $\Pr[X_{n+2} = 1]$ for $n = 1, 2, 3$.

(e) Compute $\Pr[X_{n+2} = 1 | X_n = 0]$ for $n = 1, 2, 3$.

(f) Which values in (a) to (e) depend on n and which do not?

8. For the DiMarco problem (Example 9.7) compute the following.

(a) $\Pr[X_1 = 1]$.

(b) $\Pr[X_2 = 1 | X_0 = 1]$.

(c) $\Pr[X_2 = 1]$.

(d) $\Pr[X_2 = 1,\ X_1 = 1 | X_0 = 1]$.

(e) $\Pr[X_3 = 1 | X_2 = 1,\ X_1 = 1,\ X_0 = 1]$.

Assume here that $\mathbf{p}^{(0)} = (1, 0, 0)$. Give a physical meaning to your results in terms of DiMarco's application.

9. Give an example of a random process with a discrete state, discrete time random process that possesses the Markov property but does not have stationary transition probabilities.

10. Determine the missing elements in the following transition matrix.

$$
\begin{array}{c}
\begin{array}{cccc} 0 & 1 & 2 & 3 \end{array}\\
P = \begin{array}{c} 0\\ 1\\ 2\\ 3 \end{array}\left[\begin{array}{cccc}
 & 1/3 & 1/3 & 1/3\\
1/10 & & 1/10 & 1/10\\
 & & & 1\\
1/4 & 3/4 & &
\end{array}\right]
\end{array}
$$

11. In Example 10.3 assume that it is dry on Monday. What is the probability that the rest of the week will be dry up to and including Sunday? What is the probability that it is dry on Wednesday, Friday, and Sunday and wet on Tuesday, Thursday, and Saturday still assuming it is dry on Monday?

12. What is the probability that it will be wet tomorrow if it was wet yesterday?

13. In Example 10.2 what is the probability that an item is successfully completed without being reworked?

14. In Example 10.2 compute the following.

(a) $\Pr[X_0 = 0]$.
(b) $\Pr[X_1 = 3 | X_1 = 2]$.
(c) $\Pr[X_2 = 3]$.
(d) $\Pr[X_4 = 3 | X_2 = 2]$.

15. In Example 10.2 compute the state probabilities $\mathbf{p}^{(1)}$, $\mathbf{p}^{(2)}$, and $\mathbf{p}^{(3)}$.

16. In Example 10.3 compute $\mathbf{p}^{(1)}$, $\mathbf{p}^{(2)}$, and $\mathbf{p}^{(3)}$ assuming $\mathbf{p}^{(0)} = (\frac{1}{4}, \frac{3}{4})$.

17. In Example 10.3 compute

$$f_{ij}^{(k)} \quad \text{for } i = 0, 1, \ j = 0, 1, \quad \text{and } k = 1, 2$$

18. Suppose that we have a two-state Markov chain with

$$
\begin{array}{c}
\begin{array}{cc} 0 & 1 \end{array}\\
P = \begin{array}{c} 0\\ 1 \end{array}\left[\begin{array}{cc}
p_{00} & p_{01}\\
p_{10} & p_{11}
\end{array}\right]
\end{array}
$$

Argue that

$$f_{00}^{(k)} = p_{01} p_{11}^{k-2} p_{10}, \quad k \geq 2$$

Determine analogous expressions for $f_{01}^{(k)}$ and $f_{10}^{(k)}$ for $k = 1, 2, 3$, and 4.

19. In Example 10.2 compute $f_{20}^{(k)}$ and $f_{21}^{(k)}$ for $k = 1, 2, 3$, and 4.

20. A missile is launched and, as it is tracked, a sequence of course correction signals are sent to it. Suppose that the system has four states that are labeled as follows.

state 0: *on-course*, no further correction necessary
state 1: *minor deviation*
state 2: *major deviation*
state 3: *abort*, off-course so badly a self-destruct signal is sent

Let X_n represent the state of the system after the nth course correction and assume that the behavior of $\{X_n\}$ can be modeled by a Markov chain that has the following one-step transition matrix.

$$P = \begin{pmatrix} 1 & 0 & 0 & 0 \\ 1/2 & 1/4 & 1/4 & 0 \\ 0 & 1/2 & 1/4 & 1/4 \\ 0 & 0 & 0 & 1 \end{pmatrix}$$

(a) Show that states 0 and 3 are absorbing states and that states 1 and 2 are transient states.

(b) If the missile starts initially in state 1, compute the probability that it eventually gets on-course.

(c) If the missile has a 50–50 chance of starting in either state 1 or state 2, compute the probability that is eventually gets on-course.

(d) If the missile is equally likely to start in any of the states 0, 1, 2, or 3, compute the probability that it eventually gets on-course.

21. A Markov chain has state space $\{0,1,2,3,4\}$ and one-step transition matrix

$$P = \begin{pmatrix} 1/2 & 1/2 & 0 & 0 & 0 \\ 1/3 & 2/3 & 0 & 0 & 0 \\ 1/4 & 0 & 1/4 & 1/4 & 1/4 \\ 0 & 1/4 & 1/4 & 1/4 & 1/4 \\ 0 & 0 & 0 & 0 & 1 \end{pmatrix}$$

(a) Determine the closed communicating classes and the transient states.

(b) For each closed communicating class find the probabilities t_i that the chain will enter that class, given that it starts in the transient state i.

22. In Exercise 20 compute the probability that the second course correction puts the missile on course by assuming initially that it has a 50–50 chance of starting in state 1 or state 2.

23. A gambler plays a "fair game" in which the odds are 2 to 1. In other words, at each play he has probability $\frac{1}{3}$ of winning and $\frac{2}{3}$ of losing. If he wins, he wins $2. If he loses, he loses $1. Assume that the total dollar resources of the gambler and his opponent are N dollars. If the capital of either player falls below the point where he could pay up if he lost on the next play, the game terminates. Let X_n represent the first player's capital after n plays.

(a) Set up the one-step transition matrix of the Markov chain $\{X_n\}$.

(b) Suppose that the two players agree that if the capital of either one falls to $1 they will perform the next play at even odds—win or lose $1 with equal probability. Set up the one-step transition matrix for this case.

24. In Exercise 6 compute $p_{ij}^{(3)}$ for $i, j = 0, 1, 2$.

25. Let $\{X_n\}$ be a Markov chain with state space $\{0, 1, 2, 3, 4\}$ and one-step transition matrix $\mathbf{P}$. For each of the following cases, determine the closed communicating classes and absorbing states.

(a)
$$\mathbf{P} = \begin{pmatrix} 1/2 & 0 & 0 & 1/2 & 0 \\ 1/2 & 1/2 & 0 & 0 & 0 \\ 1/4 & 1/2 & 0 & 0 & 1/4 \\ 0 & 0 & 0 & 1 & 0 \\ 0 & 0 & 1/2 & 1/4 & 1/4 \end{pmatrix}$$

(b)
$$\mathbf{P} = \begin{pmatrix} 0 & 1/3 & 0 & 1/3 & 1/3 \\ 1/3 & 1/3 & 0 & 1/3 & 0 \\ 0 & 0 & 2/3 & 0 & 1/3 \\ 1/4 & 1/4 & 0 & 1/4 & 1/4 \\ 0 & 0 & 1/3 & 0 & 2/3 \end{pmatrix}$$

26. Discuss the limiting behavior as $n \to \infty$ of the nonirreducible Markov chain in Exercise 25(a).

27. A certain factory has two machines and one repair crew. Assume that the probability of any one machine breaking down in a given day is α. Assume that if the repair crew is working on a machine, the probability that they will complete the repair in one more day is β. For simplicity, ignore the possibility of a repair completion or a breakdown taking place except at the end of a day. Let X_n represent the number of machines in operation at the end of the nth day. Assume that the behavior of $\{X_n\}$ can be modeled by a Markov chain.

(a) Set up the one-step transition matrix $\mathbf{P}$.

(b) If the system starts out with both machines operating, what is the probability that both will be operating 2 days later?

28. The owner of the local one-chair barber shop is thinking of expanding because there seem to be too many people waiting. Observations indicate that in the time required to cut one person's hair there may be zero, one, or two new arrivals with probability .3, .4, and .3, respectively. The shop has a fixed capacity of six people, which includes the person whose hair is being cut. Let X_n be the number of people in the shop at the completion of the nth customer's haircut. $\{X_n\}$ is a Markov chain assuming independent, identically distributed arrivals. Find its one-step transition probability matrix. Determine the "long-run" proportion of time that the shop has six people in it; that it has five people in it.

29. Suppose that you have conducted a series of tests on a manual dexterity training procedure and find that the following probability matrix describes the course of "correct" and "incorrect" responses.

		$(j+1)$st trial	
		correct	incorrect
jth trial	correct	0.95	0.05
	incorrect	0.01	0.99

(a) What proportion of correct responses would one expect of a "fully trained" trainee?

(b) What proportion of correct responses would one expect of a trainee after five repetitions of the procedure if the initial response is equally likely to be correct or incorrect?

(c) How many repetitions (trials) of this procedure must one perform before the trainee is within 1% of that individual's ultimate performance?

(d) What is the probability that a correct response is obtained for the first time exactly four trials after an incorrect response?

30. Suppose that we have a container with a membrane that separates it into compartments A and B. Initially there are j molecules in compartment A and $a - j$ in compartment B. A *transition* is said to occur whenever a molecule crosses through the membrane (either from A to B or from B to A). Let X_n = the number of molecules in compartment A after n transitions. At each transition, X_n increases or decreases by exactly one molecule. Suppose the probability that a molecule changes compartments is proportional to the number of molecules in the compartment that the molecule leaves. Set up a Markov chain model of this diffusion process. The model discussed here is called the Ehrenfest diffusion model.

31. Suppose that on a final assembly line for automobiles we have the following set of rules.

(1) Two convertibles can never follow each other down the line because the work content would unbalance the line.

(2) A station wagon must be followed by a sedan to balance the line.

(3) A sedan can be followed by either a station wagon or a convertible, but not by a sedan. (Only those three models are produced on this line.)

Set up a possible transition matrix for these rules by inserting the letters $a, b, c, d, \ldots$, et cetera, for the values that are not numerically defined by the above rules.

(a) Is the chain irreducible?

(b) What is the probability that after one convertible the next one occurs on the line one space apart (separated by one other vehicle)? Use your letters $a, b, c, d, \ldots$ if they are needed.

(c) If **P** is the one-step transition matrix, what interpretation would you give the (ij)th element of $\mathbf{P}^n$ for large n?

(d) If we had a production schedule that was given by a vector of the proportion of each type of auto to be made in a month, how might we determine a probability sequence that met the schedule and the rules of sequencing given in this problem?

(e) Suppose that we had contracted to deliver 10,000 autos at the end of a month. Of these 10,000, approximately 50% had to be sedans, 10% had to be station wagons, and 40% had to be convertibles. Set up some probability schedule rule that ensures this production and that still meets constraints (1), (2), and (3) of this problem.

32. A power conveyor is a materials-handling device. Material is placed on carriers that are equally spaced along a constant speed-powered chain. As a loaded carrier passes by a worker, the worker removes the material from the conveyor, performs some work on it, and returns it to the conveyor so that it can be moved to the next work station. Because of the constant speed and equal spacing of carriers, it is convenient to measure time in terms of the number of carriers that pass and, hence, to have a discrete variable to measure time.

Suppose that a particular work station has work brought to it by a power conveyor on which every carrier has a single part to be worked on by this station. Suppose further that, in the time between two consecutive carriers, the work station can complete zero, one, or two parts with probability $\frac{3}{8}$, $\frac{1}{2}$, and $\frac{1}{8}$, respectively. Suppose further that this work station can accumulate (as temporary storage) at most one part in addition to a part that is being worked on. If a carrier passes by and the station is empty or has just one part in operation, then the carrier is unloaded and the part is laid aside to be operated on when the one now in progress is completed. If a carrier passes by and the station already has one part in operation and one in storage, then that carrier load bypasses the station and is lost.

Let the states of the work station be 0, 1, and 2, representing an empty station, a station with exactly one part in progress, and a station with one part in progress and one in storage, respectively. Let the nth step occur just *before* the arrival of the nth carrier.

(a) Show that the one-step transition probability matrix is given by

$$\mathbf{P} = \begin{pmatrix} 5/8 & 3/8 & 0 \\ 1/8 & 1/2 & 3/8 \\ 1/8 & 1/2 & 3/8 \end{pmatrix}$$

(that is, give a logical argument based on the system description and probabilities that will support the use of **P** as the transition probabilities for this problem).

(b) Find the proportion of time that the work station spends in each state for the long run.

(c) If the station starts the day empty, what is the probability that the third carrier that arrives will not be able to discharge its load and, hence, bypasses the station.

(d) Find the average length of time between idle times.

(e) What is the probability that the station will go from state 2 to state 0 for the first time in two steps? How can we interpret this probability?[1]

33. Suppose that we run an (s, S) inventory system as follows. At fixed equally spaced instants $t_1 < t_2 < t_3 \ldots$ we make a decision whether to purchase a resupply of inventory. This resupply is immediately delivered. How much is ordered depends on the level of inventory at the instant t_i. If X_i is the on-hand inventory at t_i, then we reorder if $X_i \le s$ and do not reorder if $s < X_i \le S$. Whenever an order is placed, it is always for an amount that brings the on-hand inventory up to S.

Hence X_i takes values $S, S-1, S-2, \ldots 0, -1, -2, \ldots$ where negative values indicate that demand is back-ordered. Backlogged demand is satisfied immediately on receipt of a new supply.

Suppose that d_i represents the total demand that occurs in (t_{i-1}, t_i) and that d_i is Poisson distributed, with the demand in consecutive periods independent, identically distributed random variables. Thus,

$$X_{i+1} = X_i - d_{i+1}, \qquad \text{if } s < X_i \le S$$
$$= S - d_{i+1}, \qquad \text{if } X_i \le s$$

Show that the process $\{X_i\}$ is a Markov process and determine its transition probability matrix **P**.

34. Suppose that we try to model a sequence of decisions in the following manner.

In decision-making one collects and interprets data. For many reasons, including collecting the wrong data, we might interpret the data correctly, with probability p, or incorrectly, with probability q, irrespective of the decision. Certainly $p + q = 1$. Having the data we then make a decision. The decision can be "correct" or "incorrect." Which decision occurs depends, of course, on whether the data were interpreted correctly or not. Suppose also that whether a decision is correct or not depends on whether the previous decision was correct or not. For example, if a correct decision is made, then

---

[1]In this problem we have "discretized" time and in the process have created some anomalies. For example, if a part is completed at any time within the interval between two carrier passings, the worker is considered to be idle for the entire time. Hence estimates, based on the probabilities obtained, may be in error in any real-life problem in which time is continuous.

we would expect that, at the next decision point, the process being controlled was near its control level. Hence, the next decision would be routine and would have a large chance of being correct. On the other hand, an incorrect decision might put the process so far from its control level that drastic action was required that had a large chance of being incorrect.

Let $X_n = 0$ if a correct decision is made at the nth decision point and $X_n = 1$ if an incorrect decision is made at the nth decision point.

Determine the transition probabilities of this decision process.

$$p_{ij} = \Pr[X_n = j | X_{n-1} = i], \qquad i, j = 0, 1$$

Suppose the probability p that the data are interpreted correctly is $\frac{8}{9}$ and the probability q that they are interpreted incorrectly is $\frac{1}{9}$. Suppose that

$$\Pr[X_n = 0 | X_{n-1} = 0 \text{ and the data for the } n\text{th}$$
$$\text{decision are interpreted correctly}] = .95$$

$$\Pr[X_n = 1 | X_{n-1} = 1 \text{ and the data for the } n\text{th}$$
$$\text{decision are interpreted incorrectly}] = .99$$

$$\Pr[X_n = 0 | X_{n-1} = 0 \text{ and the data are interpreted}$$
$$\text{incorrectly}] = 0.5$$

$$\Pr[X_n = 1 | X_{n-1} = 1 \text{ and the data are interpreted}$$
$$\text{correctly}] = 0.25$$

From this description determine the Markov chain transition probability matrix **P**.

(a) Suppose that the "mission" is successful if the fourth decision is correct. ("A missile mission is successful if the fourth corrective signal puts it on course.") If the initial decision is correct, what is the probability that the mission is successful for the above calculated probabilities?

(b) If a decision-maker is a "good one" if in the long run he or she is correct 75% of the time, is the above decision-maker a good one?

(c) If the decision process is to be improved (increase the "long-run" probability of a correct decision) should we improve the data interpretation step (that is, change p and q) or improve the decision-maker (that is, change the probabilities $\Pr[X_n = j | X_{n-1} = i$ and the data are correct (or incorrect)])? Discuss the relative importance of these two ways of improving matters.

35. Show that if a Markov chain is irreducible and if $p_{ii} > 0$ for any $i \in E$ then the chain is aperiodic. (This is a quick test for aperiodicity in many Markov chains.)

36. Suppose state j is recurrent and periodic with period δ. Compute

$$\sum_{n=1}^{\infty} f_{ii}^{(n)}$$

$$\sum_{n=1}^{\infty} f_{ii}^{(n\delta)}$$

37. Show that if i and j are two communicating recurrent states then they are either both periodic with the same period or they are both aperiodic.

38. Prove or disprove: If state j is absorbing then $\{j\}$ is a closed communicating class.

39. Prove: A Markov chain is irreducible if its only closed set is the entire state space.

11 · Continuous Parameter Markov Chains

11.1 Introduction

In the previous chapter our attention was focused on Markov chains with a discrete parameter, that is, sequences of discrete random variables with the Markov property. In this chapter we turn our attention to random processes with a continuous parameter. We will assume that these processes also possess the Markov property as discussed in Section 9.4. Thus, we consider a random process $\mathbf{X} = \{X_t : t \geq 0\}$. For each t, X_t is a discrete random variable, and this uncountable collection of random variables possesses the Markov property. We will call this process a *Markov process* or *Markov chain with a continuous parameter*. As in Chapter 10 we will continue to take the state space E as a countable set that we usually assume to be the nonnegative integers. Then if for some t, $X_t = i$, we will say "the process is in state i at the time t." Thus, we might say that $X_t = 21$ means "the number of mine accidents up to time t is 21."

In these continuous parameter random processes a single experiment now results not in a single number but in an entire function X_t. Thus, one experimental result or single sample point defines X_t for all $t \geq 0$. In the Poisson process example (Section 9.7) each observation consists of an entire sequence of accidents that defines the corresponding counts X_t for all $t \geq 0$. Each sample point will yield a function X_t—a *realization* of the process—that will have a form like the one shown in Figure 11.1. In the Poisson process example these transitions will occur at the times a mine accident occurs and X_t will proceed successively through the integer values $0, 1, 2, \ldots$.

11.2 The Markov Property

We repeat here a modified form of Definition 9.1.

DEFINITION 11.1   Let $\mathbf{X} = \{X_t : t \geq 0\}$ be a discrete state random process. $\mathbf{X}$ is a Markov process if

$$\Pr\left[X_{t_{n+1}} = k_{n+1} | X_{t_n} = k_n, \ldots, X_{t_1} = k_1\right] = \Pr\left[X_{t_{n+1}} = k_{n+1} | X_{t_n} = k_n\right] \quad (11.1)$$

for every $0 < t_1 < t_2 < \cdots < t_n < t_{n+1}$ and $k_j \in E$.

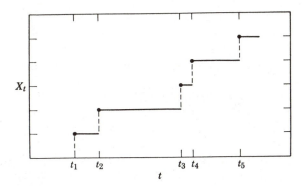

FIGURE 11.1 One realization of a Poisson process X_i = number of accidents up to time t.

Now, using Definition 11.1 it is an easy matter to compute any probability of the form

$$\Pr\left[X_{t_n} = k_n, X_{t_{n-1}} = k_{n-1}, \ldots, X_{t_1} = k_1 | X_0 = k_0 \right]$$

simply by multiplication of the conditional probabilities on the right-hand side of the equation in that Definition. Then by summing on the appropriate k_j one can determine all joint distribution functions for **X** for a given X_0. If we also know

$$\Pr[X_0 = j] = p_j \tag{11.2}$$

then all finite distribution functions are known. Thus, given $\Pr[X_{t_{n+1}} = k_{n+1} | X_{t_n} = k_n]$ and p_j for every $t_n \le t_{n+1}$ and k_n, k_{n+1}, $j \in E$ we have the complete probability structure of **X**. But now notice that life is not quite as simple as it was in the discrete parameter Markov chain. Essentially our argument above says that to have the complete probability structure of a continuous parameter Markov chain we need a collection of *functions*, say, $f_{i,j}(t_n, t_{n+1})$, where

$$f_{ij}(t_n, t_{n+1}) = \Pr\left[X_{t_{n+1}} = j | X_{t_n} = i \right]$$

for all $i, j \in E$, $0 \le t_n \le t_{n+1}$, $n = 0, 1, 2, \ldots$. We also need $\Pr[X_0 = j]$ for $j \in E$. These latter probabilities are usually easy to get, but the $f_{ij}(t_n, t_{n+1})$ can be troublesome.

Let us rewrite the transition probabilities in the form

$$p_{ij}(s, t + s) = \Pr[X_{t+s} = j | X_s = i] \tag{11.3}$$

and call each of these a *transition function*, or the i, jth transition function. This change of notation doesn't help much. We still need to know these for every $i, j \in E$ and $t, s \ge 0$.

As in Chapter 10 we make one more simplifying assumption because it makes messy calculations less messy and, more important, it gives a very important class of processes for modeling real phenomena. If $p_{ij}(s, t + s)$ does not depend on s

(but may depend on t), then we say that the Markov process, **X**, has the *time homogeneity property* or is *time homogeneous*. These Markov processes with the time homogeneity property are the only ones we shall study.

Notice that for Markov processes with the time homogeneity property, if $p_{ij}(t) = p_{ij}(s, t + s)$ is given for every i, $j \in E$ and $t \geq 0$ and if $\Pr[X_0 = j]$ is given for every $j \in E$, then the probability structure of **X** is completely determined. Thus, to study continuous parameter Markov processes we need two pieces of information: a matrix $\mathbf{P}(t)$ whose elements are the $p_{ij}(t)$ and a vector $\mathbf{p}(0)$ whose elements are the $\Pr[X_0 = j] = p_j$. The result then is reminiscent of the discrete parameter Markov chain of Chapter 10. The new idea is that the matrix of one-step transition probabilities that was needed in Chapter 10 is now a matrix of transition functions, $p_{ij}(t)$. For now we will look at a few of the properties of a Markov process when $\mathbf{P}(t)$ and $\mathbf{p}(0)$ are known. In Section 11.4 we will see that in most applications we do not even need to know $\mathbf{P}(t)$ since it can be obtained from more basic information.

11.3   Notation and Basic Properties

We now assume that **X** is a time-homogeneous Markov process and introduce the following notation for the transition probability.

$$p_{ij}(t) = \Pr[X_{s+t} = j | X_s = i] \tag{11.4}$$

for i, $j = 0, 1, 2, \ldots$ and $t \geq 0$ (and for any initial time s for which $\Pr[X_s = i] \neq 0$). Let us also introduce the notation

$$p_j(t) = \Pr[X_t = j] \tag{11.5}$$

for $j = 0, 1, 2, \ldots$ and $t \geq 0$, for the state probabilities at time t.

Clearly the functions $p_{ij}(t)$ and $p_j(t)$ must satisfy

$$\sum_j p_{ij}(t) = 1 \tag{11.6}$$

for each i and t and

$$\sum_j p_j(t) = 1 \tag{11.7}$$

for each $t \geq 0$, since at any given time the process must be in *some* state.

By using the Markov property (11.1) and the basic formulas (2.5) and (2.10) for conditional probabilities, any desired individual or joint probability of the process can be expressed in terms of the transition probabilities $p_{ij}(t)$ and the *initial probabilities*

$$p_i = p_i(0) = \Pr[X_0 = i]$$

For example,

$$p_j(t) = \Pr[X_t = j] = \sum_i \Pr[X_t = j \text{ and } X_0 = i]$$

$$= \sum_i \Pr[X_0 = i] \Pr[X_t = j | X_0 = i]$$

$$= \sum_i p_i p_{ij}(t) \tag{11.8}$$

by using formula (2.10). Formula (11.8) can be expressed more compactly by using vector and matrix notation. Introduce the vectors $\mathbf{p}(t)$, $\mathbf{p}$, and the matrix $\mathbf{P}(t)$ as follows:

$$\mathbf{p}(t) = (p_0(t), p_1(t), p_2(t), \ldots) \tag{11.9}$$

$$\mathbf{p} = \mathbf{p}(0) = (p_0, p_1, p_2, \ldots) \tag{11.10}$$

$$\mathbf{P}(t) = (p_{ij}(t)) = \begin{pmatrix} p_{00}(t) & p_{01}(t) & \cdots \\ p_{10}(t) & p_{11}(t) & \cdots \\ \vdots & \vdots & \end{pmatrix} \tag{11.11}$$

If we observe that the right side of (11.8) is simply a term obtained by multiplying a vector and a matrix, this formula can be written in the form

$$\mathbf{p}(t) = \mathbf{p}\mathbf{P}(t) \tag{11.12}$$

Similarly, joint probabilities can be computed. For example, if $0 < t_1 < t_2$ and if j_1 and j_2 are integers,

$$\Pr[X_{t_1} = j_1 \text{ and } X_{t_2} = j_2] = \Pr[X_{t_1} = j_1] \Pr[X_{t_2} = j_2 | X_{t_1} = j_1]$$

$$= p_{j_1}(t_1) p_{j_1 j_2}(t_2 - t_1)$$

$$= \left[\sum_i p_i p_{ij_1}(t_1) \right] p_{j_1 j_2}(t_2 - t_1)$$

If $0 < t_1 < t_2 < t_3$ and if j_1, j_2, and j_3 are integers,

$$\Pr[X_{t_1} = j_1, X_{t_2} = j_2, X_{t_3} = j_3] = \Pr[X_{t_1} = j_1, X_{t_2} = j_2] \Pr[X_{t_3} = j_3 | X_{t_1} = j_1, X_{t_2} = j_2]$$

$$= \Pr[X_{t_1} = j_1, X_{t_2} = j_2] \Pr[X_{t_3} = j_3 | X_{t_2} = j_2]$$

by using the Markov property (11.2). Thus,

$$\Pr[X_{t_1} = j_1, X_{t_2} = j_2, X_{t_3} = j_3] = \left[\sum_i p_i p_{ij_1}(t_1) \right] p_{j_1 j_2}(t_2 - t_1) p_{j_2 j_3}(t_3 - t_2) \tag{11.13}$$

Formula (11.13) extends to give joint probabilities of any order: For $0 < t_1 < t_2 < \cdots < t_n$,

$$\Pr[X_{t_1} = j_1, X_{t_2} = j_2, \ldots, X_{t_n} = j_n]$$

$$= \left[\sum_i p_i p_{ij_1}(t_1) \right] p_{j_1 j_2}(t_2 - t_1) \cdots p_{j_{n-1} j_n}(t_n - t_{n-1}) \tag{11.14}$$

The analogy with formula (10.8) in the discrete parameter case is clear.

EXAMPLE 11.1

The Poisson process, discussed in Section 9.7, has been mentioned previously as an example of a continuous parameter Markov chain. In Section 9.7 we showed that when events occurred randomly in time (for instance, mine accidents) under certain precisely stated hypotheses the number of events (accidents) in a time interval of length t had a Poisson distribution, with parameter proportional to t:

$$\Pr[n \text{ accidents in time } t] = e^{-\lambda t}(\lambda t)^n/n!, \qquad n = 0, 1, 2, \ldots$$

If X_t represents the number of events in the time interval from 0 to t, then we can easily observe that $\mathbf{X}$ is a Markov chain with infinite state space $0, 1, 2, \ldots$ and with transition probabilities given by

$$p_{ij}(t) = \begin{cases} e^{-\lambda t}\dfrac{(\lambda t)^{j-i}}{(j-i)!}, & \text{if } j \geq i \\ 0, & \text{if } j < i \end{cases} \qquad (11.15)$$

for $i, j = 0, 1, 2, \ldots$. [Obviously, $p_{ij}(t) = 0$ for $j < i$, since the total number of calls that have come in cannot *decrease* over a positive time interval.]

We also found that for such a process the time *between* events (accidents) has a negative exponential distribution with probability density function $f(t) = \lambda e^{-\lambda t}$, $t > 0$. This fact is closely connected with the "forgetfulness" property of the negative exponential discussed in Section 7.7.

EXAMPLE 11.2

Suppose a store has a single phone. Let $X_t = 0$ if the phone is being used and $X_t = 1$ otherwise. Assume that $\{X_t : t \geq 0\}$ is a Markov process with transition matrix

$$\mathbf{P}(t) = \begin{array}{c} \\ 0 \\ 1 \end{array} \begin{bmatrix} \dfrac{1 + 7e^{-8t}}{8} & \dfrac{7 - 7e^{-8t}}{8} \\ \dfrac{1 - e^{-8t}}{8} & \dfrac{7 + e^{-8t}}{8} \end{bmatrix} \begin{array}{c} 0 \qquad\qquad 1 \end{array}$$

Notice that each row sums to one for every $t \geq 0$. Suppose that the initial distribution is

$$\mathbf{p}(0) = \left(\frac{1}{10}, \frac{9}{10} \right)$$

and that we wish to compute

$$\Pr[X_{.5} = 0, X_{.7} = 1 \mid X_0 = 1]$$

and

$$\Pr[X_{.5} = 0, X_{.7} = 1]$$

The first probability is simply

$$\Pr[X_{.5}=0, X_{.7}=1|X_0=1] = p_{10}(.5)\,p_{01}(.2)$$
$$= (.123)(.698)$$
$$= .085$$

To compute $\Pr\{X_{.5} = 0, X_{.7} = 1\}$ we use equation (11.14) yielding

$$\Pr[X_{.5}=0, X_{.7}=1] = p_0 p_{00}(.5)\,p_{01}(.2) + p_1 p_{10}(.5)\,p_{01}(.2)$$
$$= \frac{1}{10}(.141)(.698) + \frac{9}{10}(.123)(.698)$$
$$= .086$$

11.4   Properties of Transition Probabilities

The transition probabilities $p_{ij}(t)$ of a Markov process clearly have the properties

$$0 \le p_{ij}(t) \le 1 \tag{11.16}$$

for each i, j, and $t \ge 0$,

$$\sum_j p_{ij}(t) \equiv 1 \tag{11.17}$$

for each j and $t \ge 0$ [see (10.6)], and

$$p_{ij}(0) = \begin{cases} 1, & \text{if } i=j \\ 0, & \text{if } i \ne j \end{cases} \tag{11.18}$$

The last condition simply states that if the process is in state i at any given time, it is also in state i zero time units later.

In terms of the matrix $\mathbf{P}(t) = (p_{ij}(t))$, (11.17) states that every row adds up to 1, while (11.18) is equivalent to

$$\mathbf{P}(0) = \mathbf{I} \tag{11.18'}$$

$\mathbf{I}$ representing the identity matrix with ones on the main diagonal and zeros elsewhere.

If $s \ge 0$ and $t \ge 0$ are two times, an extremely important identity can be derived by expressing $p_{ij}(t+s)$ in terms of the sets of transition probabilities $\{p_{ij}(t)\}$ and $\{p_{ij}(s)\}$. We notice that to pass from state i to state j in time $t + s$, at time t the process must be in *some* state, say, k. Thus,

$$p_{ij}(t+s) = \sum_k \Pr[X_{t+s} = j, X_t = k | X_0 = i] \tag{11.19}$$

The Markov property (11.1) can then be applied as follows:

$$\Pr[\,X_{t+s}=j,\,X_t=k\,|\,X_0=i\,] = \frac{\Pr[\,X_0=i,\,X_t=k,\,X_{t+s}=j\,]}{\Pr[\,X_0=i\,]}$$

$$= \frac{\Pr[\,X_0=i,\,X_t=k\,]}{\Pr[\,X_0=i\,]}\,\frac{\Pr[\,X_0=i,\,X_t=k,\,X_{t+s}=j\,]}{\Pr[\,X_0=i,\,X_t=k\,]}$$

$$= \Pr[\,X_t=k\,|\,X_0=i\,]\,\Pr[\,X_{t+s}=j\,|\,X_0=i,\,X_t=k\,]$$

$$= \Pr[\,X_t=k\,|\,X_0=i\,]\,\Pr[\,X_{t+s}=j\,|\,X_t=k\,]$$

and by using homogeneity property

$$= p_{ik}(t)\,p_{kj}(s)$$

Thus (11.19) yields

$$p_{ij}(t+s) = \sum_k p_{ik}(t)\,p_{kj}(s) \tag{11.20}$$

for all states i, j, and $t \geq 0$, $s \geq 0$. This identity (11.20) is termed the *Chapman–Kolmogorov Equation*. It can be interpreted as stating that to make a transition from i to j in time $t + s$, one must first make a transition to *some* state k in time t and then a transition from k to j in the remaining time s. This identity allows us to analyze the behavior of the transition probabilities over *long* time intervals in terms of their behavior over *short* time intervals.

Notice that the sum on the right side of (11.20) is of the type obtained when two matrices are multiplied. Thus, (11.20) can be written in matrix form as

$$\mathbf{P}(t+s) = \mathbf{P}(t)\mathbf{P}(s) \tag{11.21}$$

To simplify our discussion we also assume that, considered as functions of t, the functions $p_{ij}(t)$ are sufficiently "smooth." More precisely, we assume that the functions $p_{ij}(t)$ are *continuous* and *differentiable* for $t \geq 0$ with *uniformly bounded derivatives*

$$\left| \frac{dp_{ij}(t)}{dt} \right| < K$$

for all i, j, and $t \geq 0$. This assumption is much stronger than we actually need and much of our discussion would hold true under weaker hypotheses.

Graphically speaking, the functions $p_{ij}(t)$ all start at 0 or 1, depending on whether $i \neq j$ or $i = j$, by (11.18), and they always lie between 0 and 1 by (11.16) (see Figure 11.2). We also notice that the *slope* (derivative) of $p_{ij}(t)$ at $t = 0$ must be ≥ 0 if $i \neq j$ and ≤ 0 if $i = j$.

We state without proof two important properties (but see Section 9.7).

Theorem 11.1

The probability of two or more transitions between states in time Δt is $o(\Delta t)$ as $\Delta t \to 0$.

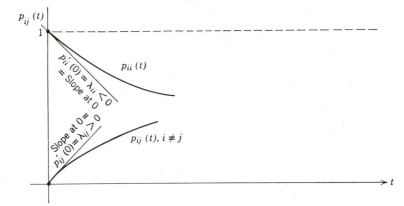

FIGURE 11.2 Typical graphs for transition probability functions.

Note. A function $f(t)$ is termed $o(\Delta t)$ if

$$\lim_{\Delta t \to 0} \frac{f(\Delta t)}{\Delta t} = 0$$

In other words, $f(\Delta t)$ approaches zero so rapidly as $\Delta t \to 0$ that, even when divided by the small quantity Δt, the ratio *still* approaches zero.

Theorem 11.2

The time that the process stays in a given state (time between transitions) is a random variable with a negative exponential distribution.

A proof of this theorem can be constructed by showing that the Markov property implies that the length of time the process stays in a given state must have the "forgetfulness" property of the negative exponential discussed in Section 7.7.

11.5   The Rate Matrix and the Kolmogorov Differential Equations

Let us examine more closely the behavior of the transition probability functions $p_{ij}(t)$ for small values of t. If $i \neq j$, we have noted that $p_{ij}(t)$ starts at 0 when $t = 0$ and initially moves with an upward slope while, if $i = j$, it starts at 1 and initially moves with a downward slope. Let us denote this initial slope by

$$\lambda_{ij} = \frac{dp_{ij}(t)}{dt}\bigg|_{t=0} \tag{11.22}$$

where $\lambda_{ij} \geq 0$ if $i \neq j$ and $\lambda_{ii} \leq 0$. The quantities λ_{ij} can be collected into a matrix Λ that is termed the *rate matrix* of the process or the *generator* of the process.

$$\Lambda = (\lambda_{ij}) = \begin{pmatrix} \lambda_{00} & \lambda_{01} & \lambda_{02} & \cdots \\ \lambda_{10} & \lambda_{11} & \lambda_{12} & \cdots \\ \lambda_{20} & \lambda_{21} & \lambda_{22} & \cdots \\ \vdots & \vdots & \vdots & \end{pmatrix} \tag{11.23}$$

This matrix turns out to be most important in describing the properties of the process.

In case $i \neq j$, the definition of a derivative yields

$$\lambda_{ij} = \left. \frac{dp_{ij}(t)}{dt} \right|_{t=0} = \lim_{\Delta t \to 0} \frac{p_{ij}(\Delta t) - 0}{\Delta t}$$

Hence, for small values of Δt we have approximately

$$p_{ij}(\Delta t) \approx \lambda_{ij} \Delta t \tag{11.24}$$

or, more precisely,

$$p_{ij}(\Delta t) = \lambda_{ij} \Delta t + o(\Delta t) \tag{11.25}$$

where again $o(\Delta t)$ refers to a term so small that, even when divided by Δt, it still approaches zero as $\Delta t \to 0$,

$$\lim_{\Delta t \to 0} \frac{o(\Delta t)}{\Delta t} = 0$$

Note. This $o(\Delta t)$ notation is very useful to describe a term of *smaller order* than Δt. Since addition or multiplication by a scalar does not change the property of approaching zero, even when divided by Δt, this small-o symbol satisfies useful identities such as $o(\Delta t) + o(\Delta t) = o(\Delta t)$ and $ao(\Delta t) = o(\Delta t)$ for all constants a.

If $i = j$, we must modify our procedure slightly:

$$\lambda_{ii} = \left. \frac{dp_{ii}(t)}{dt} \right|_{t=0} = \lim_{\Delta t \to 0} \frac{p_{ii}(\Delta t) - 1}{\Delta t}$$

Hence

$$p_{ii}(\Delta t) \approx 1 + \lambda_{ii} \Delta t \tag{11.26}$$

for small Δt or, more precisely,

$$p_{ii}(\Delta t) = 1 + \lambda_{ii} \Delta t + o(\Delta t) \tag{11.27}$$

Formulas (11.25) and (11.27) are useful in determining the values of the λ_{ij}'s in applications. They are illustrated in Figure 11.3.

Recall formula (11.17):

$$\sum_j p_{ij}(t) = 1$$

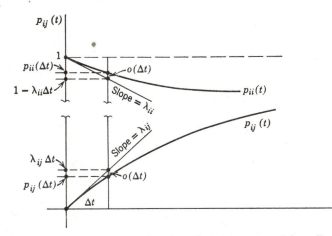

FIGURE 11.3   An approximation of $p_{ij}(t)$ near $t = 0$ by a linear function.

If we differentiate each side of this identity with respect to t and set $t = 0$, we obtain

$$\sum_j \lambda_{ij} = 0 \qquad (11.28)$$

for each i. This derivation is valid if the state space is finite and, under the assumptions that we have made, it can also be justified in the infinite case with the sum in (11.28) being an infinite series.

Since $\lambda_{ij} \geq 0$ for $i \neq j$, it follows from (11.28) that the one negative term λ_{ii} must cancel all the positive terms. Thus, the matrix Λ has the following properties.

(i)   The off-diagonal terms are ≥ 0 and the diagonal terms are ≤ 0.

(ii)   In each row the sum of the off-diagonal terms is equal in magnitude to the diagonal term.

Hence, in applications, if all the λ_{ij} terms with $i \neq j$ are known, properties (i) and (ii) allow one to compute the diagonal terms λ_{ii}.

Now let us return to the Chapman–Kolmogorov equation (11.20) and let us differentiate each side with respect to s:

$$\frac{\partial p_{ij}(t+s)}{\partial s} = p'_{ij}(t+s) = \sum_k p_{ik}(t) \frac{dp_{kj}(s)}{ds}$$

This term-by-term differentiation is valid even in the infinite state case under our assumptions. If we then set $s = 0$, we obtain

$$p'_{ij}(t) = \frac{dp_{ij}(t)}{dt} = \sum_k p_{ik}(t)\lambda_{kj} \qquad (11.29)$$

for $i, j = 0, 1, 2, \ldots$. Formula (11.29) represents a system of linear differential equations that, theoretically at least, can be solved for the unknown functions $p_{ij}(t)$. The initial conditions necessary to the solution are given by (11.18):

$$p_{ij}(0) = \begin{cases} 1, & \text{if } i = j \\ 0, & \text{if } i \neq j \end{cases}$$

The equations (11.29) are termed the *Kolmogorov Forward System of Differential Equations* of the Markov process. By studying their solutions one can determine the complete behavior of the transition probabilities $p_{ik}(t)$ in terms of the *rate constants* λ_{ij}.

If the Chapman–Kolmogorov equation is differentiated with respect to t instead of s, and then t is set equal to 0, another system of differential equations with the same initial conditions is obtained.

$$\frac{dp_{ij}(s)}{ds} = \sum_k \lambda_{ik} p_{kj}(s) \tag{11.30}$$

for $i, j = 0, 1, 2, \ldots$. This is termed the *Kolmogorov Backward System of Differential Equations*. In all the cases in which we are interested both of these systems have the same solution (although this is not always true under more general hypotheses). We use whichever system is simplest in each application that we consider.

The two Kolmogorov systems (11.29) and (11.30) can be written in matrix form as

$$\frac{d\mathbf{P}(t)}{dt} = \mathbf{P}(t)\mathbf{\Lambda} \tag{11.29'}$$

$$\frac{d\mathbf{P}(s)}{d} = \mathbf{\Lambda}\mathbf{P}(s) \tag{11.30'}$$

while the initial condition is

$$\mathbf{P}(0) = \mathbf{I}$$

The analogy is striking between (11.29′) or (11.30′) and the simple ordinary differential equation $dy/dt = \lambda y$, with $y(0) = 1$. This equation of exponential growth has solution $y = e^{\lambda t}$, which suggests that we seek a solution of (11.29′) or (11.30′) in the form

$$\mathbf{P}(t) = e^{\mathbf{\Lambda} t} \tag{11.31}$$

This indeed *is* the solution to these systems, under our assumptions, provided that we define properly what is meant by an exponential with a matrix exponent. In analogy with the series expansion $e^{\lambda t} = \sum_{n=0}^{\infty} \lambda^n t^n / n!$, the proper *definition* of $e^{\mathbf{\Lambda} t}$ turns out to be

$$e^{\mathbf{\Lambda} t} = \mathbf{I} + \sum_{n=1}^{\infty} \frac{\mathbf{\Lambda}^n t^n}{n!} \tag{11.32}$$

where the sum of a series of matrices is the matrix having entries equal to the sum of the corresponding entries in the terms.

Formulas (11.31) and (11.32) provide a formal solution to the problem of finding the transition matrix $\mathbf{P}(t)$ in terms of the rate matrix Λ. In practice, this method is often cumbersome in direct calculations and alternate methods must be sought. Note that the initial distribution $\mathbf{p}$ and the rate matrix Λ characterize a Markov process since the transition matrix $\mathbf{P}(t)$ can be found from Λ by using (11.31).

11.6   Examples

EXAMPLE 11.3

The *Poisson process* discussed in Sections 9.7 and 11.3 is a continuous parameter Markov process with state space $\{0, 1, 2, \ldots\}$. By differentiating formula (11.15) and setting $t = 0$, we find that the rate constants λ_{ij} are given by

$$
\begin{aligned}
\lambda_{i\,i+1} &= \lambda, &&\text{for } i = 0, 1, 2, \ldots \\
\lambda_{i\,i} &= -\lambda, &&\text{for } i = 0, 1, 2, \ldots \\
\lambda_{i\,j} &= 0, &&\text{for all other } i, j
\end{aligned}
$$

$$
\Lambda = \begin{pmatrix}
-\lambda & \lambda & 0 & 0 & \cdots \\
0 & -\lambda & \lambda & 0 & \cdots \\
0 & 0 & -\lambda & \lambda & \cdots \\
\vdots & \vdots & \vdots & \vdots
\end{pmatrix} \tag{11.33}
$$

Alternately the λ_{ij}'s can be determined by using (11.25). We observed in Chapter 7 that the time between events in such a Poisson process has a negative exponential distribution with density function $\lambda e^{-\lambda t}$. Since $p_{i\,i+1}(\Delta t)$ is the probability of an event occurring within time Δt,[1] we have

$$
p_{i\,i+1}(\Delta t) = \int_0^{\Delta t} \lambda e^{-\lambda t}\, dt = 1 - e^{-\lambda \Delta t}
$$

$$
= 1 - \left(1 - \lambda\, \Delta t + \frac{\lambda^2 (\Delta t)^2}{2} \mp \cdots \right)
$$

$$
= \lambda\, \Delta t + (\Delta t)^2 \left(-\frac{\lambda^2}{2} \mp \cdots \right)
$$

$$
= \lambda\, \Delta t + o(\Delta t) \tag{11.34}
$$

where the last term is $o(\Delta t)$, since it has a factor of $(\Delta t)^2$ and, hence, approaches zero as $\Delta t \to 0$ even when divided by Δt.

By comparison with (11.25) we find that $\lambda_{i\,i+1} = \lambda$.

Since $p_{ij}(\Delta t)$ is the probability of $j - i$ events in time Δt, we notice that $p_{ij}(\Delta t) = 0$ if $i > j$ and that $p_{ij}(\Delta t) = o(\Delta t)$ if $j - i \geq 2$ (by Theorem 11.1).

---

[1] We have ignored the possibility of two or more events in time Δt. By Theorem 11.1 this would have probability $o(\Delta t)$ and, hence, would not affect the final result (11.34).

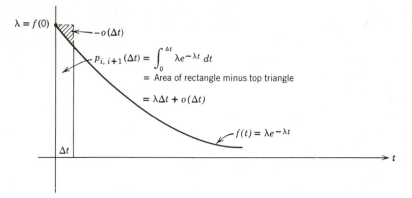

FIGURE 11.4 Pr[one event in time $\Delta t] \approx \lambda \Delta t$ for negative exponential interevent times.

Comparison with (11.25) gives $\lambda_{ij} = 0$ in both these cases. This determines all the elements of the matrix Λ with the exception of the diagonal elements λ_{ii}, which are obtained by using the fact that each row sums to zero.

Formula (11.34) is illustrated in Figure 11.4.

In Example 6.2 the mean of a random variable that has the negative exponential probability density function of the form $\lambda e^{-\lambda t}$ was shown to be $1/\lambda$. Thus, (11.33) can be written in the form

$$p_{i\,i+1}(\Delta t) = \frac{1}{\beta}\Delta t + o(\Delta t) \tag{11.35}$$

where β is the mean time *between* events.

By using (11.33), the Kolmogorov Forward System of differential equations becomes

$$p'_{ii}(t) = -\lambda p_{ii}(t)$$
$$p'_{ij}(t) = -\lambda p_{ij}(t) + \lambda p_{ij-1}(t), \qquad j = i+1, i+2, \ldots$$

This system is essentially the same as the system studied in Section 9.7 but uses a slightly different notation. It can be solved in the same way, with initial conditions given by (11.18). The solution is (11.15).

EXAMPLE 11.4 THE TWO-STATE PROCESS

Consider the simplest possible nontrivial example of a random process as one that has only two states, designated 0 and 1. For instance, a machine could be considered as either in state 1 ("operating" state) or in state 0 ("under repair" state). As we have mentioned previously, if the length of stay in each state has a negative exponential distribution, independent of the lengths of previous stays, then the "forgetfulness" property of the negative exponential can be used to show that the process is Markovian.

Suppose that the length of an operating period has density function $\mu e^{-\mu t}$, $t > 0$, and that the length of a repair period has density function $\lambda e^{-\lambda t}$, $t > 0$. The parameters λ and μ can be interpeted by noting that $1/\lambda$ is the *mean* length of a repair period and that $1/\mu$ is the *mean* length of an operating period. Thus, these quantities can be estimated by observing the *average* operating times and repair times over a lengthy period of observation.

Now $p_{01}(t)$ represents the probability of a machine, which initially is under repair, being in operation t units later. Thus, $p_{01}(\Delta t)$ is the probability of a repair being completed in time Δt, ignoring the possibility of two or more transitions in time Δt as a term of smaller order, $o(\Delta t)$ (Theorem 11.1). Thus, the same argument that we used in deriving formula (11.34) applies:

$$p_{01}(\Delta t) = \int_0^{\Delta t} \lambda e^{-\lambda t}\, dt = 1 - e^{-\lambda \Delta t}$$

$$= \lambda\,\Delta t - \frac{\lambda^2 (\Delta t)^2}{2} \pm \cdots$$

$$= \lambda\,\Delta t + o(\Delta t)$$

By comparison with (11.25), we have $\lambda_{01} = \lambda$. Similarly, $p_{10}(\Delta t) = \mu\,\Delta t + o(\Delta t)$ and $\lambda_{10} = \mu$. Since there are only two states Λ is a 2×2 matrix and the diagonal terms λ_{00} and λ_{11} are determined, since each row sums to 0:

$$\Lambda = \begin{pmatrix} -\lambda & \lambda \\ \mu & -\mu \end{pmatrix}$$

The Kolmogorov Forward Equations are

$$p_{i0}'(t) = -\lambda p_{i0}(t) + \mu p_{i1}(t)$$

$$p_{i1}'(t) = \lambda p_{i0}(t) - \mu p_{i1}(t) \tag{11.36}$$

$i = 0, 1$. Since $p_{i0}(t) + p_{i1}(t) = 1$, $p_{i1}(t) = 1 - p_{i0}(t)$, and the first of these equations can be written as a simple first-order linear differential equation

$$p_{i0}'(t) + (\lambda + \mu) p_{i0}(t) = \mu$$

which has general solution

$$p_{i0}(t) = \frac{\mu}{\lambda + \mu} + C e^{-(\lambda + \mu)t}$$

The constant C can be found by using the fact that $p_{i0}(0) = 1$ if $i = 0$ and $= 0$ if $i = 1$. The final solutions for $p_{i0}(t)$ and $p_{i1}(t) = 1 - p_{i0}(t)$ are

$$p_{00}(t) = \frac{\mu + \lambda e^{-(\lambda + \mu)t}}{\lambda + \mu}, \qquad p_{01}(t) = \frac{\lambda - \lambda e^{-(\lambda + \mu)t}}{\lambda + \mu}$$

$$p_{10}(t) = \frac{\mu - \mu e^{-(\lambda + \mu)t}}{\lambda + \mu}, \qquad p_{11}(t) = \frac{\lambda + \mu e^{-(\lambda + \mu)t}}{\lambda + \mu} \tag{11.37}$$

[Notice that the substitution $p_{i1}(t) = 1 - p_{i0}(t)$ changes the second of the equations (11.36) into a form identical to the first and, hence, the solution that uses the second equation would be the same.] The graphs of the transition probabilities $p_{ij}(t)$ are sketched in Figure 11.5.

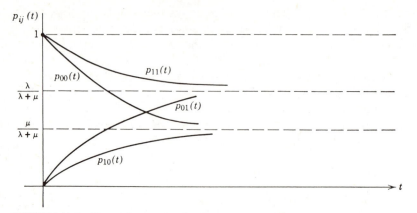

FIGURE 11.5 Transition probabilities for a two-state Markov process.

EXAMPLE 11.5   MACHINERY BREAKDOWN

The situation of the previous example can be generalized. Suppose that we have N identical machines, operating independently and serviced by a single repair crew. If a machine breaks down while another is being repaired it must wait its turn before repairs can start. As before, assume that *each* machine has an operating time that is a negative exponentially distributed random variable with mean $1/\mu$. Also assume that the time required to repair a machine is negative exponentially distributed with mean $1/\lambda$.

Let X_t be the number of machines in operating condition at time t. Thus, **X** is a Markov chain with state space $0, 1, 2, \ldots, N$.

Next we must determine the transition rates λ_{ij}.

First notice the probability that two breakdowns (or a breakdown and a repair completion) will occur exactly simultaneously is clearly 0. It then follows from Theorem 11.1 that $p_{ij}(\Delta t) = o(\Delta t)$ whenever the subscripts i and j differ by more than 1. Thus, $\lambda_{ij} = 0$ when $j \neq i - 1, i, i + 1$.

Let us consider the breakdown and repair sides of the problem separately. $p_{i,i+1}(\Delta t)$ represents the probability of one machine being repaired during a time interval of length Δt (and, hence, the number of operating machines increases from i to $i + 1$). The same argument we used in the previous example applies. Again, by ignoring the possibility of more than one transition (for example, two repairs and one further breakdown in time Δt) as a term of order $o(\Delta t)$ we have

$$p_{ii+1}(\Delta t) = \int_0^{\Delta t} \lambda e^{-\lambda t}\, dt = 1 - e^{-\lambda \Delta t}$$

$$= \lambda\, \Delta t - \lambda^2 \frac{(\Delta t)^2}{2} \mp \cdots$$

$$= \lambda\, \Delta t + o(\Delta t)$$

for $i = 0, 1, 2, , \ldots, N - 1$. Thus,

$$\lambda_{i\,i+1} = \lambda, \qquad \text{for } i = 0, 1, 2, \ldots, N - 1 \tag{11.38}$$

To determine $\lambda_{i,\,i-1}$ we must estimate $p_{i,i-1}(\Delta t)$. This is the probability of one machine breaking down when i are operating. The probability of any given machine breaking down is $\int_0^{\Delta t} \mu e^{-\mu t}\,dt = 1 - e^{-\mu \Delta t} = \mu \Delta t + o(\Delta t)$. The probability of exactly one out of i machines breaking down is i times this, $i[\mu \Delta t + o(\Delta t)]$, *minus* the probability of two or more breaking down. By Theorem 11.1 this latter probability is $o(\Delta t)$. Thus,

$$p_{i\,i-1}(\Delta t) = i[\mu \Delta t + o(\Delta t)] + o(\Delta t)$$
$$= i\mu \Delta t + o(\Delta t)$$

and

$$\lambda_{i\,i-1} = i\mu, \qquad \text{for } i = 1, 2 \ldots, N \tag{11.39}$$

Since $\lambda_{ij} = 0$ unless $j = i - 1$, i, or $i + 1$ and the rows of Λ sum to zero, the terms λ_{ii} can be calculated from (11.38) and (11.39):

$$\lambda_{ii} = -(\lambda + i\mu), \qquad \text{for } i = 0, 1, 2, \ldots, N - 1$$

$$\lambda_{NN} = -N\mu$$

$$\Lambda = \begin{pmatrix} -\lambda & \lambda & 0 & 0 & \ldots & 0 \\ \mu & -(\lambda + \mu) & \lambda & 0 & \ldots & 0 \\ 0 & 2\mu & -(\lambda + 2\mu) & \lambda & \ldots & 0 \\ \vdots & \vdots & \vdots & & \vdots & \\ 0 & 0 & 0 & 0 & \ldots & -N\mu \end{pmatrix} \tag{11.40}$$

The explicit solution of the Kolmogorov differential equations for this Λ is somewhat more complicated.

EXAMPLE 11.6   A SIMPLE QUEUEING MODEL

Consider a situation in which individuals or demands arrive at a certain service point in random sequence, line up, and wait their turn at this point. Then, successively, they have some service operation performed on them and they depart. Such a situation is termed a single-server queue. Examples readily come to mind that involve supermarket checkout lines, telephone traffic, computer demands, factory production lines, and inventories, in addition to many others. Assume that interarrival times and service times are all independent random variables and that the times between arrivals have negative exponential distributions with mean $1/\lambda$ (or, equivalently, that the arrivals form a Poisson process with parameter λ). Assume likewise that the service times also have negative exponential distributions with mean $1/\mu$.

Now let X_t represent the length of the queue or waiting line at time t. Under our assumptions **X** is a Markov process with state space $0, 1, 2, \ldots$. The probability of an arrival in time Δt is $\int_0^{\Delta t} \lambda e^{-\lambda t}\,dt = \lambda \Delta t + o(\Delta t)$ and the probability of a service completion in time Δt is $\mu \Delta t + o(\Delta t)$ (assuming that some individual is actually being served at the beginning of this time interval). Thus, we have $p_{ii+1}(\Delta t) = \lambda \Delta t + o(\Delta t)$, $i = 0, 1, 2, \ldots$, and $p_{ii-1}(\Delta t) = \mu \Delta t + o(\Delta t)$, $i = 1, 2, 3, \ldots$, and hence

$$\lambda_{ii+1} = \lambda, \qquad i = 0, 1, 2, \ldots$$

$$\lambda_{ii-1} = \mu, \qquad i = 1, 2, 3, \ldots \tag{11.41}$$

By Theorem 11.1, $\lambda_{ij} = 0$ otherwise for $i \neq j$, and consequently we have necessarily

$$\lambda_{00} = -\lambda$$

$$\lambda_{ii} = -(\lambda + \mu), \qquad i = 1, 2, 3, \ldots$$

in order that the rows of Λ sum to 0.

$$\Lambda = \begin{pmatrix} -\lambda & \lambda & 0 & 0 & \ldots \\ \mu & -(\lambda + \mu) & \lambda & 0 & \ldots \\ 0 & \mu & -(\lambda + \mu) & \lambda & \\ \vdots & \vdots & \vdots & \vdots & \end{pmatrix} \tag{11.42}$$

The Kolmogorov Forward Equations are

$$p'_{i0}(t) = -\lambda p_{i0}(t) + \mu p_{i1}(t)$$

$$p'_{ij}(t) = \lambda p_{i,j-1}(t) - (\lambda + \mu) p_{ij}(t) + \mu p_{i,j+1}(t), \qquad j = 1, 2, \ldots \tag{11.43}$$

for $i = 0, 1, 2, \ldots$. The solution of this system requires certain special techniques.

EXAMPLE 11.7   THE COMPOUND POISSON PROCESS

A Markov process **X** of wide applicability is one that stays constant between points of change that occur at times given by a Poisson process with parameter λ, but where the amount of each change is given by a positive random variable, independent of the change times, and not always 1 as in the simple Poisson process. These situations occur frequently—for example, orders to a factory (orders are not always for the same dollar amount), claims on an insurance company, certain radioactive disintegrations, and many others.

Given that an increase in X_t occurs at time t, assume that the amount of the increase is one of the integers $1, 2, 3, \ldots$ with probabilities $q_1, q_2, q_3, \ldots, \Sigma q_i = 1$, with the increases being independent of each other and of the times of increase.

Under these circumstances X_t is termed a *compound Poisson process*.

The probability $p_{ij}(\Delta t)$ of a change from state i to state j, $i < j$, in time Δt is obtained by multiplying the probability of any change, $\lambda \Delta t + o(\Delta t)$, by the probability q_{j-i} that the change is of the amount $j - i$. Thus,

$$p_{ij}(\Delta t) = q_{j-i} \lambda \Delta t + o(\Delta t), \qquad \text{for } i < j$$

and

$$\lambda_{ij} = \lambda q_{j-i}, \quad \text{for } i < j$$
$$\lambda_{ij} = 0, \quad \text{for } i > j$$
$$\lambda_{ii} = -\lambda \sum_j q_{j-i} = -\lambda \tag{11.44}$$

$$\Lambda = \begin{pmatrix} -\lambda & \lambda q_1 & \lambda q_2 & \cdots \\ 0 & -\lambda & \lambda q_1 & \cdots \\ 0 & 0 & -\lambda & \cdots \\ \vdots & \vdots & \vdots & \end{pmatrix}$$

The Kolmogorov Forward Equations are

$$p'_{ij}(t) = -\lambda \left[p_{ij}(t) - \sum_{k=i}^{j-1} q_{j-k} p_{ik}(t) \right] \tag{11.45}$$

for $j = i, i+1, i+2, \dots$ and for each fixed i.

A further analysis of the functions $p_{ij}(t)$ is facilitated by using the concept of a probability-generating function that is discussed in Section 6.13. For each t introduce the generating function

$$\psi = \psi_i(t, z) = \sum_j p_{ij}(t) z^j \tag{11.46}$$

for $i = 0, 1, 2, \dots, t \geq 0, |z| \leq 1$. Also define the generating function for the q_k distribution as

$$q(z) = \sum_k q_k z^k$$

Now

$$\frac{\partial \psi}{\partial t} = \sum_j p'_{ij}(t) z^j$$

$$= -\lambda \left[\sum_{j=i}^{\infty} p_{ij}(t) z^j - \sum_{j=i+1}^{\infty} \sum_{k=i}^{j-1} q_{j-k} p_{ik}(t) z^j \right]$$

by using (11.45). The first summation yields ψ again. The double summation adds over all values of the indices k and i that satisfy $i \leq k < j < \infty$, with i fixed. If the order of summation is reversed, we obtain

$$\frac{\partial \psi}{\partial t} = -\lambda \left[\psi - \sum_{k=i}^{\infty} \sum_{j=k+1}^{\infty} q_{j-k} p_{ik}(t) z^j \right]$$

$$= -\lambda \left[\psi - \sum_{k=1}^{\infty} p_{ik}(t) z^k \sum_{j=k+1}^{\infty} q_{j-k} z^{j-k} \right]$$

$$= -\lambda \left[\psi - \psi \sum_{n=1}^{\infty} q_n z^n \right]$$

$$= -\lambda [\psi - \psi q(z)]$$

$$\frac{\partial \psi}{\partial t} = -\lambda \psi [1 - q(z)] \tag{11.47}$$

Now (11.47) is a simple linear differential equation of the form $\partial\psi/\partial t = \text{constant} \cdot \psi$. (It is only formally a partial differential equation.) If $t = 0$, ψ has initial condition

$$\psi_i(0, z) = \sum_j p_{ij}(0) z^j = z^i$$

since $p_{ii}(0) = 1$ and $p_{ij}(0) = 0$ for $i \neq j$. The solution of (11.47), subject to this initial condition, is easily found to be

$$\psi = \psi_i(t, z) = z^i e^{-\lambda t[1 - q(z)]} \tag{11.48}$$

For any given $q(z)$, (11.48) can be expanded in powers of z and the functions $p_{ij}(t)$ can be determined as the coefficients of these powers by using (11.46).

For the compound Poisson process $\{X_t : t \geq 0\}$, it is possible to determine $E[X_t]$ from (11.48). Let m denote the expected size of an increase, that is,

$$m = q_1 + 2q_2 + 3q_3 + \cdots$$

From the results in Section 6.13 we know that

$$m = q'(z)|_{z=1}$$

Let us assume that $X_0 = i$. In that case

$$E[X_t] = \left. \frac{\partial \psi_0(t, z)}{\partial z} \right|_{z=1}$$

$$= \left. \frac{\partial [e^{-\lambda t[1 - q(z)]}]}{\partial z} \right|_{z=1}$$

$$= e^{-\lambda t[1 - q(z)]} \lambda t q(z) q'(z)|_{z=1}$$

$$= \lambda t m$$

Thus, the expected value of X_t is simply the expected number of increase points times the expected size of each increase.

In case each increase of X_t is one unit, then $q_1 = 1$, $q_k = 0$ for $k > 1$, $q(z) = z$, and

$$\psi = e^{-\lambda t(1 - z)}$$

provided that the initial state is assumed to be $i = 0$. By comparison with (7.24) we notice that, in this case, X_t has a Poisson distribution with mean λt. Under these circumstances, the compound Poisson process specializes to an ordinary Poisson process.

Sometimes one encounters compound processes where at an arrival point of the Poisson process there may be no increase. For example, suppose that customers arriving at a roadside vegetable stand form a Poisson process with rate λ' and that each customer purchases i ears of corn with probability q'_i, $i = 0, 1, \dots$. Note that if $q_0 > 0$, a customer might not purchase any corn. Let X_t denote the number of ears of corn sold between time 0 and time t. Even though this situation does not fit into the description of the model since $q'_0 > 0$, by a slight modification we can use the results. It can be shown that the sequence of arrival times at which customers do buy corn, that is, the jump points of X_t, forms a Poisson process with rate $\lambda'(1 - q_0')$. Furthermore, the conditional probability that a customer buys k ears of corn given that the customer buys some corn is

$$\frac{q_k'}{1 - q_0'}, \qquad k = 1, 2, \ldots$$

Hence X_t can be considered as a compound Poisson process with arrival rate

$$\lambda = \lambda'(1 - q_0')$$

and with jump probabilities

$$q_k = \frac{q_k'}{1 - q_0'}, \qquad k = 1, 2, \ldots$$

Note that if

$$m' = q_1' + 2q_2' + 3q_3' + \cdots$$
$$m = q_1 + 2q_2 + 3q_3 + \cdots$$

then $E[X_t]$ can be expressed as either

$$E[X_t] = \lambda m$$
$$E[X_t] = \lambda' m'$$

11.7 Birth-and-Death Processes

All of the examples in the previous section, except the last, shared the property that changes in X_t occurred only in unit steps, up or down; each change in the value of X_t was either $+1$ or -1. Only in the compound Poisson process, among the examples discussed in Section 11.6, could X_t change by 2 or more. In that process changes could be $1, 2, 3, \ldots$ with probabilities $q_1, q_2, q_3, \ldots$. If X_t can only change by $+1$ or -1 at any instant, the process is said to be a *birth-and-death* process. These processes have many applications and an extensively worked-out theory. They are characterized as continuous parameter Markov processes with state space $0, 1, 2, \ldots$ and with rates that satisfy $\lambda_{ij} = 0$ if i and j differ by 2 or more. Thus, the rate matrix has the form

$$\Lambda = \begin{pmatrix} -\lambda_0 & \lambda_0 & 0 & 0 & \cdots \\ \mu_1 & -(\lambda_1 + \mu_1) & \lambda_1 & 0 & \cdots \\ 0 & \mu_2 & -(\lambda_2 + \mu_2) & \lambda_2 & \cdots \\ 0 & 0 & \mu_3 & -(\lambda_3 + \mu_3) & \cdots \\ \vdots & \vdots & \vdots & \vdots & \end{pmatrix} \qquad (11.49)$$

The Kolmogorov Forward Equations are

$$p_{i0}'(t) = -\lambda_0 p_{i0}(t) + \mu_1 p_{i1}(t)$$
$$p_{ij}'(t) = \lambda_{j-1} p_{ij-1}(t) - (\lambda_j + \mu_j) p_{ij}(t) + \mu_{j+1} p_{ij+1}(t), \qquad j = 1, 2, \ldots$$

Historically this structure of transition probabilities probably arose in connection with bacterial studies. One interpretation with regard to these studies is the following.

Suppose that we are observing a colony of bacteria. At each instant the colony can increase by one member. The probability of an increase depends on how many members there are in the colony, and, in an interval of length Δt,

$$p_{i\,i+1}(\Delta t) = \lambda_i \Delta t + o(\Delta t)$$

is the probability of an increase, given that the population is of size i at the start of the interval. This increase is called a *birth*.

Furthermore, suppose that members of the colony die. At each instant the colony can decrease by one because of a death. The probability of a death in an interval of length Δt depends on how many are present at the start of the interval, and it is given by

$$p_{i\,i-1}(\Delta t) = \mu_i \Delta t + o(\Delta t)$$

The biological questions involve finding the transition and state probabilities as functions of t, or finding the probability that the colony eventually dies out, or finding the probability that its size exceeds some limit or, if it does eventually die out, estimating how long it takes to do so.

Questions like the ones asked of the bacterial population are often asked of other systems that have nothing to do with biological populations, such as the birth-and-death examples 11.3 to 11.6 in Section 11.6.

Exercises 11

1. Let X_t be the Poisson process in Example 11.1. By using (11.15) show the following.

   (a) $p_{ij}(\Delta t) = o(\Delta t)$, whenever $|j - i| > 1$.

   (b) $\sum_{k=i}^{j} p_{ik}(t) p_{kj}(s) = p_{ij}(t + s)$, whenever $i < j$, $t \geq 0$, and $s \geq 0$.

2. Consider the process $\{ X_t : t \geq 0 \}$ described in Example 11.2.

   (a) Compute the following: $\Pr[X_2 = 0]$, $\Pr[X_2 = 0 | X_0 = 0]$, $\Pr[X_1 = 0, X_{.6} = 1, X_{1.1} = 1 | X_0 = 0]$, and $\Pr[X_{1.1} = 0, X_{.6} = 1, X_1 = 0]$.

   (b) Compute $\mathbf{p}(0)$. Compare the result with (11.18).

   (c) Sum each row of $\mathbf{P}(t)$. Compare the result with (11.17).

   (d) Compute $\mathbf{P}(t)\mathbf{P}(s)$ and $\mathbf{P}(t + s)$. Compare the results with (11.21).

   (e) Compute $\mathbf{P}'(t)$. Now using (11.22) determine Λ.

   (f) Compute $\mathbf{P}(t)\Lambda$ and $\Lambda\mathbf{P}(t)$. Compare $\mathbf{P}'(t)$, $\mathbf{P}(\Lambda)$, and $\Lambda\mathbf{P}(t)$ with (11.29') and (11.30').

   (g) Sum the rows of Λ.

3. Determine the missing elements of Λ.

$$\begin{array}{cc} & \begin{array}{ccc} 0 & 1 & 2 \end{array} \\ \Lambda = \begin{array}{c} 0 \\ 1 \\ 2 \end{array} & \left[\begin{array}{ccc} -5 & & 3 \\ & -6 & 6 \\ & & 0 \end{array} \right] \end{array}$$

4. Two communications satellites are placed in orbit. The lifetime of a satellite is negative exponential with mean $1/\mu$. If one fails a replacement is sent up. The time necessary to prepare and send up a replacement is negative exponential with mean $1/\lambda$. Let X_t = the number of satellites in orbit. Assume that X_t is a Markov process with state space $\{0, 1, 2\}$.

   (a) Show that the rate matrix Λ is given by

$$\Lambda = \begin{pmatrix} -\lambda & \lambda & 0 \\ \mu & -(\lambda + \mu) & \lambda \\ 0 & 2\mu & -2\mu \end{pmatrix}$$

   (b) Set up the Kolmogorov Forward Equations for this process.

   (c) Set up the Kolmogorov Backward Equations and compare to (b).

5. Write and solve the Kolmogorov Backward Equations for the Poisson process, Example 11.3. [*Hint*: Remember that $p_{ij}(t) = 0$ if $i > j$.]

6. Show for Example 11.4 of Section 11.6 that $p_{10}(\Delta t) = \mu \Delta t + o(\Delta t)$.

7. For Example 11.4 find the Kolmogorov Backward Equations and solve for $p_{ij}(t)$.

8. Write the Kolmogorov Forward and Backward Equations for Example 11.5.

9. In Example 11.5 set up the Forward Equations assuming that there are three identical repair crews, each with the same service rate λ. Each crew services at most one machine at a time and the crews do not cooperate with each other. That is, if i machines are down for repairs, then for $i < 3$, $3 - i$ crews stand idle awaiting a new breakdown and i crews perform service on the down machines. Assume $N = 7$.

10. In the birth-and-death process of Section 11.7 suppose that the "birth rate" λ_i is inversely proportional to one more than the number in the population and the "death rate" μ_i is constant, $\mu_i = \mu$. Write the Kolmogorov Forward Equations and determine Λ.

11. In the simple queueing model of Example 11.6 suppose that the waiting line can be at most K individuals (including the one in service). Write the Kolmogorov Forward Equations and determine Λ.

12. Suppose that a company has four operators serving a single telephone number, that is, up to four different people can have dialed the same number and be talking to the operators; however, anybody who calls while all four

operators are busy will receive a busy signal. Let X_i denote the number of busy operators. Assume arrival and individual service processes are Poisson with rates λ and μ. Determine the Kolmogorov Forward Equations and Λ.

13. Suppose that arrivals to a service center are composed of exactly two units. Each arrival unit is serviced one at a time. With this single change, write the Kolmogorov Forward Equations for Example 11.6 and determine Λ.

14. The birth-and-death process of Section 11.7 is called a *pure birth process* if $\mu_i = 0$ for every i. The Poisson process is an example of this process. Suppose that $\lambda_i = i\lambda$ for $i = 0, 1, 2, \ldots$. This process would occur in a system, such as a biological population, in which the probability of a birth is proportional to the number of individuals present to act as parents. Assume that initially $X_0 = n$. Show that

$$\Pr[\,X_t = i\,] = p_{ni}(t) = \binom{i-1}{n-1} e^{-n\lambda t}(1 - e^{-\lambda t})^{i-n}, \qquad \text{for } i \geq n$$

15. A birth-and-death process is called a *pure death process* if $\lambda_i = 0$ for every i. Suppose that $\mu_i = i\mu$ for $i = 1, 2, 3, \ldots$ and that initially $X_0 = n$. Show that $\Pr[\,X_t = i\,] = p_{ni}(t)$ has a binomial distribution with parameter $p = e^{-\mu t}$, for $i = 0, 1, 2, \ldots, n$.

16. An insurance company might be faced with the following problem. The time between accidents is a negative exponentially distributed random variable with parameter λ_1. The dollar cost to the insurance company of each accident is a random variable with a Poisson density function with parameter λ_2. Let $\mathbf{X}$ denote the total claims in $[0, t]$. Assume that $\mathbf{X}$ is a Markov chain.

(a) Determine the generating function of X_t.

(b) Compute $E[\,X_t\,]$.

17. Find the transition probabilities $p_{ij}(t)$ for the two-state process, discussed in Example 11.4, by using the methods of Section 11.8. [Use linear algebra techniques to determine a 2×2 matrix $\mathbf{S}$ so that $\Lambda = \mathbf{S}^{-1}\mathbf{DS}$, where $\mathbf{D}$ is diagonal.] Compare with (11.37).

18. Use the methods of Exercise 17 to determine the transition probabilities $p_{ij}(t)$ for the machinery repair in Example 11.5 in the case $\lambda = 2$, $\mu = 1$, and $N = 2$.

19. Answer Exercise 18 for general λ and μ in the case $N = 2$.

20. Consider the following queueing system with reneging. Customers arrive according to a Poisson process with rate λ. The service times are exponentially distributed with mean $1/\mu$. There is a single-server and an infinite waiting area. Each customer is willing to wait for a random length of time that is exponentially distributed with mean $1/\theta$. If the customer has not started service within this time the customer reneges, that is, leaves the queueing system. Determine Λ. [*Hint*: $\lambda_{32} = (\mu + 2\theta)$.]

21. Develop a simple model of the behavior of people at the checkout counter in a supermarket. Assume that there are two checkout counters and two lines. In addition, assume arrival and individual service processes are Poisson with rates λ and μ, and that each arriving customer joins the shortest queue. After joining a queue the customer does not switch to the other queue. If both lines are the same length, the customer is equally likely to enter either line. Let $X_t = (i, j)$ if there are i customers in one line and j in the other. Determine $\lambda_{ij,kl}$. [*Hint*: If $i > j > 0$, the possible transitions are to states $(i, j + 1)$, $(i - 1, j)$, and $(i, j - 1)$.]

22. Suppose that a man operates a small auto collision shop. The arrival process of cars needing repair is a Poisson process with rate λ. The man first bumps a car, then paints the car, and then starts on the next car. The length of time to bump a car is an exponentially distributed random variable with mean $1/\mu_1$ and the length of time to paint a car is an exponentially distributed random variable with mean $1/\mu_2$. Model the shop as a Markov process $\{X_t : t \geq 0\}$, where $X_t = (i, j)$ if there are $i > 0$ cars in the shop and the car being repaired is in state j. Let $j = 0$ mean that the car is in the bumping stage and $j = 1$ that the car is in the painting stage. If the shop is empty, let $X_t = 0$. Determine Λ.

23. In the mine accident example (see Section 9.7) suppose that, measured in units of $100,000, the amount of damage done by an accident is a random variable Y that is a geometrically distributed random variable with $p = .2$. Let Z_t be the total amount of damage up to time t. Determine the probability-generating function, the mean, and the variance of $\{Z_t : t \geq 0\}$. For $t = 100$ give an estimate of the probability that $Z_t \geq 30$.

12 · Limiting Distribution of Continuous Parameter Markov Processes

12.1 Introduction

In Chapter 11 we derived a system of differential equations, the Kolmogorov equations, that, as we observed, completely defined the transition probabilities of the continuous parameter Markov process (with the initial conditions $\mathbf{P}(0) = \mathbf{I}$). Our job in that chapter was to show the basic probability structure of the Markov process and to present several examples of how these processes arise naturally in real-life processes. Except for the simplest examples, such as the Poisson process (Example 11.3) and the two-state process (Example 11.4), we did not attempt to solve the differential equations explicitly. This was not an oversight on our part for two reasons. First, the solution of the Kolmogorov equations for all but the simplest cases often requires methods that are beyond the scope of this book. Second, in many applied problems one's primary interest is to study the behavior of the solutions for large values of t. It turns out that simple methods are available for studying this behavior that do not require solving the differential equations themselves explicitly. In this chapter we shall explore and exploit these methods to get limiting values of the probabilities as $t \to \infty$, and, hence, approximate values for large t.

In effect, our approach follows rather closely the approach of Section 10.8, where we studied Markov chains with a discrete parameter. We recall from our discussion that there were cases where the limiting probabilities exist but that there were also cases in which these limits did not themselves form a probability distribution. Thus, in Theorems 10.4, 10.5, and 10.6 we observed that if the Markov chain was finite, irreducible, and aperiodic, then there always existed a limiting distribution with the property that

$$\lim_{n \to \infty} p_{ij}^{(n)} = v_j, \text{ not depending on } i$$

$$\mathbf{v} = \mathbf{vP}$$

$$0 \le v_j \le 1$$

$$\sum_j v_j = 1$$

Our job in this chapter is to study the behavior of the $p_{ij}(t)$ for $t \to \infty$, to try to understand under what conditions these limits exist as probabilities (that is, under what conditions a limiting distribution exists for the continuous parameter Markov chain), and to find ways of determining these distributions. Of course one way to find the limiting distribution is to solve the Kolmogorov differential equations in Chapter 11 and then to compute $\lim_{t \to \infty} p_{ij}(t)$. It is this method that we want to avoid, because of the difficulty of determining $p_{ij}(t)$. Proofs of some of the assertions in this chapter require analysis well beyond the scope of this book. In these cases we will state the results without proof. The interested student can consult one of the listed references for more details.

Thus, in this chapter we wish to explore the following problem.

If $p_j(t) = \Pr[X_t = j]$, where X_t is a continuous, parameter Markov process:

(a) Do the limits

$$q_j = \lim_{t \to \infty} p_j(t) \tag{12.1}$$

exist? If so, how do they depend on the initial probability distribution

$$\{ p_i = p_i(0) \}$$

(b) How can these limits be computed?

(c) Do the quantities $\{q_j\}$ form a probability distribution? That is, does Σq_j equal 1?

12.2 The Existence of the $\{q_i\}$

It can be shown that, under the assumptions of Chapter 11, all of the limits q_j defined by (12.1) do, in fact, exist and that $0 \le q_j \le 1$ for each j.

In general these limits will depend on the initial value, X_0 (or on the probability distribution of X_0). For example, if X_t represents the number of bacteria in a sealed culture at time t, then the long-run behavior of X_t will depend crucially on whether the initial value of X_0 is 0 or not. In this case, 0 is termed an *absorbing state*, since if the process ever enters this state it never leaves it (see Sections 10.7 and 10.9).

There is one important situation in which the initial distribution does not affect the limiting probabilities. This occurs when the process is *irreducible*.

In analogy with the discrete parameter case, a continuous parameter Markov process is termed *irreducible* if, for each i and j, the transition probability $p_{ij}(t)$ is strictly positive for some t. That is, a chain is irreducible if there is a positive probability of eventually reaching any state j from any state i.

Theorem 12.1

Under our assumptions, if the Markov process **X** is irreducible, then the limits q_j exist and always have the same values, regardless of the initial distribution at time 0.

Since $p_{ij}(t)$ can be interpreted as simply $p_j(t) = \Pr[X_t = j]$ provided that the initial value X_0 is assumed to be equal to i ($p_i = 1$, $p_k = 0$ for $k \neq i$), it follows that the statement that the $\{q_j\}$ exist and are independent of the initial distribution is equivalent to the formula

$$\lim_{t \to \infty} p_{ij}(t) = q_j, \qquad \text{for all } i \tag{12.2}$$

12.3   Determination of $\{q_i\}$—The Normal Equations

It seems reasonable to hope that if the functions $p_{ij}(t)$ approach the constants q_j as $t \to \infty$, then the derivatives of these functions should approach 0 (since the derivative of a constant is 0). This is indeed true, although a rigorous proof is quite deep.

Theorem 12.2

Under our assumptions

$$\lim_{t \to \infty} \frac{dp_{ij}(t)}{dt} = 0 \tag{12.3}$$

for each i and j.

If one lets $t \to \infty$ in the Kolmogorov Forward Equations (11.29) and formally uses (12.2) and (12.3), the following system of linear homogeneous equations for the probabilities q_i is obtained (the *normal equations*):

$$\sum_k q_k \lambda_{kj} = 0, \qquad \text{for } j = 0, 1, 2, \ldots \tag{12.4}$$

In matrix form this system can be written as

$$\mathbf{q}\Lambda = 0 \tag{12.5}$$

where $\mathbf{q} = (q_0, q_1, q_2, \ldots)$. Even for infinite state spaces (12.4) can be justified rigorously. (The proof that the limit of the infinite series is the series of limits requires great care here.)

Thus, we observe that the limits q_j do, in fact, exist and are independent of the initial conditions if the Markov chain is irreducible. We can also hope that in these cases $\{q_j\}$ can be determined by solving the system (12.4) or (12.5). This gives a partial answer to (a) and (b) of Section 12.1. Question (c) must still be explored.

12.4   The Existence of a Limiting Probability Distribution

A homogeneous system of equations like (12.4) or (12.5) always has the trivial solution $\mathbf{q} = (0, 0, 0, \ldots)$ as one possible solution. If another solution exists, then an infinite number of solutions can be found by multiplying by scalars. Thus, some additional information is needed to uniquely determine the q_j's. This additional information is provided in the following theorem.

If **X** is an irreducible Markov process, then it can be shown that *either* the limiting probabilities are identically 0, $q_j \equiv 0$ for all j, *or* that they are all positive and form a probability distribution, $q_j > 0$ for all j and $\Sigma q_j = 1$. In the second case, we say that the process is *positive recurrent*.

Combining the results stated in this section with the results of the previous section, we can show the following.

Theorem 12.3

Let **X** be an irreducible Markov process that satisfies our assumptions.

(i)   If the system (12.4) or (12.5) has the trivial solution $(0, 0, 0, \ldots)$ as the only solution for which $\Sigma |q_j| < \infty$, then $q_j \equiv 0$ for all j.

(ii) If the system (12.4) or (12.5) has exactly one solution

$$\mathbf{q} = (q_0, q_1, q_2, \ldots)$$

in which each $q_j > 0$ and $\Sigma q_j = 1$, then the system is positive recurrent, and this solution gives the limiting probabilities.

(iii) If the state space is finite, case (ii) must occur.

Suppose that the process **X** does have a limiting distribution $\{q_j\}$ that is itself a probability distribution. If that limiting distribution is then taken to be the initial distribution at $t = 0$, $p_i = p_i(0) = q_i$, it follows that

$$p_j(t) \equiv q_j$$

for every $t \geq 0$. That is, the state probability functions remain constant for all t.

We notice, of course, that (iii) is just a restatement of Theorem 10.6 for the continuous parameter case, and that (i) and (ii) are restatements of Theorem 10.7.

12.5   Irreducible Examples

Examples 11.4, 11.5, and 11.6, as well as the general birth-and-death model in Section 11.7, are irreducible, provided that all the parameters involved are nonzero. The student should convince him or herself of the truth of this assertion. We shall illustrate how the results of the preceding section can be applied to obtain the limiting stationary distribution in these cases. The following examples refer to the example numbers in Section 11.6. Notice that Examples 11.4 and 11.5 have finite state spaces and hence are known to be positive recurrent by the theorem of the preceding section.

EXAMPLE 12.1   THE MACHINE OPERATION CYCLE   (see Example 11.4)

In this example the limiting probabilities can be computed explicitly by using formulas (11.37) for the $p_{ij}(t)$.

$$q_0 = \lim_{t \to \infty} p_{i0}(t) = \frac{\mu}{\lambda + \mu}$$

$$q_1 = \lim_{t \to \infty} p_{i1}(t) = \frac{\lambda}{\lambda + \mu} \qquad\qquad (12.6)$$

These results could also be found by solving the system (12.4), which in this case becomes

$$-\lambda q_0 + \mu q_1 = 0$$
$$\lambda q_0 - \mu q_1 = 0$$

Each equation is equivalent to $q_1 = (\lambda/\mu)q_0$, and hence $q_0 + q_1 = [(\lambda + \mu)/\mu]q_0$. Thus, (12.6) is easily seen to be the only solution of this system that satisfies $q_i > 0$, $q_0 + q_1 = 1$. By (ii) of the preceding theorem the system is positive recurrent with these limiting probabilities. The approach of the $p_{ij}(t)$ to their limiting values is illustrated in Figure 11.5.

EXAMPLE 12.2   THE MACHINERY BREAKDOWN EXAMPLE WITH SEVERAL MACHINES   (see Example 11.5)

In this case the system of normal equations (12.4) becomes

$$-\lambda q_0 + \mu q_1 = 0$$
$$\lambda q_{i-1} - (\lambda + \mu i) q_i + \mu(i+1) q_{i+1} = 0, \qquad \text{for } i = 1, 2, \dots, N-1$$
$$\lambda q_{N-1} - \mu N q_N = 0$$

From the first equation we have

$$q_1 = \left(\frac{\lambda}{\mu}\right) q_0$$

The second equation (with $i = 1$) can be written as

$$q_2 = \frac{1}{2\mu}\left[(\lambda + \mu)q_1 - \lambda q_0\right]$$

$$= \frac{1}{2\mu}\left[(\lambda + \mu)\left(\frac{\lambda}{\mu}\right)q_0 - \lambda q_0\right]$$

$$= \frac{1}{2\mu}\left[\left(\frac{\lambda^2}{\mu}\right)q_0\right]$$

$$= \frac{1}{2}\left(\frac{\lambda}{\mu}\right)^2 q_0$$

This can now be substituted in the equation with $i = 2$, et cetera, and the terms

$q_3, q_4, \ldots$ solved for recursively in terms of q_0. The general result is

$$q_i = \frac{1}{i!}\left(\frac{\lambda}{\mu}\right)^i q_0, \qquad \text{for } i = 0, 1, 2, \ldots, N \tag{12.7}$$

(The student should check this.) To find q_0 we simply set

$$\sum q_i = \left[\sum \frac{1}{i!}\left(\frac{\lambda}{\mu}\right)^i\right] q_0 = 1$$

and find

$$q_0 = \frac{1}{\displaystyle\sum_{i=0}^{N} \frac{1}{i!}\left(\frac{\lambda}{\mu}\right)^i} \tag{12.8}$$

By the theorem in the preceding section, (12.7) and (12.8) do, in fact, give the limiting probabilities for this positive recurrent system.

EXAMPLE 12.3 THE SIMPLE QUEUEING MODEL (see Example 11.6)

In this case the normal equations (12.4) take the form

$$-\lambda q_0 + \mu q_1 = 0$$

$$\lambda q_{i-1} - (\lambda + \mu) q_i + \mu q_{i+1} = 0, \qquad \text{for } i = 1, 2, 3, \ldots \tag{12.9}$$

As in the preceding example, we can solve recursively for $q_1, q_2, q_3, \ldots$ in terms of q_0. The result is

$$q_i = (\lambda/\mu)^i q_0, \qquad \text{for } i = 0, 1, 2, \ldots \tag{12.10}$$

(The student should check this.) In order that $\sum q_i = 1$, we must have

$$q_0 \sum_{i=0}^{\infty} \left(\frac{\lambda}{\mu}\right)^i = 1 \tag{12.11}$$

Two cases must now be considered.

(i) $\lambda < \mu$. In this case the series in (12.11) converges and can be evaluated by using the formula for the sum of a geometric series (Appendix A.3):

$$q_0 \cdot \frac{1}{1 - \lambda/\mu} = 1$$

$$q_0 = 1 - \frac{\lambda}{\mu}$$

$$q_i = \left(1 - \frac{\lambda}{\mu}\right)\left(\frac{\lambda}{\mu}\right)^i, \qquad \text{for } i = 0, 1, 2, \ldots \tag{12.12}$$

Formula (12.12) gives the only solution of the system (12.9) that satisfies $q_i > 0$, $\sum q_i = 1$, and hence by (ii) of the theorem of the preceding section the system is positive recurrent with limiting probabilities given by (12.12).

(ii) $\lambda \geq \mu$. In this case the series in (12.11) diverges, and thus equation (12.11) cannot be satisfied for any choice of q_0. The series Σq_j will diverge for any choice of $q_0 > 0$. By (i) of the theorem of the preceding section it follows that

$$q_j \equiv 0, \qquad \text{for all } j$$

and thus that the system is not positive recurrent.

By formula (12.12) we observe that in the positive recurrent case, $\lambda < \mu$, the limiting distribution of X_t is *modified geometric* with common ratio $p = \lambda/\mu$ (see Section 7.4). It will have mean and variance

$$\frac{p}{1-p} = \frac{\lambda}{\mu - \lambda} \qquad \text{and} \qquad \frac{p}{(1-p)^2} = \frac{\lambda\mu}{(\mu - \lambda)^2}$$

EXAMPLE 12.4 THE GENERAL BIRTH-AND-DEATH MODEL

This process defined in Section 11.7 is irreducible only if all parameters are nonzero. For example, if $\lambda_0 = 0$, then the state 0 becomes an absorbing state that once entered can never be left, $p_{0j}(t) = 0$ for every $j > 0$ (no "birth" from state 0 is possible). Let us assume that all the parameters λ_i and μ_i are nonzero.

For this case the system (12.4) becomes

$$-\lambda_0 q_0 + \mu_1 q_1 = 0 \tag{12.13}$$

$$\lambda_{j-1} q_{j-1} - \left(\lambda_j + \mu_j\right) q_j + \mu_{j+1} q_{j+1} = 0, \qquad \text{for } j = 1, 2, 3, \ldots \tag{12.14}$$

Equation (12.14) can be written as

$$\mu_{j+1} q_{j+1} - \lambda_j q_j = \mu_j q_j - \lambda_{j-1} q_{j-1} \tag{12.15}$$

By (12.13), $\mu_1 q_1 - \lambda_0 q_0 = 0$. If this is substituted into (12.15) with $j = 1$, we find that $\mu_2 q_2 - \lambda_1 q_1 = 0$. Substituting again, we obtain by successive substitution

$$\mu_{j+1} q_{j+1} - \lambda_j q_j = 0, \qquad \text{for each } j = 0, 1, 2, \ldots$$

or

$$q_{j+1} = \frac{\lambda_j}{\mu_{j+1}} q_j \tag{12.16}$$

Successive application of (12.16) yields

$$q_n = \frac{\lambda_0 \lambda_1 \ldots \lambda_{n-1}}{\mu_1 \mu_2 \ldots \mu_n} q_0, \qquad n = 1, 2, \ldots \tag{12.17}$$

If the series

$$S = 1 + \frac{\lambda_0}{\mu_1} + \frac{\lambda_0 \lambda_1}{\mu_1 \mu_2} + \cdots + \frac{\lambda_0 \lambda_1 \ldots \lambda_{n-1}}{\mu_1 \mu_2 \ldots \mu_n} + \cdots \tag{12.18}$$

is convergent, then by (ii) of the theorem of the preceding section the system is positive recurrent with limiting probabilities given by (12.17) with

$$q_0 = \frac{1}{S}$$

If the series is divergent, then case (i) applies and the limiting probabilities are

$$q_i = 0, \qquad \text{for } i = 0, 1, 2, \ldots$$

From this example we obtain the following very important and useful theorem.

Theorem 12.4

In an irreducible birth-and-death process with $\lambda_i > 0$, $i = 0, 1, 2, \ldots$ and $\mu_i > 0$, $i = 1, 2, \ldots$, a limiting stationary distribution exists if the sum (12.18) converges.

12.6   An Interpretation of the $\{q_i\}$

In many practical problems for which a limiting distribution exists, we interpret q_j as the probability that the system will be in state j a "long time" after $t = 0$. As a consequence, we talk of q_j as the *long-run probability* of being in state j, or the *steady-state probability* for state j. In many cases, we can show that $p_{ij}(t)$ approaches q_j at a fairly rapid rate as $t \to \infty$. Thus, in many practical applications we assume that these limiting probabilities are, in fact, good measures of a system's behavior even if the system is observed for modest t-values instead of for large t. Of course, from our study, we do not know the rate at which a particular process approaches its steady-state probabilities, and hence care must be exercised in this interpretation of the q_j.

There is another interpretation of the $\{q_j\}$ that make them useful in application. Our approach to the $\{q_j\}$ shows that they are limiting *state* probabilities. That is, for large t, q_j is the probability of *being in state j* (or alternatively the $\{q_j\}$ are the limiting transition probabilities). With a deeper study of the theory of stochastic processes q_j can also be interpreted as *the proportion of time that the system spends in state j* over the long run, assuming the process is irreducible. This is an interesting relation. What we are asserting is that state probabilities can be interpreted as proportions of time. The limiting probability of being in state j is equal to the proportion of time spent in the state. Clearly these two concepts are quite different, and we might not expect, in general, that probabilities with respect to states were in any way related to the length of time that a state is occupied. It is of practical value (and theoretically verifiable) that the relation exists. As one consequence, the relation implies that to estimate empirically the probabilities q_j we can proceed in one of two ways. First, we could observe many different realizations of the Markov process, long after their commencements. From these observations we could form a relative frequency ratio (as we did in Chapter 2) of the number of observations in state j relative to the total number of observations. The resulting proportion of observations of j would then serve as an estimate of q_j. Alternatively, we could observe only *one* realization of the Markov process over a long period of time. From this observation we could form the relative frequency ratio of the total time spent in state j relative to the total time of observation. The resulting *proportion* of time in state j would also serve as an estimate of q_j. The importance of this result is that in many applied problems we have available a reasonably long series of observations on a single realization of the process. But many different series of observations are not available, because of the expense of time consumed in collecting this amount of data. Nonetheless, from our remark, we can still estimate the q_j from this one series of observations by using the proportion of time the process spends in state j.

EXAMPLE 12.5

Suppose that a particular production line can be modeled as a two-state Markov process as in Example 11.4, in which state 0 means that the production line is stopped and state 1 means that the production line is working. Assume that the company loses money at a rate of \$$a$ per hour that the line is stopped. We wish to determine the average cost per hour due to line stoppages. Assuming $\lambda, \mu > 0$, q_j can be interpreted as the long-run proportion of time that the system is in state j, since the process is irreducible. Hence, the production line is stopped q_0 of the time. The average long-run cost per hour is

$$aq_0 + 0q_1 = \frac{a\mu}{\mu + \lambda}$$

12.7   Some Nonirreducible Processes

One way in which a process can fail to be irreducible is to have one or more *absorbing states* [that is, states i for which $p_{ij}(t) = 0$ for every $j \neq i$] or *absorbing classes* of states. An absorbing class C is a class of states such that $p_{ij}(t) > 0$ for each $i, j \in C$, and some $t > 0$, while $p_{ij}(t) = 0$ for $i \in C$, $j \notin C$. If such states or classes exist, then the existence of a unique limiting distribution cannot be guaranteed, since the behavior of the process as $t \to \infty$ might depend crucially on whether or not it started in an absorbing state.

For a nonirreducible process, by a deeper analysis one can show that as $t \to \infty$ one of two things must happen to X_t: either it eventually enters one of the absorbing states or classes (is *absorbed*) or it eventually grows larger and larger, $|X_t| \to \infty$. Of course, if the state space is finite only the first possibility exists.

EXAMPLE 12.6

Suppose that we consider the birth-and-death model (Section 11.7), with

$$\lambda_0 = 0$$
$$\lambda_i = i\lambda, \qquad \text{for } i = 1, 2, 3, \ldots$$
$$\mu_i = i\mu, \qquad \text{for } i = 1, 2, 3, \ldots$$

In this case, X_t could be interpreted as the number of bacteria or individuals present in a certain sealed environment and λ and μ are the per unit birth and death rates. The terms $i\lambda$ and $i\mu$ indicate that the probability of a birth or death is proportional to the number present. We observe that state 0 is an absorbing state. Thus, as $t \to \infty$, either X_t eventually enters state 0 (the population becomes extinct) or $X_t \to \infty$. The important question here is: Which of these possibilities happen, given that the population starts with i individuals? Thus, we wish to compute

$$\alpha_i = \Pr[\, X_t \text{ is absorbed in } 0 \,|\, X_0 = i \,]$$

$$= \Pr[\text{population becomes extinct} | \text{initially } i \text{ individuals}]$$

Clearly $\alpha_0 = 1$.

For absorption into 0 to occur starting with state i, in a small time interval Δt the process will either proceed from state i to state $i + 1$ and then eventually be absorbed, or it will proceed from state i to state $i - 1$ and then eventually be absorbed, or it will stay in state i and eventually be absorbed. Thus,

$$\alpha_i = i\lambda\,\Delta t\,\alpha_{i+1} + i\mu\,\Delta t\,\alpha_{i-1} + [1 - i(\lambda + \mu)\,\Delta t]\,\alpha_i + o(\Delta t)$$

for $i = 1, 2, 3, \ldots$. It follows that

$$i(\lambda + \mu)\,\Delta t\,\alpha_i = i\lambda\,\Delta t\,\alpha_{i+1} + i\mu\,\Delta t\,\alpha_{i-1} + o(\Delta t)$$

or, dividing by $i\,\Delta t$ and letting $\Delta t \to 0$,

$$(\lambda + \mu)\alpha_i = \lambda\alpha_{i+1} + \mu\alpha_{i-1}, \qquad \text{for } i = 1, 2, 3, \ldots$$

$$\alpha_i = \frac{\lambda}{\lambda + \mu}\alpha_{i+1} + \frac{\mu}{\lambda + \mu}\alpha_{i-1}, \qquad \text{for } i = 1, 2, 3, \ldots$$

This system of equations is identical with the system discussed in the gambler's ruin example, Example 10.1 (continued), with $p = \lambda/(\lambda + \mu)$, $q = \mu/(\lambda + \mu)$, and $N \to \infty$. The solution is

$$\alpha_i = \begin{cases} (\mu/\lambda)^i, & \text{if } \mu < \lambda \\ 1, & \text{if } \mu \geq \lambda \end{cases} \tag{12.19}$$

Thus, extinction of the population is *certain* if the death rate exceeds *or is equal to* the birth rate. If the birth rate exceeds the death rate, extinction is not *certain*, but is *possible* with probability given by (12.19).

EXAMPLE 12.7

In the Poisson process no absorbing states exist, yet the process is not irreducible since only transitions to higher states are possible. Thus, necessarily, $X_t \to \infty$ as $t \to \infty$ and $p_{ij}(t) \to 0$ for every i, j. (This can be seen directly by using formula (11.15):

$$q_j = \lim_{t \to \infty} p_{ij}(t) = \lim_{t \to \infty} e^{-\lambda t}\frac{(\lambda t)^{j-1}}{(j-i)!} = 0$$

for all $j \geq i$. Notice that $p_{ij}(t) \equiv 0$ for $j < i$, so that $q_j = \lim_{t \to \infty} p_{ij}(t) = 0$ in this case as well.) This agrees with our intuition. As $t \to \infty$, the number of calls coming in would increase without bound, so the probability of any fixed number of calls coming in would approach zero.

Exercises 12

1. Find the limiting probabilities for the process of Exercise 11.4. For what proportion of time in a year will the communications system be out of action with no satellites in operation?

2. Find the limiting probabilities for the machinery repair problem discussed in Exercise 11.9.

3. Consider the process discussed in Exercise 11.10. This can be considered as a queueing process in which arrivals are discouraged by a long queue (arrival rate inversely proportional to $i + 1$, where i is number present). Show that this process is always positive recurrent, unlike the simple queueing example, and determine its long-run probabilities.

4. Find the limiting probabilities for $\{ X_t : t \geq 0 \}$ if

$$
\Lambda = \begin{array}{c} 0 \\ 1 \\ 2 \end{array} \begin{array}{ccc} 0 & 1 & 2 \\ \left[\begin{array}{ccc} -10 & 6 & 4 \\ 1 & -3 & 2 \\ 8 & 1 & -9 \end{array} \right] \end{array}
$$

5. Find the limiting probabilites in Exercise 11.4.

6. Consider the queueing process with batch arrivals (two at a time) discussed in Exercise 11.13.

   (a) Under what conditions is this positive recurrent?
   (b) Compute the limiting probabilities.

7. In the machinery breakdown Example 12.2, show that the mean and variance of the limiting distribution are

$$
\left(\frac{\lambda}{\mu} \right) (1 - q_N) \quad \text{and} \quad \left(\frac{\lambda}{\mu} \right) (1 - q_N) \left(\frac{\lambda}{\mu} q_N + 1 \right) - \left(\frac{\lambda}{\mu} \right)^2 q_{N-1}
$$

8. Compute the long-run average cost per hour in the queueing model in the following two cases:

   (a) The cost per hour of being in state i is proportional to i, that is, $c_i = ci$.
   (b) The cost per hour of being in state i is proportional to i^2, that is, $c_i = ci^2$.

9. Determine the optimal number of repair crews R in the machinery breakdown example with $N = 3$ machines. Assume that each repair crew is paid $50 per hour whether they are working or waiting for a machine to fail. Assume that each broken machine costs the company $200 per hour.

10. Determine the limiting distribution of the queueing system with reneging in Exercise 11.20.

13 · Introduction to Queueing Theory

13.1 Introduction

One of the useful areas of application for the theory of Markov processes is in the study of waiting lines. Queues (waiting lines) abound in practical situations. The history of the development of queueing theory is found in fields as seemingly diverse as hospital management and time-shared computer system design. The earliest use of queueing theory was in the design of telephone systems. In a queueing system one supposes that there are "customers" arriving at a "server" for service. The time between arrivals is a random variable and the time to perform service is a random variable. Because of this random character, it is impossible to guarantee that service completions coincide exactly with customer arrivals. Consequently, there are times when the server completes his service on a customer and finds no one else available to work on. Thus, he is idle until the next customer arrives. But there are also times when a customer arrives and finds the server already occupied with some previous arrival. This customer must then wait his turn or leave. In telephone studies, physical storage devices for letting calls wait are not always available and, hence, in these studies one assumes that customers who arrive during the time when a server (trunk line) is busy are lost. They leave without waiting. In other applications (e.g., computer networks), there are physical storage devices that are large and, hence, one can reasonably suppose that jobs that arrive when the server is busy wait their turn for service. They queue up and form a waiting line or backlog of work to be done.

There are four basic problems of interest in the study of waiting lines. We can ask how long an arriving customer will wait. This *waiting time*, of course, is a random variable, and we would be interested in its density or distribution function. The sequence of waiting times of the successive customers forms a stochastic process with a discrete parameter space and continuous state space. A second stochastic process is called the *busy period*. Here one is interested in determining the density function for the time from the commencement of service of a customer who terminates an idle time until the server becomes free for the first time after this. The third problem is the *queue-length problem*. In this we are concerned with the number of customers waiting for service (including the ones that are being served) at any given time. In this chapter we are primarily concerned with this third problem. A fourth problem would be the study of the *departure process*, which describes the behavior of the stream of customers that are leaving the system after completing service. This would be of particular .

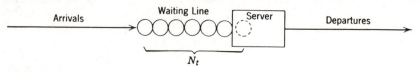

FIGURE 13.1   A schematic queueing system.

interest if these individuals were to go to another queueing system for another service operation.

Many of the examples discussed in Chapters 9, 10, 11, and 12 can be considered as queueing processes. The machine repair Examples 11.4 and 11.5 can be interpreted as queueing problems in which the queue consists of machines awaiting repair and the servers are the repair crews. Other situations involve the demands on a computer system, the demands for long-distance telephone lines, the customers at banks and toll booths, the scheduling of patients at hospitals and clinics, and the arrival of airplanes at terminal facilities, et cetera. (See Figure 13.1.)

13.2   Definitions

The description of any queueing system requires a discussion of three parts: the *arrival process*, which describes the way in which individuals join the queue, the *queue discipline*, which describes the way in which choices are made for service (first come–first served, last come–first served, random choice of an individual to be served, et cetera), and the *server mechanism*, which describes the length of time each service operation requires. Since in this chapter we are concerned primarily with queue lengths rather than with waiting times of individuals, we can ignore the effects of different types of queue disciplines.

Consider the sequence $X_1, X_2, X_3, \ldots$ of *interarrival times*. Thus, X_1 is the time until the first individual arrives at the system and X_2 is the time between the first and second arrivals, et cetera. This random sequence constitutes a discrete parameter random process, each realization of which defines an entire sequence of arrivals to the system. The time T_n of the nth arrival is given by

$$T_n = X_1 + X_2 + \cdots + X_n \tag{13.1}$$

Instead of describing the arrivals in terms of their interarrival times $\{X_n\}$, they can also be described by the process $Y_t = $ the total number of arrivals up to time t, $t \geq 0$. Thus, Y_t is a continuous parameter random process termed the *arrival process from time* 0. The event "n arrivals before time t" is equivalent to "the nth arrival is before time t but the $(n + 1)$st arrival is after time t." In symbols:

$$[Y_t = n] = [T_n \leq t < T_{n+1}]$$

Thus, the Y_t process is determined once the sequence of interarrival times $X_1, X_2, \ldots$ is known, and conversely. The Y_t process proceeds successively through the states $0, 1, 2, \ldots$, and hence is termed a *counting process*.

More generally, let $Y_{\tau,t}$ represent the number of arrivals *between* times τ and t, $0 \leq \tau \leq t$. Thus, $Y_t = Y_{0,t}$; $Y_{\tau,t}$ is termed the *arrival process from time τ*.

The service mechanism is specified by giving the random sequence $Z_1, Z_2, \ldots$ of service times of the successive individuals served. In a way similar to that used in defining the arrival process Y_t in terms of the interarrival times, we can also define a continuous parameter process by describing the service mechanism. Some care is needed, however, since the server will be idle for certain time intervals when the queue is empty. Thus, for each $\tau \geq 0$, we define the *service process from time τ*, $Z_{\tau,t}$, as giving the number of service completions between time τ and time t, $\tau \leq t$, *with the clock being stopped whenever the server is idle*. Thus, $Z_{\tau,t}$ counts the number of service completions between times τ and t, assuming that an individual is always available to enter the service system immediately after each service completion.

The random sequences $\{X_n\}$ and $\{Z_n\}$ or, equivalently, the processes $Y_{\tau,t}$ and $Z_{\tau,t}$ determine the behavior of the waiting line.

Let N_t represent the queue length at time t, that is, the total number of individuals either waiting for service or being served at time t. Thus, **N** is a continuous parameter random process with state space $\{0, 1, 2, \ldots\}$. This is the random process that we wish to discuss.

In a complete discussion of queueing systems one should classify systems according to the *number of servers* making up the service system and the maximum number of individuals that can be accommodated in the queue (*storage size*).

13.3   The Simple Markovian Queue (M/M/1)

In general, the process **N** is not a Markov process. In certain important special cases, however, it does have the Markov property and the discussion in Chapters 11 and 12 can be applied.

Let us consider a single-server, unlimited-storage queueing system with the following properties:

(i)   The probability of an arrival in a time interval of length Δt is $\lambda \, \Delta t + o(\Delta t)$.

(ii)   The probability of a service completion in a time interval of length Δt is $\mu \, \Delta t + o(\Delta t)$, subject only to the condition that the queue is not empty at the start of the interval, $N_t > 0$.

(iii)   The probability of more than one arrival or service completion in a time interval of length Δt is $o(\Delta t)$.

(iv)   The events in (i) and (ii) are independent of each other and of the past history of the system.

From properties (i) to (iv) we can show rather readily that the interarrival times $\{X_n\}$ and service times $\{Z_n\}$ satisfy the "forgetfulness" property that characterizes the negative exponential distribution. The sequence $\{X_n\}$ is a sequence of independent random variables that each have density $\lambda e^{-\lambda x}$, $x > 0$; the sequence $\{Z_n\}$ is a sequence of independent random variables that each have density $\mu e^{-\mu z}$, $z > 0$; and these two sequences are independent of each other. The processes $Y_{\tau,t}$ and $Z_{\tau,t}$ are Poisson processes of the type discussed in Section 9.7.

In view of the "forgetfulness" property of the exponential, the probability of an arrival or a departure in some future time interval is unaffected by the past history of the process, given the queue length at some fixed time t, $N_t = n$. Thus, N represents a continuous parameter Markov process, termed a *simple Markovian queue*. Such a system is particularly easy to analyze and serves as a good approximation to many real-life queueing situations.

From properties (i) to (iii) it follows that the rate matrix for this Markov chain has the form

$$\Lambda = \begin{pmatrix} -\lambda & \lambda & 0 & 0 & \cdots \\ \mu & -(\lambda + \mu) & \lambda & 0 & \cdots \\ 0 & \mu & -(\lambda + \mu) & \lambda & \cdots \\ \vdots & \vdots & \vdots & \vdots & \end{pmatrix}$$

Thus, this process is the one given in Example 11.6, with limiting properties discussed in Example 12.3.

If $\lambda < \mu$ and $t \to \infty$, then by (12.12) the state probabilities approach those of a random variable that has the modified geometric distribution discussed in Section 7.4, with parameter $p = \lambda/\mu$,

$$p_n(t) = \Pr[N_t = n] \to q_n = \left(1 - \frac{\lambda}{\mu}\right)\left(\frac{\lambda}{\mu}\right)^n \tag{13.2}$$

for $n = 0, 1, 2, \ldots$. The mean and variance of N_t approach those of the modified geometric given by (7.17) and (7.18):

$$E[N_t] \to \frac{\lambda/\mu}{1 - \lambda/\mu} = \frac{\lambda}{\mu - \lambda} \tag{13.3}$$

$$\mathrm{Var}[N_t] \to \frac{\lambda/\mu}{(1 - \lambda/\mu)^2} = \frac{\lambda\mu}{(\mu - \lambda)^2} \tag{13.4}$$

If $\lambda \geq \mu$, then the system is not positive recurrent. The queue length N_t grows beyond any bound as $t \to \infty$, with probability 1. Thus, no limiting probability distribution exists. It is important to notice that this happens even when $\lambda = \mu$. This implies that any attempt to exactly balance the queue by making the service rate exactly equal to the arrival rate will inevitably lead to an unstable queue. Basically this is because random downward fluctuations in queue length are bounded, since the queue length cannot be less than zero, while no such bound exists on random upward fluctuations.

13.4   The Multiple-Server Markovian Queue (M/M/r)

Let us assume that the same situation as in the previous section holds true except that the system has r identical serving units operating in parallel, rather than just 1. Properties (i), (iii), and (iv) still hold true; however, property (ii) must be modified somewhat. If the number of individuals in the system does not exceed r, $N_t = i \le r$, then the probability of a departure in time Δt is approximately proportional to i, as well as to Δt. This follows since, by neglecting $o(\Delta t)$ terms, the probability of a service completion in time Δt if two individuals are being served is twice what it would be if only one individual were to be served, et cetera. Thus, (ii) must be replaced by the following.

(ii)′ Given that $N_t = i$, the probability of a service completion in the time interval $(t, t + \Delta t)$ is $i\mu \Delta t + o(\Delta t)$ if $i \le r$, and is $r\mu \Delta t + o(\Delta t)$ if $i > r$.

Now the departures after time t clearly depend on the value of N_t, but not on the past history. Thus, the Markovian property still holds true. From (i) and (ii)′ the rate matrix Λ is found to be

$$\Lambda = \begin{pmatrix} -\lambda & \lambda & 0 & 0 & \cdots \\ \mu & -(\lambda+\mu) & \lambda & 0 & \cdots \\ 0 & 2\mu & -(\lambda+2\mu) & \lambda & \cdots \\ & & \ddots & \ddots & \ddots \\ & & & r\mu & -(\lambda+r\mu) & \lambda \\ & & & & r\mu & -(\lambda+r\mu) & \lambda \\ & & & & & & \ddots \end{pmatrix}$$

We see that N_t is a birth-and-death process of the type that is discussed in Example 12.4, with $\lambda_i = \lambda$ and $\mu_i = i\mu$ for $i \le r$ and $\mu_i = r\mu$ for $i > r$. The series (12.18) becomes

$$S = 1 + \frac{\lambda}{\mu} + \frac{\lambda^2}{2\mu^2} + \cdots + \frac{\lambda^r}{r!\mu^r} + \frac{\lambda^r}{r!\mu^r} \sum_{i=1}^{\infty} \frac{\lambda^i}{(r\mu)^i} \tag{13.5}$$

This infinite series will converge if, and only if, $\lambda < r\mu$. In this case, the sum can be evaluated by using the formula for the sum of a geometric series, although the resulting expression is somewhat complicated. By (12.17) and the theorem of Section 12.4, we conclude the following.

If $\lambda < r\mu$,

$$p_n(t) = \Pr[N_t = n] \rightarrow \begin{cases} \dfrac{\lambda^n}{n!\mu^n S}, & \text{if } n \le r \\[3mm] \dfrac{\lambda^n}{r!r^{n-r}\mu^n S}, & \text{if } n > r \end{cases} \tag{13.6}$$

as $t \rightarrow \infty$ where S is given by (13.5). Limiting formulas for the mean and variance can also be derived, but they are rather complex.

If $\lambda \geq r\mu$ the queue length grows beyond any bound with probability 1 as $t \to \infty$.

13.5   The Simple Markovian Queue with Bounded Storage (M/M/1/K)

Consider a simple one-server Markovian queue of the type discussed in Section 13.3, with the additional restriction that there is space for only K individuals in the system. If the queue reaches this size, then additional arrivals are turned away until the number waiting drops below K. The assumptions made in Section 13.3 still apply with the single exception that property (i) must be modified as follows.

(i)′   Given that $N_t < K$, the probability of an addition to the queue in the time interval $(t, t + \Delta t)$ is $\lambda \, \Delta t + o(\Delta t)$. Given that $N_t = K$, the probability of an addition in this interval is $o(\Delta t)$.

Properties (i)′ and (ii) to (iv) imply that N_t is a Markov chain having the finite state space $\{0, 1, 2, \ldots, K\}$ and transition rate matrix

$$
\Lambda = \begin{pmatrix}
-\lambda & \lambda & 0 & 0 & \ldots & 0 & 0 & 0 \\
\mu & -(\lambda+\mu) & \lambda & 0 & \ldots & 0 & 0 & 0 \\
0 & \mu & -(\lambda+\mu) & \lambda & \ldots & 0 & 0 & 0 \\
\vdots & \vdots & \vdots & \vdots & & \vdots & \vdots & \vdots \\
0 & 0 & 0 & 0 & \ldots & \mu & -(\lambda+\mu) & \lambda \\
0 & 0 & 0 & 0 & \ldots & 0 & \mu & -\mu
\end{pmatrix}
$$

Since the chain is irreducible and finite, it must be positive recurrent for all values of λ and μ. The limiting probabilities are easily found, by using (12.17), to be

$$
p_n(t) = \Pr[N_t = n] \to \left(\frac{1 - \lambda/\mu}{1 - (\lambda/\mu)^{K+1}} \right) (\lambda/\mu)^n \tag{13.7}
$$

for $n = 0, 1, 2, \ldots, K$.

13.6   Independent Interarrival and Service Times

Assume that the sequence $\{X_n\}$ of interarrival times forms a sequence of independent, identically distributed random variables. We assume also that there is only one server and that the sequence $\{Z_n\}$ of service times also forms a sequence of independent, identically distributed random variables. Finally, we assume that the terms of the $\{X_n\}$ sequence are independent of the terms of the $\{Z_n\}$ sequence. Under these conditions, we say that the process **N** is a *general independent input and service queueing process*. The notation GI/GI/1 is often used to describe such a process. (GI stands for *general independent input and service* and 1 for the *number of servers*.) For r servers the notation GI/GI/r is used. However, in this case, the independent, identically distributed sequence of service times refers to each individual server, not to the overall system. This notation allows us to specify a whole hierarchy of different queueing systems by specializing the probability distribution of the interarrival times $\{X_n\}$ or the service times $\{Z_n\}$. Many real-life queueing systems fall into these categories.

The simple Markovian queue discussed in Section 13.3 has this form. The interarrival times $\{X_n\}$ are independent and each has a negative exponential density $\lambda e^{-\lambda x}$ for $x > 0$. Similarly, the service times $\{Z_n\}$ are independent and each also has a negative exponential density $\mu e^{-\mu z}$ for $z > 0$. This situation is symbolized by the notation M/M/1. Here M stands for *Markovian* and refers to negative exponential interarrival times and service times (or, equivalently, to a Poisson arrival process and a Poisson service process). Similarly, the *r*-server case discussed in Section 13.4 can be symbolized by M/M/r.

It can be shown that the M/M/1 and M/M/r queues are the *only* general independent input and service queueing processes for which N is Markovian.

13.7   The M / GI / 1 Queue. Embedded Chains

Let us now illustrate how we might handle a non-Markov queueing process of the type discussed in the previous section. Assume that we have a single-server queue in which the arrivals form a Poisson process, that is, the interarrival times are negative exponential with density $\lambda e^{-\lambda x}$ for $x > 0$. Suppose that the service times $\{Z_n\}$ each have some fixed density function $f(z), z > 0$. Again, we also assume that all interarrival and service times are mutually independent. If $f(z)$ is not specified, this queue is said to be of type M/GI/1.

Such a queueing system results if arrivals come in essentially at random. For example, telephone calls arrive at an exchange, but the length of each call has a density, illustrated in Figure 13.2, which might appear related more to a gamma density than to a negative exponential.

Suppose that N_t is observed for some fixed time t, $N_t = n$. If $n > 0$, then some individual is being serviced at this time. Clearly, information about when this service operation began assists us in predicting when service will be completed on this individual (unless the service time is negative exponential and has the

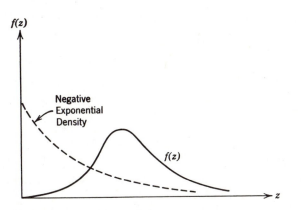

FIGURE 13.2   A possible service time density.

"forgetfulness" property). This information can be obtained by looking at the past history of the process—particularly, at the time that the previous individual completed his service. Thus, the future behavior is not independent of the past; **N** does not satisfy the Markov property.

However, *if we happened to observe N_t at the instant after a departure has taken place from the queue*, then we know that the next service operation is just starting, and no other information about the past will tell us any more about that operation. Thus, the length of this service operation *is* independent of past history. Since the interarrival times are negative exponential, prior information does not help with them either. These observations suggest the following construction.

Let $\tau_1, \tau_2, \tau_3, \ldots$ represent the random sequence of *departure times*. If we consider the queue length N_t only at these times, we obtain a *discrete parameter random sequence*

$$\{ N_{\tau_k} \} = \{ N_{\tau_1}, N_{\tau_2}, N_{\tau_3}, \ldots \}^1$$

This sequence is termed an *embedded chain* since it is extracted from the continuous parameter process N_t by observing it only at a discrete sequence of times. The observations in the preceding paragraph imply that this embedded chain $\{ N_{\tau_k} \}$ is a Markovian sequence. Thus, instead of working with the entire continuous parameter process **N**, which is non-Markovian and complicated, we extract from it a discrete parameter process $\{ N_{\tau_k} \}$, which is Markovian, and study this simpler process.

$\{ N_{\tau_k} \}$ has state space $\{0, 1, 2, \ldots\}$. We need to determine the one-step transition probabilities

$$p_{ij} = \Pr\left[N_{\tau_{k+1}} = j \mid N_{\tau_k} = i \right]$$

If $j < i - 1$ then clearly $p_{ij} = 0$, since the queue cannot decrease by more than 1 in an interval having only one departure.

If $j \geq i - 1$ and $i > 0$, then p_{ij} is the probability of the queue length increasing from i to j during one service time. Since one individual departs, $j - i + 1$ must arrive during this interval. The number of arrivals during a *fixed* time interval has a Poisson distribution. Thus, the conditional density of the total number M of arrivals during the $(k + 1)$st service time Z_{k+1}, given that $Z_{k+1} = z$, is

$$f(m|z) = \Pr[m \text{ arrivals in time } z]$$

$$= \Pr[M = m \mid Z_{k+1} = z] = e^{-\lambda z} \frac{(\lambda z)^m}{m!}$$

for $m = 0, 1, 2, \ldots$. The joint density of M and Z_{k+1} is

$$f(m, z) = f(m|z) f(z) = e^{-\lambda z} \frac{(\lambda z)^m}{m!} \cdot f(z)$$

for $m = 0, 1, 2, \ldots$ and $z > 0$. This is a mixed situation that combines a discrete

---

[1]N_{τ_k} represents the number left behind after the departure at time τ_k. The departing individual is not counted.

and a continuous random variable, but the analysis of Section 4.8 still applies. The density g_m of M alone is obtained by integrating $f(m, z)$ with respect to z [see (4.20)]:

$$g_m = \Pr[M = m] = \int_0^\infty e^{-\lambda z} \frac{(\lambda z)^m}{m!} f(z) \, dz \tag{13.8}$$

It follows that

$$p_{ij} = g_{j-i+1} = \int_0^\infty e^{-\lambda z} \frac{(\lambda z)^{j-i+1}}{(j-i+1)!} f(z) \, dz$$

for $j \geq i - 1$ and $i > 0$.

If $i = 0$, then we must modify the argument slightly. This means that after time τ_k the queue is empty. One must wait for an individual to arrive before the service mechanism can start up again, and this first individual to arrive will be the $(k+1)$st departure. He will leave behind a queue that contains all the arrivals during his service time. Thus, if $i = 0$,

$$p_{0j} = g_j = \int_0^\infty e^{-\lambda z} \frac{(\lambda z)^j}{j!} f(z) \, dz \tag{13.9}$$

for $j = 0, 1, 2, \ldots$, and hence the zeroth row of the transition matrix coincides with the first row.

We conclude that the one-step transition matrix for the embedded chain $\{N_{\tau_k}\}$ has the form

$$\mathbf{P} = \begin{pmatrix} g_0 & g_1 & g_2 & g_3 & \cdots \\ g_0 & g_1 & g_2 & g_3 & \cdots \\ 0 & g_0 & g_1 & g_2 & \cdots \\ 0 & 0 & g_0 & g_1 & \cdots \\ \vdots & \vdots & \vdots & \vdots & \end{pmatrix} \tag{13.10}$$

where g_m is given by (13.8).

The chain $\{N_{\tau_k}\}$ is irreducible and its distribution can be found by solving the system $\mathbf{v} = \mathbf{vP}$ to find the stationary vector $\mathbf{v}$ (see Section 10.8). We will show that a stationary probability vector exists provided that

$$\lambda m < 1 \tag{13.11}$$

where $m = E[Z_k] = \int_0^\infty z f(z) \, dz$. By Theorem 10.7 the chain is positive recurrent if, and only if, (13.11) holds true. If $\lambda m \geq 1$, then $N_{\tau_k} \to \infty$ (and hence $N_t \to \infty$) with probability 1.

Most required information about the behavior of the original process $\mathbf{N}$ as $t \to \infty$ can be obtained by studying the behavior of the embedded chain $\{N_{\tau_k}\}$ as $k \to \infty$.

The equation $\mathbf{v} = \mathbf{vP}$ for the stationary vector $\mathbf{v}$ can be written in component form by using (13.10),

$$v_i = v_0 g_i + \sum_{k=1}^{i+1} v_k g_{i+1-k} \tag{13.12}$$

for $i = 0, 1, 2, \ldots$.

To solve (13.12), we use the method of generating functions. Define the function

$$V(\xi) = \sum_{i=0}^{\infty} v_i \xi^i \tag{13.13}$$

If we can determine this function and show it to be convergent for $\xi = 1$, with $V(1) = \Sigma v_i = 1$ and coefficients satisfying (13.12), this will assure us that the system is positive recurrent. The limiting probabilities v_i can be found as the coefficients of the ξ^i-terms in the series expansion of $V(\xi)$.

To find $V(\xi)$, multiply each side of (13.12) by ξ^i and sum, for $i = 0, 1, 2, \ldots$.

$$V(\xi) = \sum_{i=0}^{\infty} v_i \xi^i = v_0 \sum_{i=0}^{\infty} g_i \xi^i + \sum_{i=0}^{\infty} \sum_{k=1}^{i+1} v_k g_{i+1-k} \xi^i$$

$$= v_0 \sum_{i=0}^{\infty} g_i \xi^i + \xi^{-1} \sum_{k=1}^{\infty} v_k \xi^k \sum_{i=k-1}^{\infty} g_{i+1-k} \xi^{i+1-k} \qquad 1$$

$$= v_0 \sum_{i=0}^{\infty} g_i \xi^i + \xi^{-1} [V(\xi) - v_0] \sum_{j=0}^{\infty} g_j \xi^j$$

$$= [v_0(1 - \xi^{-1}) + \xi^{-1} V(\xi)] G(\xi) \tag{13.14}$$

where

$$G(\xi) = \sum_{i=0}^{\infty} g_i \xi^i \tag{13.15}$$

If we solve (13.14) for $V(\xi)$ we obtain

$$V(\xi) = \frac{v_0(\xi - 1) G(\xi)}{\xi - G(\xi)} \tag{13.16}$$

By the definition (13.9) of the terms g_i,

$$G(\xi) = \sum_{i=0}^{\infty} g_i \xi^i = \sum_{i=0}^{\infty} \xi^i \int_0^{\infty} e^{-\lambda z} \frac{(\lambda z)^i}{i!} f(z) \, dz$$

$$= \int_0^{\infty} e^{-\lambda z} \sum_{i=0}^{\infty} \frac{(\lambda z \xi)^i}{i!} f(z) \, dz$$

$$= \int_0^{\infty} e^{-\lambda z} \cdot e^{\lambda z \xi} f(z) \, dz$$

$$= \int_0^{\infty} e^{\lambda(\xi - 1) z} f(z) \, dz$$

$$= M[\lambda(\xi - 1)]$$

---

[1] The double summation here is over all indices i and k that satisfy $1 \leq k \leq i+1 < \infty$, and we have rearranged the order of summation.

where

$$M(\theta) = E[e^{\theta z_k}] = \int_0^\infty e^{\theta z} f(z)\, dz$$

is the moment-generating function of the service distribution. Thus, (13.16) can be written in the form

$$V(\xi) = \frac{v_0(\xi - 1) M[\lambda(\xi - 1)]}{\xi - M[\lambda(\xi - 1)]} \tag{13.17}$$

To find v_0 we let $\xi \to 1$, and hence $V(\xi) \to V(1) = 1$ in (13.17). Now $M[\lambda(\xi - 1)] \to M(0) = 1$. To compute the limit of the right side of (13.17) we must use l'Hôpital's rule.

$$\lim_{\xi \to 1} V(\xi) = 1 = v_0 \lim_{\xi \to 1} \frac{\xi - 1}{\xi - M[\lambda(\xi - 1)]}$$

$$= v_0 \cdot \frac{1}{1 - \lambda M'(0)}$$

Thus, $v_0 = 1 - \lambda M'(0)$. But

$$M'(0) = \int_0^\infty z f(z)\, dz$$

$$= E[Z_k] = m$$

We conclude that

$$v_0 = 1 - \lambda m \tag{13.18}$$

Clearly v_0 will be a probability, and $V(\xi)$ will not be identically zero, only if $1 - \lambda m > 0$. But this is precisely condition (13.11). If (13.11) holds true, it can be shown by simple analytic methods that (13.17) exists and that it defines a convergent series for all $|\xi| \leq 1$. The final solution for $V(\xi)$ is

$$V(\xi) = \frac{(1 - \lambda m)(\xi - 1) M[\lambda(\xi - 1)]}{\xi - M(\lambda(\xi - 1))} \tag{13.19}$$

for $|\xi| \leq 1$ and $\lambda m < 1$. In any given special case $M(\theta)$ can be computed and hence $V(\xi)$ can be determined by (13.19). The limiting probabilities v_i can then be found by expanding in powers of ξ.

Formula (13.19) can be used to derive a useful formula for the long-run mean queue length $\lim_{k \to \infty} E[N_{\tau_k}]$. This is termed the *Pollaczek–Khintchine* formula:

$$\lim_{k \to \infty} E[N_{\tau_k}] = \lambda m + \frac{(\lambda m)^2 + \lambda^2 \operatorname{Var}[Z_k]}{2(1 - \lambda m)} \tag{13.20}$$

The proof is left for Exercise 13.4.

13.8   Examples

EXAMPLE 13.1

Customers arrive at a certain airline counter in a terminal at a mean rate of one per minute. The manager wants to adjust his staff so that the probability of more than five customers waiting for service is no more than 5%. At what rate should he be prepared to service customers?

For convenience, assume that arrivals are completely random, so that the interarrival time is negative exponential, and that the service times are also negative exponential. Thus, we have a simple Markovian queue, as discussed in Section 13.3. Arrivals form a Poisson process with mean rate λ per unit time. The hypothesis is that $\lambda = 1$. We wish to determine the mean service rate μ so that the long-run probability of a queue > 5 does not exceed .05.

$$\sum_{n=6}^{\infty} q_n \le .05$$

By (13.2) this becomes

$$\left(1 - \frac{1}{\mu}\right) \sum_{n=6}^{\infty} \left(\frac{1}{\mu}\right)^n \le .05$$

or

$$\frac{1}{\mu^6} \le .05$$

$$\mu^6 \ge 20$$

$$\mu \ge 20^{1/6} = 1.65$$

Thus, the facility must be prepared to handle customers at the mean rate of at least 1.65 per minute (or with mean service time $1/1.65 = 0.61$ minutes per customer) to meet the requirement. This rate is substantially greater than the mean arrival rate. When $\lambda = 1$ and $\mu = 1.65$, $q_0 = 1 - \lambda/\mu = 1 - 1/1.65 = 0.39$, so the manager must be prepared to accept the fact that his staff will be standing idle 39% of the time for the stated requirement to be met.

By varying the length of line that he can permit and by comparing it with the cost of providing the desired service, analyses like the one above permit the manager to achieve the best balance between cost and service.

EXAMPLE 13.2

In the preceding example the manager decides to arrange his staff so that there are two service positions at the counter, each with mean rate μ. How would this affect the value of the combined service rate (which will now be 2μ) that is necessary to meet the stated condition?

The discussion of Section 13.4 now applies with $r = 2$ and $\lambda = 1$. By (13.5) and (13.6)

$$S = 1 + \frac{1}{\mu} + \frac{1}{2\mu^2} + \frac{1}{2\mu^2} \sum_{i=1}^{\infty} \left(\frac{1}{2\mu}\right)^i$$

$$= 1 + \frac{1}{\mu} + \frac{1/(2\mu^2)}{1 - 1/(2\mu)}$$

$$= 1 + \frac{1}{\mu} + \frac{1}{2\mu^2 - \mu}$$

$$= \frac{2\mu + 1}{2\mu - 1}$$

$$q_0 = \frac{1}{S} = \frac{2\mu - 1}{2\mu + 1}$$

$$q_1 = \frac{2\mu - 1}{\mu(2\mu + 1)}$$

$$q_n = \frac{2\mu - 1}{2^{n-1}\mu^n(2\mu + 1)}, \qquad \text{for } n \geq 2$$

$$\sum_{n=6}^{\infty} q_n = \frac{2\mu - 1}{2^5\mu^6(2\mu + 1)} \cdot \frac{1}{1 - 1/(2\mu)} = \frac{1}{2^4\mu^5(2\mu + 1)}$$

This probability will be $\leq .05$ provided that

$$2^4\mu^5(2\mu + 1) \geq 20$$

This inequality can be solved to give $\mu \geq 0.856$. The necessary combined service rate is $2\mu = 1.712$, which exceeds the single rate of 1.65 found previously. Thus, splitting the service system does not improve matters. A still greater service capacity would be required then for a single channel.

EXAMPLE 13.3

Answer the question raised in Example 13.1 with the system designed so that the *mean* queue length should not exceed 2.

By (13.3) the limiting mean queue length is $\lambda/(\mu - \lambda) = 1/(\mu - 1)$. We can solve the inequality

$$\frac{1}{\mu - 1} \leq 2$$

and obtain $\mu \geq 1.5$. Thus, a mean service rate of only 1.5 (less than the 1.65 found in Example 13.1) will guarantee that the mean length queue is only 2, although there will then be a probability greater than .05 that the length will actually exceed 5.

Exercises 13

1. Show that if $\{X_1, X_2, \ldots\}$ is a sequence of identically distributed, negative exponential random variables, then Y_t has a Poisson distribution for each t. (*Hint*: See Theorem of Section 7.8.)

2. Under the conditions of Exercise 1 show that $Y_{\tau, t}$ has a Poisson distribution for arbitrary $\tau \le t$.

3. Determine the limiting probabilities v_j, by using (13.19), for the special case of negative exponentially distributed service times. Compare your answer with (7.16) and infer that the queue-length probabilities are the ones given by (13.2). Compare it with the result obtained directly in Example 10.13. (*Note*: The limiting probabilities here are identical, but the two stochastic processes N_t and N_{τ_k} are clearly different.)

4. Prove the *Pollaczek–Khintchine* formula (13.20) using (13.19). [*Hint*: The mean $\Sigma i v_i = V'(1) = \lim_{\xi \to 1} V'(\xi)$ and l'Hôpital's rule must be used in evaluating this limit.]

5. By using the formula in Exercise 4, show that the mean queue-length is minimized for fixed λm whenever the variance of the service time is 0. Describe the service process for which this holds true and give the mean queue length.

6. Cars arrive for repair at a one-man body shop during business hours in the form of a Poisson process with mean arrival rate $\lambda = \frac{1}{4}$, that is, mean time between arrivals $= 4$ hours. Service consists of bumping followed by painting. Assume bumping time has a negative exponential density with mean 2 hours, painting time has a negative exponential density with mean 1 hour, and that these two times are independent. Compute the service time density $f(z)$ and the terms of the transition matrix $\mathbf{P}$ of the embedded queueing process.

7. In the bounded storage example (Section 13.5), if a customer arrives when the queue is at full capacity K that customer is turned away (lost). One can show that with probability $\lambda/(\lambda + \mu)$ the next event in this queueing system will be an arrival, while with probability $\mu/(\lambda + \mu)$ the next event will be a departure. Determine the probability that the next arriving customer will be turned away.

8. In Example 13.1 it was shown that the probability of the server being idle was .39 if one wished to have a probability of less than .05 that the queue length exceeded 5. Suppose that in that example the server was a computer that had to be kept running and that one could tolerate no more than 5% idle time resulting from the lack of work. By using the values given in Example 13.1, determine the queue-length probabilities and the mean and variance of the queue length for this system.

9. In Exercise 6 use the Pollaczek–Khintchine formula (13.20) to determine the long-run mean number of cars in the system.

10. In Example 13.2 it was argued that we could change the performance of the system by adding another server. Clearly, for fixed λ, say, $\lambda = 1$ as in that example, we could increase the service rate of one server or add another server. Suppose that we could speed up the servicing process so that the mean service time was one-half of its previous value or that we could add another identical server. Compare q_0 and $\sum_{n=6}^{\infty} q_n$ for these two possible systems. Also compare the means and variances of the queue length for the two systems.

11. *The GI/M/1 Queue.* Consider a single-server queue with negatively exponentially distributed service times, for which the interarrival times $\{X_n\}$ have some fixed density function $g(x)$, $x > 0$. The process N_t is not Markovian for this queue. Discuss why.

12. (Exercise 11 continued) Define $T_1, T_2, \ldots, T_n$ to be the times at which *arrivals* occur. Verify that $\{N_{T_k}\}$ is a Markov chain with a discrete parameter space and a countable state space.

13. (Exercise 11 continued) Let $p_{ij} = \Pr[N_{T_{k+1}} = j \mid N_{T_k} = i]$ be the one-step transition probabilities for the *embedded Markov chain* $\{N_{T_k}\}$. Notice that $p_{ij} = 0$ if $j > i + 1$. Determine p_{ij} for $j \leq i + 1$ in terms of $g(x)$ and the mean service rate μ.

APPENDIX A ·
Some Useful Formulas

The formulas discussed here occur frequently in probability theory. For convenience they are combined here and referred to at various points in the text. Some proofs are given or sketched.

A.1 Arithmetic Series

$$\sum_{i=1}^{m} i = 1 + 2 + 3 + \cdots + m = \frac{m(m+1)}{2}$$

PROOF Denote the sum by S. Thus,

$$S = 1 + 2 + 3 + \cdots + m$$

or, reversing it,

$$S = m + (m-1) + (m-2) + \cdots + 1$$

Adding, we obtain

$$2S = (m+1) + (m+1) + (m+1) + \cdots + (m+1) = m(m+1)$$

and the result follows.

A.2 Finite Geometric Series

$$\sum_{i=0}^{m} x^i = 1 + x + x^2 + \cdots + x^m = \frac{1 - x^{m+1}}{1 - x}, \qquad x \neq 1$$

PROOF Denote the sum by S. Thus,

$$S = 1 + x + x^2 + \cdots + x^m$$
$$xS = x + x^2 + \cdots + x^m + x^{m+1}$$

Subtraction gives

$$(1 - x)S = 1 - x^{m+1}$$

from which the result follows.

A.3   Infinite Geometric Series

$$\sum_{i=0}^{\infty} x^i = 1 + x + x^2 + \cdots = \frac{1}{1-x}, \qquad |x| < 1$$

PROOF   Let $m \to \infty$ in A.2. $x^{m+1} \to 0$ if $|x| < 1$.

A.4   Taylor's Formula

$$f(x) = \sum_{i=0}^{\infty} f^{(i)}(0)\frac{x^i}{i!} = f(0) + f'(0)\frac{x}{1!} +$$

$$f''(0)\frac{x^2}{2!} + \cdots + f^{(i)}(0)\frac{x^i}{i!} + \cdots$$

for all x in some interval about $x = 0$, provided $f(x)$ has continuous uniformly bounded derivatives of all orders in that interval.

$$\left(f^{(i)}(x) \text{ stands for the } i\text{th derivative, } \frac{d^i f(x)}{dx^i} \right)$$

PROOF   See any standard calculus text.

A.5   Exponential Series

$$e^x = 1 + \frac{x}{1!} + \frac{x^2}{2!} + \cdots + \frac{x^i}{i!} + \cdots = \sum_{i=0}^{\infty} \frac{x^i}{i!}$$

for all x.

PROOF   In some calculus developments this series is taken as the definition of e^x. Alternately, use A.4 with $f(x) = e^x$ and note that $f'(x) = f''(x) = \cdots = e^x$, $f'(0) = f''(0) = \cdots = 1$, and that e^x is bounded in every finite interval about 0.

A.6   Binomial Theorem

$$(a+b)^n = a^n + \frac{n}{1!}a^{n-1}b + \frac{n(n-1)}{2!}a^{n-2}b^2 + \cdots + b^n$$

$$= \sum_{i=0}^{n} \binom{n}{i} a^{n-i}b^i, \qquad \text{for every positive integer } n$$

where

$$\binom{n}{i} = \frac{n!}{i!(n-i)!} = \frac{n(n-1)\ldots(n-i+1)}{i!}, \qquad i = 0,1,2,\ldots,n$$

PROOF

$$(a+b)^n = a^n\left(1 + \frac{b}{a}\right)^n = a^n f(b/a)$$

where

$$f(x) = (1+x)^n$$

Now $f'(x) = n(1+x)^{n-1}$, $f''(x) = n(n-1)(1+x)^{n-2}$, and, by induction, $f^{(i)}(x) = n(n-1)\ldots(n-i+1)(1+x)^{n-i}$. Set $x = 0$ in each of these and use Taylor's Formula (A.4), (noting that $f^{(i)}(x) = 0$ if $i > n$):

$$f(x) = 1 + n\frac{x}{1!} + n(n-1)\frac{x^2}{2!} + \cdots = \sum_{i=0}^{n} \binom{n}{i} x^i$$

and the result follows.

A.7   Formula

$$\sum_{i=0}^{n} \binom{n}{i} = 2^n$$

PROOF   Set $a = b = 1$ in the Binomial Theorem (A.6).

A.8   Normal Density Constant

$$\int_{-\infty}^{\infty} e^{-x^2/2}\, dx = \sqrt{2\pi}$$

and hence $(1/\sqrt{2\pi})e^{-x^2/2}$ is a density function,

$$\int_{-\infty}^{\infty} \frac{1}{\sqrt{2\pi}} e^{-x^2/2}\, dx = 1$$

PROOF   Let $I = \int_{-\infty}^{\infty} e^{-x^2/2}\, dx$. Note that the integral converges since $x^2/2 > |x|$ for $|x| > 2$, and $\int_{-\infty}^{\infty} e^{-|x|}\, dx = 2\int_0^{\infty} e^{-x}\, dx$ converges.

Since the letter used for the variable is irrelevant in a definite integral, we can also write

$$I = \int_{-\infty}^{\infty} e^{-y^2/2}\, dy$$

Thus,

$$I^2 = \int_{-\infty}^{\infty} e^{-x^2/2}\, dx \cdot \int_{-\infty}^{\infty} e^{-y^2/2}\, dy = \int_{-\infty}^{\infty} \int_{-\infty}^{\infty} e^{-(x^2+y^2)/2}\, dx\, dy$$

Now change from rectangular coordinates (x, y) to polar coordinates (r, θ). By standard calculus techniques the element of volume $dx\, dy$ is changed to $r\, dr\, d\theta$ and $x^2 + y^2$ becomes r^2. To sweep out the entire plane r must range from 0 to ∞

and θ from 0 to 2π. Thus,

$$I^2 = \int_0^{2\pi} \int_0^\infty e^{-r^2/2} r\, dr\, d\theta$$

$$= \int_0^{2\pi} -e^{-r^2/2}\Big|_0^\infty d\theta$$

$$= \int_0^{2\pi} 1\, d\theta = 2\pi$$

and the result follows.

A.9   Formula

$$\int_0^\infty e^{-at} dt = \frac{1}{a}, \qquad a > 0$$

PROOF

$$\int_0^\infty e^{-at} dt = \lim_{k \to \infty} -\frac{1}{a} e^{-at}\Big|_0^k = 0 - \left(-\frac{1}{a}\right) = \frac{1}{a}$$

A.10   Gamma Function

The gamma function $\Gamma(\alpha)$ is defined by

$$\Gamma(\alpha) = \int_0^\infty x^{\alpha-1} e^{-x}\, dx, \qquad \text{for } \alpha > 0$$

(1) $\Gamma(\alpha)$ satisfies the recursion formula

$$\Gamma(\alpha + 1) = \alpha\Gamma(\alpha)$$

(2) $\Gamma(n) = \int_0^\infty x^{n-1} e^{-x}\, dx = (n-1)!, \qquad$ for each integer $n \geq 1$

(3) $\Gamma(n + \frac{1}{2}) = \Gamma\left(\dfrac{2n+1}{2}\right) = \sqrt{\pi}\, \dfrac{1 \cdot 3 \cdot 5 \cdot \cdots \cdot (2n-1)}{2^n}$, for each integer $n \geq 0$

PROOF   The convergence of the integral defining $\Gamma(\alpha)$ can be shown by splitting the integral into $\int_0^1 + \int_1^\infty$ and treating the cases $0 < \alpha < 1$ and $\alpha \geq 1$ separately.

To show (1), integrate by parts setting $u = x^\alpha$ and $dv = e^{-x}\, dx$.

$$\Gamma(\alpha + 1) = \int_0^\infty x^\alpha e^{-x}\, dx = \left[\int u\, dv\right]_{x=0}^\infty$$

$$= \left[uv - \int v\, du\right]_{x=0}^\infty$$

$$= -x^\alpha e^{-x}\Big|_0^\infty + \alpha \int_0^\infty x^{\alpha-1} e^{-x}\, dx$$

$$= 0 + \alpha\Gamma(\alpha)$$

Note that (2) holds for $n = 1$ by A.9. Thus, using (1),

$$\Gamma(2) = 1 \cdot \Gamma(1) = 1 \cdot 0! = 1!$$

$$\Gamma(3) = 2 \cdot \Gamma(2) = 2 \cdot 1! = 2!$$

$$\vdots$$

and the result follows in general by induction. To show (3), first set $n = 0$:

$$\Gamma(\tfrac{1}{2}) = \int_0^\infty x^{-1/2} e^{-x} \, dx$$

Now change variables, $x = y^2/2$, $dx = y\,dy$, and $x^{-1/2} = \sqrt{2}/y$.

$$\Gamma(\tfrac{1}{2}) = \int_0^\infty \frac{\sqrt{2}}{y} e^{-y^2/2} y \, dy = \sqrt{2} \int_0^\infty e^{-y^2/2} \, dy$$

$$= \sqrt{2} \cdot \frac{\sqrt{2\pi}}{2}, \qquad \text{by A.8}$$

$$= \sqrt{\pi}$$

Using (1),

$$\Gamma(\tfrac{3}{2}) = \sqrt{\pi} \cdot \frac{1}{2}$$

$$\Gamma(\tfrac{5}{2}) = \sqrt{\pi} \cdot \frac{1 \cdot 3}{2 \cdot 2}$$

and again the result follows in general by induction.

A.11 Mean of Standard Normal Density

$$\int_{-\infty}^\infty x e^{-x^2/2} \, dx = 0$$

PROOF   The integrand is an *odd* function, $f(-x) = -f(x)$, so provided the integral converges it will have the value 0 since the part with $x > 0$ will cancel the past with $x < 0$. Convergence follows by setting $t = x^2/2$ and noting that, by A.9,

$$\int_0^\infty x e^{-x^2/2} \, dx = \int_0^\infty e^{-t} \, dt = 1$$

with a similar evaluation for $\int_{-\infty}^0 x e^{-x^2/2} \, dx$.

A.12 Interchange of Summation

If $a(n, k) \geq 0$ for all nonnegative integers n and k, then

$$\sum_{n=0}^\infty \sum_{k=0}^n a(n, k) = \sum_{k=0}^\infty \sum_{n=k}^\infty a(n, k)$$

provided either side converges.

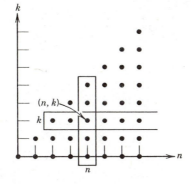

PROOF One can simply note that each side represents the sum of the terms $a(n, k)$ for all indices n and k satisfying the inequality $0 \leq k \leq n < \infty$. In the left sum one starts out holding u fixed and summing on k. In the right sum one starts out holding k fixed and summing on n.

Graphically, the indices of summation can be represented by the lattice points in the preceding graph. In the left sum, summation is first over a vertical strip of indices, and then the strips are summed from left to right. In the right sum, summation is first over a horizontal strip of indices, and then the strips are summed from the lowest level up.

APPENDIX B ·
Some Useful Tables

TABLE B.1[a]
Standard Normal Distribution

	.00	.01	.02	.03	.04	.05	.06	.07	.08	.09
0.0	.5000	.5040	.5080	.5120	.5160	.5199	.5239	.5279	.5319	.5359
0.1	.5398	.5438	.5478	.5517	.5557	.5596	.5636	.5675	.5714	.5753
0.2	.5793	.5832	.5871	.5910	.5948	.5987	.6026	.6064	.6103	.6141
0.3	.6179	.6217	.6255	.6293	.6331	.6368	.6406	.6443	.6480	.6517
0.4	.6554	.6591	.6628	.6664	.6700	.6736	.6772	.6808	.6844	.6879
0.5	.6915	.6950	.6985	.7019	.7054	.7088	.7123	.7157	.7190	.7224
0.6	.7257	.7291	.7324	.7357	.7389	.7422	.7454	.7486	.7517	.7549
0.7	.7580	.7611	.7642	.7673	.7704	.7734	.7764	.7794	.7823	.7852
0.8	.7881	.7910	.7939	.7967	.7995	.8023	.8051	.8078	.8106	.8133
0.9	.8159	.8186	.8212	.8238	.8264	.8289	.8315	.8340	.8365	.8389
1.0	.8413	.8438	.8461	.8485	.8508	.8531	.8554	.8577	.8599	.8621
1.1	.8643	.8665	.8686	.8708	.8729	.8749	.8770	.8790	.8810	.8830
1.2	.8849	.8869	.8888	.8907	.8925	.8944	.8962	.8980	.8997	.9015
1.3	.9032	.9049	.9066	.9082	.9099	.9115	.9131	.9147	.9162	.9177
1.4	.9192	.9207	.9222	.9236	.9251	.9265	.9279	.9292	.9306	.9319
1.5	.9332	.9345	.9357	.9370	.9382	.9394	.9406	.9418	.9429	.9441
1.6	.9452	.9463	.9474	.9484	.9495	.9505	.9515	.9525	.9535	.9545
1.7	.9554	.9564	.9573	.9582	.9591	.9599	.9608	.9616	.9625	.9633
1.8	.9641	.9649	.9656	.9664	.9671	.9678	.9686	.9693	.9699	.9706
1.9	.9713	.9719	.9726	.9732	.9738	.9744	.9750	.9756	.9761	.9767
2.0	.9772	.9778	.9783	.9788	.9793	.9798	.9803	.9808	.9812	.9817
2.1	.9821	.9826	.9830	.9834	.9838	.9842	.9846	.9850	.9854	.9857
2.2	.9861	.9864	.9868	.9871	.9875	.9878	.9881	.9884	.9887	.9890
2.3	.9893	.9896	.9898	.9901	.9904	.9906	.9909	.9911	.9913	.9916
2.4	.9918	.9920	.9922	.9925	.9927	.9929	.9931	.9932	.9934	.9936
2.5	.9938	.9940	.9941	.9943	.9945	.9946	.9948	.9949	.9951	.9952
2.6	.9953	.9955	.9956	.9957	.9959	.9960	.9961	.9962	.9963	.9964
2.7	.9965	.9966	.9967	.9968	.9969	.9970	.9971	.9972	.9973	.9974
2.8	.9974	.9975	.9976	.9977	.9977	.9978	.9979	.9979	.9980	.9981
2.9	.9981	.9982	.9982	.9983	.9984	.9984	.9985	.9985	.9986	.9986
3.0	.9987	.9987	.9987	.9988	.9988	.9989	.9989	.9989	.9990	.9990
3.1	.9990	.9991	.9991	.9991	.9992	.9992	.9992	.9992	.9993	.9993
3.2	.9993	.9993	.9994	.9994	.9994	.9994	.9994	.9995	.9995	.9995
3.3	.9995	.9995	.9995	.9996	.9996	.9996	.9996	.9996	.9996	.9997
3.4	.9997	.9997	.9997	.9997	.9997	.9997	.9997	.9997	.9997	.9998
3.5	.9998	.9998	.9998	.9998	.9998	.9998	.9998	.9998	.9998	.9998
3.6	.9998	.9998	.9999	.9999	.9999	.9999	.9999	.9999	.9999	.9999
3.7	.9999	.9999	.9999	.9999	.9999	.9999	.9999	.9999	.9999	.9999
3.8	.9999	.9999	.9999	.9999	.9999	.9999	.9999	.9999	.9999	1.0000

SOURCE: C. H. Goulden (1954), *Methods of Statistical Analysis*, 2nd ed., Wiley, New York. Reprinted with permission of John Wiley and Sons, Inc.

[a]Areas under the normal curve in terms of the normal deviate.

TABLE B.2[a]
Kolmogorov-Smirnov D_n

SAMPLE SIZE	LEVEL OF SIGNIFICANCE (α)	
n	0.05	0.01
1	0.975	0.995
2	0.842	0.929
3	0.708	0.828
4	0.624	0.733
5	0.565	0.669
6	0.521	0.618
7	0.486	0.577
8	0.457	0.543
9	0.432	0.514
10	0.410	0.490
11	0.391	0.468
12	0.375	0.450
13	0.361	0.433
14	0.349	0.418
15	0.338	0.404
16	0.328	0.392
17	0.318	0.381
18	0.309	0.371
19	0.301	0.363
20	0.294	0.356
25	0.27	0.32
30	0.24	0.29
35	0.23	0.27
> 35	$\dfrac{1.36}{\sqrt{n}}$	$\dfrac{1.63}{\sqrt{n}}$

SOURCE: Adapted from F. J. Massey, Jr. (1951), "The Kolmogorov–Smirnov Test of Goodness of Fit," *Journal of the American Statistical Association*, **46**, pp. 68–78.

[a] Critical values D_n^x of the maximum absolute difference between sample and population cumulative distributions.

TABLE B.3
Chi-squared Probability

d.f.	LEVEL OF SIGNIFICANCE	
	0.05	0.01
1	3.84	6.64
2	5.99	9.21
3	7.82	11.34
4	9.49	13.28
5	11.07	15.09
6	12.59	16.81
7	14.07	18.48
8	15.51	20.09
9	16.92	21.67
10	18.31	23.21
11	19.68	24.72
12	21.03	26.22
13	22.36	27.69
14	23.68	29.14
15	25.00	30.58
16	26.30	32.00
17	27.59	33.41
18	28.87	34.80
19	30.14	36.19
20	31.41	37.57
21	32.67	38.93
22	33.92	40.29
23	35.17	41.64
24	36.42	42.98
25	37.65	44.31
26	38.88	45.64
27	40.11	46.96
28	41.34	48.28
29	42.56	49.59
30	43.77	50.89

SOURCE: C. H. Goulden (1954), *Methods of Statistical Analysis*, 2nd ed., Wiley, New York. Reprinted with permission of John Wiley and Sons, Inc.

INDEX